Waste Management, Reduction, and Sustainable Practices in Modern Societies

Muhammad Rizwan Yaseen
Government College University, Faisalabad, Pakistan

Qamar Ali
https://orcid.org/0000-0002-8178-7591
The Virtual University of Pakistan, Pakistan

IGI Global
Scientific Publishing
Publishing Tomorrow's Research Today

Vice President of Editorial	Melissa Wagner
Director of Acquisitions	Mikaela Felty
Director of Book Development	Jocelynn Hessler
Production Manager	Mike Brehm
Cover Design	Jose Rosado

Published in the United States of America by
IGI Global Scientific Publishing
701 East Chocolate Avenue
Hershey, PA, 17033, USA
Tel: 717-533-8845 | Fax: 717-533-7115
Website: https://www.igi-global.com E-mail: cust@igi-global.com

Library of Congress Cataloging-in-Publication Data

(CIP Data Pending)
ISBN13: 979-8-2600-1960-3
Isbn13Softcover: 979-8-2600-1961-0
EISBN13: 979-8-2600-1962-7
British Cataloguing in Publication Data
A Cataloguing in Publication record for this book is available from the British Library.

Table of Contents

Detailed Table of Contents

In the modern age, there has been a lot of problems based on the waste produced & it is not just limited to the household waste, it has become a matter of the total waste comprising solid, liquid, gaseous waste, municipal, electronic, plastic, hazardous, biodegradable waste, etc. It indicates that waste can be categorised into distinct kinds. Compilation of many kinds of waste has led to critical & hazardous problems. Fortunately, it has gained attention, not only among big authorities, government, communities or influencers, but also in general public. Waste Management has been adopted rapidly at many places, promoting the ideas of reducing, reusing and recycling waste, which is a genuinely helpful idea. The possible environmental benefits and drawbacks of a proposed recycling or waste management project would be examined by an Environmental Impact Assessment (EIA). The social impacts of formalization programs and the function of unofficial recycling industries in low-income countries are also covered in this chapter.

Electronic waste (e-waste) is the fastest-growing waste stream worldwide, posing severe environmental, economic, and health risks. Yet global awareness and coordinated policies remain limited, especially in low- and middle-income nations where informal recycling dominates. This chapter explores how English, as the world's lingua franca, plays a vital role in raising awareness of e-waste management and driving transnational policy change. Drawing on global data, case studies, and communication theory, it argues that English connects scientific research, treaties, NGO campaigns, and media coverage, enabling diffusion of best practices, policy transfer, and civil society mobilization. However, overreliance on English risks marginalizing non-anglophone populations and widening knowledge gaps. The chapter highlights opportunities and constraints, showing how successful campaigns use a cascade model: English seed documents translated and adapted into local contexts. It proposes inclusive strategies—bilingual publishing, plain language, and participatory translation.

In the densely populated urban hubs of India, the sight of brimming garbage containers and the pervasive stench of waste have become a routine aspect of daily life for many. This predicament is not limited to India alone, but is a global issue, with the projected production of municipal solid waste expected to reach 3.8 billion tonnes by 2050. Accumulations of uncollected waste can give rise to the proliferation of disease-carrying pests and severe air and water contamination, which have a severe impact on public health. Furthermore, inadequate disposal methods, such as the open incineration of waste, make a significant contribution to air pollution, releasing harmful toxins and greenhouse gases into the atmosphere. The rapid pace of urbanization exacerbates this issue, overwhelming traditional waste disposal methods and creating severe environmental and health risks. As cities expand and populations increase, the strain on conventional waste disposal systems intensifies.

The chapter addresses the dis-ruptive power of AI-based cognitive robotics to revamp autonomous multi-material sorting and high-value recovery, in the case of the e-waste industry. Based on the analysis of cur-rent academic literature (2019-2025), this paper will generalize the emergence of artificial intelligence (AI), machine learning (ML), deep learning (DL), computer vision (CV), rein-forcement learning (RL), and enhanced robotic manipulation to outline an innovative con-cept of intelligent waste management. The methodologies include the wide range of robotic architecture,

regions, particularly addressing kitchen waste disposal. Drawing from a robust dataset of 46,334 respondents in Kerala, sourced from NFHS 5, the study uncovers a compelling commitment to responsible waste management. Both rural and urban areas exhibit a shared dedication to sustainability, with a consistent preference for closed drain disposal and reusing kitchen waste for gardening or farming. Result: Encouragingly, urban regions show higher levels of awareness and engagement in sustainable waste management, reflected in the prevalence of manual collection. However, nuanced variations in less eco-friendly methods require vigilant attention. This research underscores Kerala's resolute commitment to responsible waste disposal, buoyed by the success of eco-conscious initiatives, and highlights the necessity of tailoring strategies to address local variations for a more sustainable future for all residents.

Chapter 9
Sustainable E-Waste Disposal Methods: The Case of Discarded Mobile Phones in Zimbabwe 209
Paul Mukucha, Bindura University of Science Education, Zimbabwe

There is an exponential increase in e-waste material at most dump sites in Zimbabwe. The absence of proper guidelines for dealing with domestic e-waste material perhaps might account for its rampant presence at most dumpsites in Zimbabwean urban areas. This led to the need to empirically determine how individuals dispose of obsolete electronic gadgets in general and, more specifically, mobile phones. A phenomenological approach of qualitative research design was adopted, and 30 participants were recruited for in-depth interviews. Data was analysed thematically. The results indicated that most of the participants dispose of their obsolete mobile phones through toyification, incineration, landfilling, and using sanitary bins. The study therefore recommended the relevant statutory bodies tasked with environmental management provide comprehensive guidelines on how to domestically dispose of e-waste material.

Chapter 10
Prakriti Parv: Nature-Friendly Festivals to Combat Environmental Pollution .. 239
Ginu George, Westford University College, UAE
K. Sudheesh, Christ (Deemed to be) University, India

Indian festivals are vibrant expressions of culture, tradition, and community that significantly contribute to the nation's social and economic fabric. They foster tourism, boost local economies, and strengthen cultural identity. However, these celebrations often generate environmental concerns, including air and water pollution, excessive waste, and resource depletion. This chapter examines the dual impact of festivals—celebrating their cultural and economic importance while addressing their ecological footprint. It explores sustainable solutions such as eco-friendly materials,

Preface

Electronic waste, commonly known as e-waste, has emerged as one of the fastest-growing waste streams in the modern world. Rapid technological advancement, shortened product life cycles, and the increasing demand for new electronic devices have significantly contributed to the rising volume of discarded equipment. While technological innovation has improved human life in countless ways, it has also introduced serious environmental and social challenges.

Improper management of e-waste leads to soil contamination, water pollution, and air toxicity, posing substantial health risks—particularly for communities involved in informal recycling practices. Addressing e-waste is a complex global issue that requires coordinated international action, as discarded electronics often contain hazardous substances such as lead, mercury, cadmium, and brominated flame retardants.

At the same time, e-waste represents a valuable resource. Electronic devices contain precious and rare earth metals, including gold, silver, palladium, and copper. Through effective recycling and recovery systems, e-waste can be transformed from an environmental burden into a key component of a circular economy. In line with the growing emphasis on sustainable development and resource efficiency, smart recycling and advanced waste management systems have gained increasing importance.

Recent advancements in artificial intelligence, blockchain technology, robotics, data analytics, smart city infrastructure, and digital tracking systems have created new opportunities to redesign the entire e-waste value chain. This book, *Waste Management, Reduction, and Sustainable Practices in Modern Societies*, brings together interdisciplinary research that examines technological, economic, behavioral, and policy-driven solutions for sustainable e-waste management.

The volume explores how emerging technologies, progressive policy frameworks, circular economy principles, and community engagement can be integrated to build resilient and environmentally sound recycling systems. The contributions presented here combine both theoretical insights and practical applications, offering valuable perspectives for researchers, practitioners, policymakers, and industry leaders.

In today's global context, e-waste management is closely tied to broader sustainability goals, including climate change mitigation, resource efficiency, sustainable urban development, and the transition toward circular economies. International initiatives, such as the United Nations Sustainable Development Goals, emphasize responsible consumption and production, climate action, and environmental protection. Effective e-waste management directly supports these objectives by reducing pollution, conserving critical raw materials, and promoting green technologies.

Many countries are revising their regulatory frameworks to incorporate concepts such as extended producer responsibility, reverse logistics, and sustainable product design. At the same time, the growth of smart cities and digital infrastructure is enabling more efficient, data-driven waste management systems that improve collection, sorting, and recycling processes.

This book situates e-waste management within this evolving global landscape, highlighting the role of digital transformation and smart technologies in enhancing the efficiency, transparency, and sustainability of recycling systems. It contributes to ongoing discussions on how societies can transition toward resource-efficient and circular economies by aligning technological innovation with policy development, community participation, and sustainable business models.

The book is designed for a broad, interdisciplinary audience. Researchers and scholars in environmental management, sustainability science, circular economy studies, urban planning, and industrial engineering will find valuable theoretical and empirical insights. Policymakers and government officials can benefit from policy analyses and case studies that showcase emerging governance frameworks and best practices. Industry professionals in recycling, electronics manufacturing, supply chain management, and waste services will gain practical perspectives on technological innovations and sustainable business strategies. Additionally, graduate students and academic institutions will find this volume useful in exploring the intersection of digital technologies and environmental sustainability.

CHAPTER OVERVIEW

Chapter 1, Environmental Impact Assessment on Recycling and Waste Management, focuses on the environmental implications of recycling processes. It evaluates the environmental costs and benefits of various waste treatment technologies and highlights the importance of sustainable recycling practices.

Chapter 2, The Role of English in Raising Global Awareness about E Waste Management, highlights the importance of global communication in promoting environmental awareness.

Chapter 3, Optimizing Waste Management in Smart Cities Tracing and Tracking Solutions, investigates how smart city infrastructures can integrate digital tracking technologies for efficient waste monitoring and management.

Chapter 4, AI Driven Cognitive Robotics for Autonomous Multi Material Sorting and High Value Recovery in Circular Supply Chains, explores advanced robotic systems that enable autonomous recovery of valuable materials from electronic waste streams.

Chapter 5, Reverse Logistics and Green Supply Chains Optimizing E Waste Collection Refurbishment and Redistribution, discusses the importance of reverse logistics systems in enabling efficient collection and redistribution of used electronic products.

Chapter 6, Cementing Sustainability Lean Manufacturing for Waste Free Systems, highlights the role of lean manufacturing principles in reducing industrial waste.

Chapter 7, Waste Reduction in Public Hospitals Integrating Digital Solutions and Sustainable Practices, examines waste management strategies in healthcare institutions.

Chapter 8, Household Waste and Everyday Practices Understanding Rural Urban Differences in Kitchen Waste Disposal in Kerala, provides insights into household waste behavior and its implications for sustainable waste management.

Chapter 9, Sustainable E Waste Disposal Methods The Case of Discarded Mobile Phones in Zimbabwe, presents a case study that illustrates the challenges and opportunities associated with mobile phone recycling in developing countries.

Chapter 10, Prakriti Parv Nature Friendly Festivals to Combat Environmental Pollution, explores cultural and community based approaches to environmental protection.

Chapter 11, Environmental Applications of Disinfection Technologies against COVID 19, discusses environmental technologies that emerged during the global pandemic.

By combining conceptual discussions with real-world case studies, this book aims to bridge the gap between theory and practice. The chapters collectively examine e-waste management from multiple dimensions, with each contribution addressing a specific aspect of the broader challenge. Together, they present a comprehensive and integrated view of sustainable e-waste systems.

Chapter 1
Environmental Impact Assessment of Recycling and Waste Management

Aryan Vasa
http://orcid.org/0009-0000-1837-8587
Ajeenkya D.Y. Patil University, India

Sunil Sankathala
http://orcid.org/0009-0004-7809-7647
Ajeenkya D.Y. Patil University, India

Krutika Patre
http://orcid.org/0009-0001-7969-4409
Ajeenkya D.Y. Patil University, India

Kritikka Kunte
Ajeenkya D.Y. Patil University, India

Susanta Das
http://orcid.org/0000-0002-9314-3988
Ajeenkya D.Y. Patil University, India

ABSTRACT

In the modern age, there has been a lot of problems based on the waste produced & it is not just limited to the household waste, it has become a matter of the total waste comprising solid, liquid, gaseous waste, municipal, electronic, plastic, hazardous, biodegradable waste, etc. It indicates that waste can be categorised into distinct kinds. Compilation of many kinds of waste has led to critical & hazardous problems. Fortunately, it has gained attention, not only among big authorities, government,

DOI: 10.4018/979-8-2600-1960-3.ch001

communities or influencers, but also in general public. Waste Management has been adopted rapidly at many places, promoting the ideas of reducing, reusing and recycling waste, which is a genuinely helpful idea. The possible environmental benefits and drawbacks of a proposed recycling or waste management project would be examined by an Environmental Impact Assessment (EIA). The social impacts of formalization programs and the function of unofficial recycling industries in low-income countries are also covered in this chapter.

1. INTRODUCTION

Environmental protection depends on effective recycling and trash management because of the steep increase in waste output brought by the rapid industrialization and urbanization. Environmental Impact Assessments (EIAs) have grown in terms of importance as planning tools that aid in predicting, assessing, and mitigating the ecological impacts of development projects, especially waste management-related ones. Recycling, recovery, and disposal practices have significant environmental effects, considering both, upstream as well as downstream, making the relationship between EIA and waste management more and more crucial. The life-cycle, thinking that underpins modern EIA methods, enables decision-makers to evaluate not just direct emissions but also indirect effects such as energy consumption, resource depletion, and health hazards associated with various waste management techniques (Unegg et al., 2023).

Waste diversion and recycling are often regarded as environmentally helpful practices, but their overall effects are not always clear-cut. According to studies contrasting landfilling, energy recovery, and recycling, each option has trade-offs that vary based on the technique, local conditions, and material. In Austria, Unegg et al. (2023) compared the CO_2 emissions from material recycling and thermal recovery using a life-cycle assessment. They discovered that both techniques can drastically lower emissions when compared to creating fuels or raw materials from scratch; material recycling could cut emissions by around 1% and thermal recovery by approximately 2.6% of the country's CO_2 emissions. The true advantages, however, rely on variables such as the composition of the waste stream and the degree to which recovered goods replace new ones. These findings show that in order to accurately assess climate benefits and possible hazards throughout the system, EIA frameworks need to consider more than just diversion rates (Unegg et al., 2023).

Waste practices' environmental evaluations must take into account problems including toxicity, acidification, and nutrient contamination in addition to greenhouse gas emissions. A life-cycle method was used by Spångberg et al. (2014) to compare improved wastewater treatment with the recycling of nutrients from human excreta into agriculture. When source-separated blackwater and urine were used as fertilizer instead of chemical fertilizers, they discovered that this reduced energy consumption and climatic impacts, but it also increased ammonia emissions, which elevated the hazards of acidification and eutrophication. This highlights the significance of thorough EIA for garbage and recycling plans, as even environmentally benign techniques can cause issues that move from one impact category to another (Spångberg et al., 2014).

Plastics serve as a reminder of how complicated environmental trade-offs are. Polymer nanocomposites derived from recycled or bioplastics were the subject of life-cycle assessment research evaluated by Carroccio et al. (2021). While adding nanoparticles can improve material performance and make products lighter, potentially saving resources, they discovered that the production of nanofillers, like titanium dioxide or carbon nanotubes, is energy-intensive and produces large amounts of greenhouse gas emissions, which can negate the benefits of using recycled or bio-based materials. The necessity of applying life-cycle analysis and environmental impact assessments to waste management at the end of a product's life as well as to new material innovations prior to their entry into the trash stream is highlighted by this (Carroccio et al., 2021).

Health impacts are a crucial component of waste management's Environmental Impact Assessment. Using air pollution dispersion models and epidemiological risk data, Forastiere et al. (2011) evaluated landfills and incinerators in Italy, Slovakia, and England. Thousands of years of life lost over several decades was projected by their study, which indicated moderate but detectable increases in cancer, birth abnormalities, and low birth weight among neighboring residents. The need of including public health issues, particularly cumulative and long-term exposures, into EIA processes for waste facilities is underscored by these dangers, which are minor in comparison to those posed by big industries or traffic (Forastiere et al., 2011).

The social aspect of recycling is equally as significant as the environmental aspect, especially in low-income nations where a large portion of the recyclable material is recovered by informal waste pickers. In order to investigate the effects of integrating informal recycling into formal systems on workers' livelihoods, safety, and health, Aparcana and Salhofer (2013) developed a social life-cycle assessment (sLCA) approach. In order to provide a more systematic assessment of social effects in addition to environmental ones, their methodology established three primary social impact categories, nine subcategories, and 26 indicators specific to the realities of waste pickers. Ensuring that recycling programs intended to improve environmental performance do not inadvertently hurt vulnerable people is made possible by including such sLCA methodologies into Environmental Impact Assessment (Aparcana & Salhofer, 2013).

Studies on food packaging demonstrate the importance of a comprehensive strategy. The life-cycle assessment techniques for new packaging solutions were studied by Bremenkamp and Gallagher (2024), who discovered that labels like "biodegradable" or "recyclable" do not always indicate a decreased total environmental impact. In order to facilitate truly sustainable design decisions and guarantee fair comparisons, they underlined the necessity of precisely defining system boundaries, functional units, and effect categories. This is indicative of a larger problem with recycling and trash management: results from EIA can be deceptive and result in ineffective or dysfunctional policy decisions if they are not conducted consistently (Bremenkamp & Gallagher, 2024).

For the beginning of this chapter, a number of important lessons emerge from these various studies. First, recycling and waste management environmental impact assessments must adopt a whole life-cycle perspective, considering a variety of environmental aspects in addition to greenhouse gas emissions. Second, sensitivity analysis and scenario testing are essential because changing materials or technology can have unanticipated negative effects. Third, rather than being considered as secondary concerns, assessments should incorporate the health and social implications, particularly for populations residing near trash dumps and for workers in informal recycling. Last but not least, generating trustworthy data and conducting equitable comparisons across various locations and waste streams depend on robust methodological standards and transparency. These problems will be further examined in the rest of this chapter. It will begin by outlining the EIA's guiding principles and procedures as they relate to recycling and waste management projects, including screening, scoping, baseline data collecting, impact prediction, mitigation, and monitoring. Then, using case studies from both high- and low-income contexts, it will examine empirical data on the effects on the environment and human health of the main waste treatment options: material recycling, energy recovery, composting, landfilling, and developing technologies. By integrating recycling and trash management into a robust Environmental Impact Assessment framework, governments, corporations, and researchers can better understand trade-offs, safeguard ecosystems and human health, and transition from linear to circular material use. Comprehensive, evidence-based assessments are urgently needed to shape the next generation of waste and recycling strategies, as demonstrated by the reviewed studies, which show both the promise and the shortcomings of current approaches (Aparcana & Salhofer, 2013; Carroccio et al., 2021; Forastiere et al., 2011; Spångberg et al., 2014; Unegg et al., 2023).

This chapter will give insights and overview over many sub-categorical points. The flow of the chapter is mentioned below (see also Figure 1):

2. Overview of Waste Management Systems

Figure 1. Conceptual flow of the chapter illustrating the progression from global waste challenges and EIA fundamentals to system assessment, case-based evidence, governance, socio-economic considerations, technological challenges, and future directions for circular and sustainable waste management.

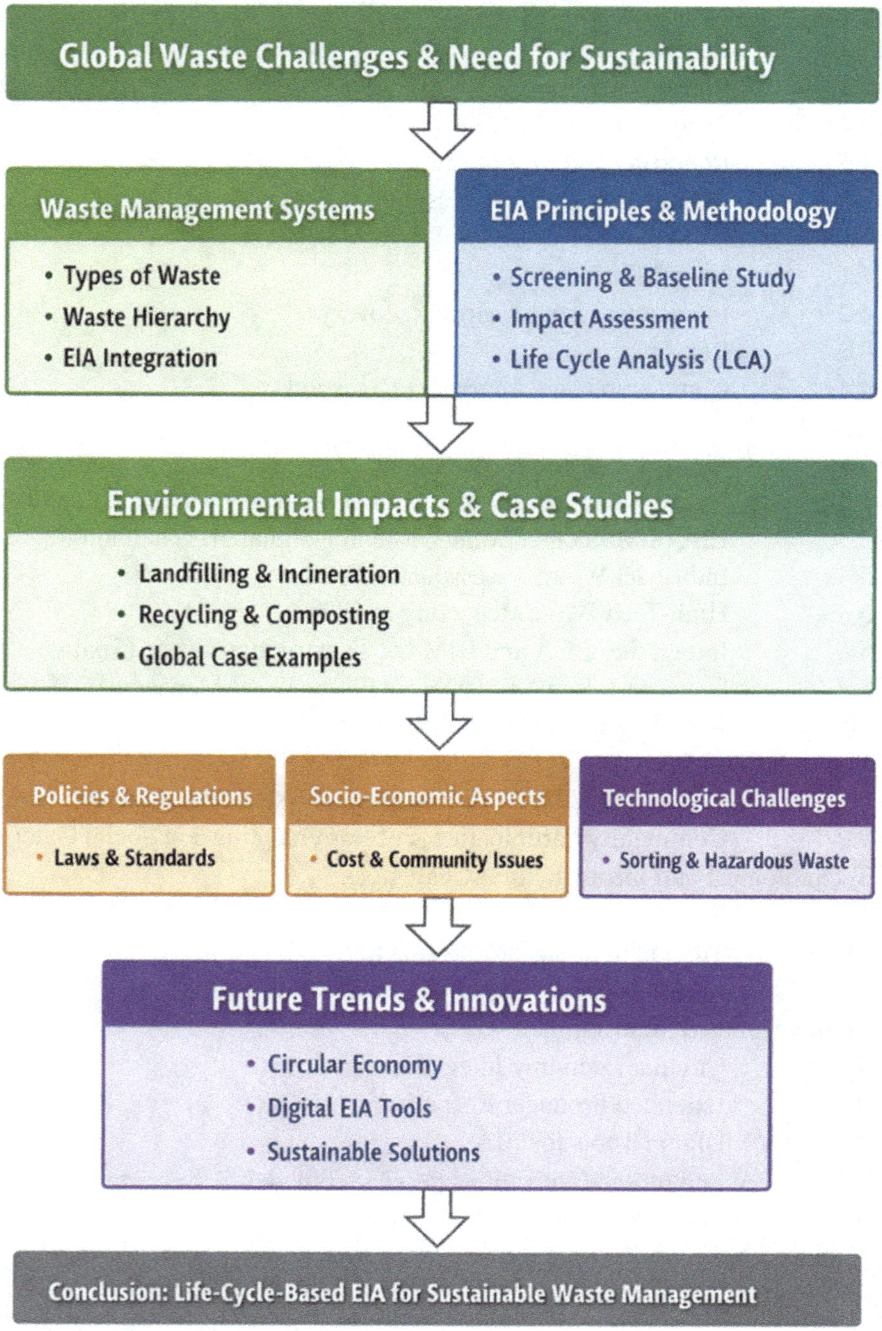

2. OVERVIEW OF WASTE MANAGEMENT SYSTEMS

2.1. Types of Waste

Since different waste kinds are mixed together, the first step in developing effective management strategies and Environmental Impact Assessments is understanding what each form of waste contains. Waste from households, businesses, and shops is known as municipal solid waste (MSW). It typically consists of food scraps, paper, plastics, glass, and metals, with more packaging and mixed materials as cities expand and people's lifestyles change (Spångberg et al., 2014). Since MSW production has increased due to rapid urbanization, local governments are under a lot of pressure to collect, treat, and dispose of it safely (Bremenkamp & Gallagher, 2024).

Power plants, manufacturers, and building sites are the sources of industrial trash. Chemical by-products, sludge, slag, and ash are examples of leftovers. Many of these wastes are filthy and require specific treatment to prevent the release of heavy metals, hazardous compounds, and greenhouse gases, while some are innocuous and may be handled with regular home waste (Unegg et al., 2023). Industrial trash can be fed into recycling or energy recovery facilities to reduce emissions and save resources if properly recognized and tracked (Carroccio et al., 2021).

Waste that is toxic, flammable, reactive, or persistent is considered hazardous waste. Medical waste, insecticides, solvents, and certain industrial sludges and residues are among its contents. Hazardous waste can damage land, water, and air if improperly handled, and it can seriously affect the health of local communities in the short and long term. People who live close to hazardous waste dumps have been found to have increased risks of cancer, birth abnormalities, and low birth weight in European studies. This highlights the importance of conducting rigorous Environmental Impact Assessments prior to licensing new treatment or disposal facilities (Forastiere et al., 2011).

E-waste, or electronic garbage, is the waste category with the fastest rate of growth worldwide. It comprises discarded electronics such as TVs, phones, computers, and other gadgets. E-waste includes hazardous materials like lead, mercury, and toxic flame retardants in addition to precious resources like gold, copper, and rare earth metals. E-waste is recycled informally in many low- and middle-income nations without the necessary safety precautions or environmental safeguards (Aparcana & Salhofer, 2013). E-waste, specifically, is a hazardous container of heavy metals and toxic chemicals, which poses key environmental risks in the lifecycle-based approach to managing the waste, extended producer responsibility, and advanced recycling technologies (Khang & Akhai, 2025). The high rate of increased electronic device consumption, which was accelerated by the COVID-19 condition, has dramatically worsened the e-waste generation, posing serious environmental and health risks and emphasizing the necessity of a circular economy strategy, and better consumer behavior as a means of sustainable e-waste management (Aggarwal et al., 2022). When this material is broken up and burned, toxic substances are released, contaminating the soil, water, and workers. To use the appropriate EIA techniques in recycling and waste operations, it is crucial to comprehend these various waste forms and their hazards (Carroccio et al., 2021).

2.2. Hierarchy of Waste Management

Contemporary waste management plans typically adhere to a "waste management hierarchy," which arranges environmental protection measures in order of best to worst. First and foremost, waste should be avoided or reduced in the first place by designing goods, services, and supply chains to produce less waste. The emphasis of Environmental Impact Assessment on preventing issues before they arise aligns with this (Bremenkamp & Gallagher, 2024). Some examples are utilizing lighter packaging, designing modular products for easy repair, and using green purchasing practices to purchase durable goods (Carroccio et al., 2021). Reuse is the next step after waste prevention, and it refers to extending the life of items or parts through refurbishing, repair, or second-hand markets. Reusing reduces emissions and energy consumption associated with recycling or creating new products. Informal reuse and repair has long been prevalent in many poor nations, helping to keep a lot of rubbish out of landfills while simultaneously giving locals jobs and cash (Aparcana & Salhofer, 2013). The waste management hierarchy, which emphasizes prevention and reuse over treatment and disposal, is presented in Figure 2.

Figure 2. Waste management hierarchy illustrating preferred waste treatment options, prioritizing prevention and reuse over recycling, energy recovery, and final disposal.

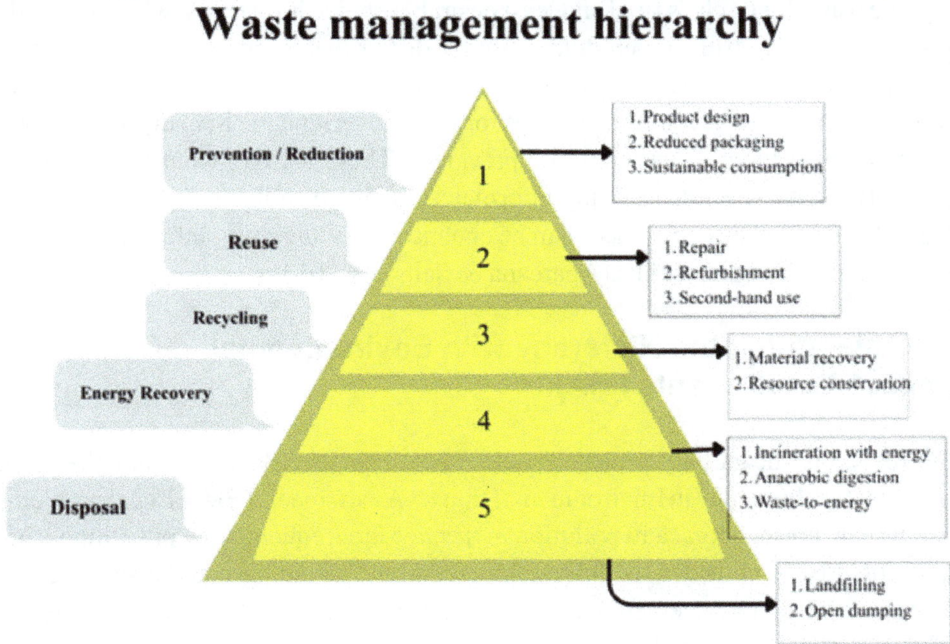

Recycling helps recover valuable materials, reduces the need for raw resources, and transforms waste into new goods. Recycling, however, can also result in emissions and energy use. According to life-cycle studies, recycling reduces greenhouse gas emissions more than creating new products (Carroccio et al., 2021; Unegg et al., 2023). However, the extent of the benefit varies depending on the waste's cleanliness, transportation distance, and ability to replace virgin materials. The trade-offs that only a comprehensive multi-impact assessment can reveal were demonstrated, discovering that recycling nutrients from human waste reduced greenhouse gas emissions when compared to manufactured fertilizers but raised the hazards of acidification and water contamination (Spångberg et al., 2014).

Waste that cannot be recycled or reused is valuable when it is recovered, primarily through energy recovery. Techniques such as burning waste to generate electricity, breaking down biological waste to create biogas, and collecting landfill gas reduce landfill volumes and partially replace the usage of fossil fuels. In comparison to existing practices, integrating heat recovery with recycling might reduce Austria's national CO_2 emissions by as much as 2.6% (Unegg et al., 2023). Strict environmental controls and health monitoring are necessary since these technologies can also discharge air pollutants and leave behind residues that require specific management (Forastiere et al., 2011).

The waste hierarchy's final and least desirable choice is disposal, which typically takes place in landfills. Open dumps are far dirtier than modern sanitary landfills with protective liners, leachate collection, and gas controls, yet they still squander resources and provide long-term issues for local governments. In many low-income nations, open dumps continue to be prevalent. These locations put neighbouring populations at risk while enabling informal recyclers to collect resources. Environmental and social circumstances can be enhanced by integrating informal recyclers into official garbage systems (Aparcana & Salhofer, 2013).

2.3. Integrating the Hierarchy with Environmental Impact Assessments (EIA):

Recycling and garbage projects require a broad perspective when using the waste management hierarchy in Environmental Impact Assessments (EIA). Local problems like traffic, noise, odors, and pollution in the area are frequently the only things that traditional EIAs consider. To fully comprehend the advantages and disadvantages of various waste management techniques, however, a comprehensive life-cycle approach is required, according to study (Carroccio et al., 2021; Unegg et al., 2023). When evaluating a new incinerator solely based on its local emissions, for instance, one may overlook the environmental benefits of substituting alternative electricity sources or the effects of alternative disposal techniques. Instead of concentrating just on the site, good environmental impact assessments should consider the entire system, including consequences that occur before and after a project. To forecast results, they must integrate health, social impact, and life-cycle assessments. They ought to look at a wide range of environmental impacts, not only greenhouse gas emissions, like toxicity, acidification, and water contamination. After the project starts, it's critical to keep an eye on the social and environmental conditions and include stakeholders in decision-making, particularly informal recyclers and local communities (Aparcana & Salhofer, 2013; Forastiere et al., 2011). In order to minimize the overall environmental impact, decision-makers can determine the optimal combination of prevention, reuse, recycling, recovery, and disposal by using this type of integrated approach. Additionally, it clarifies trade-offs, allowing the public and politicians to not only determine if a project is “good” or “bad,” but also how it compares to alternative possibilities within a broad sustainability framework (Bremenkamp & Gallagher, 2024).

Policymakers and practitioners can create initiatives that better achieve sustainability goals by precisely defining the various waste categories and adhering to the waste management hierarchy. By including these concepts into Environmental Impact

Assessments, assessments will encourage waste reduction, resource recovery, and systemic equity rather than merely approving end-of-pipe solutions. An extensive, evidence-based EIA framework is necessary to shape the next generation of recycling and waste management strategies, as demonstrated by studies on nutrient recycling, polymer nanocomposites, health risks near waste sites, and the social impacts of informal recycling (Aparcana & Salhofer, 2013; Carroccio et al., 2021; Forastiere et al., 2011; Spångberg et al., 2014; Unegg et al., 2023).

3. PRINCIPLES AND METHODS OF ENVIRONMENTAL IMPACT ASSESSMENT

A systematic method for determining, forecasting, and assessing the environmental effects of suggested activities prior to decision-making is the Environmental Impact Assessment (EIA). In addition to incorporating environmental, social, and economic factors into planning, its fundamental tenets include prudence, openness, and public involvement (Hoveidi et al., 2013; Marzouk & Azab, 2013). When it comes to recycling and waste management initiatives, EIA provides a means of going beyond end-of-pipe solutions by weighing the long-term environmental trade-offs of options including material recycling, energy recovery, and landfilling (Menikpura et al., 2012).

3.1. Screening and Scoping

In Environmental Impact Assessment (EIA), screening and scoping are the initial stages. If a recycling or waste management project is suggested, screening determines if a thorough environmental evaluation is necessary. For instance, some studies, like Yu et al. (2019) pointed out that novel techniques for recycling LCD panels have the potential to yield both advantages, like recovering indium and glass and disadvantages like acidification and health risks, which calls for a thorough investigation. By specifying which impacts to look at, the time and region covered, and the alternatives to compare, scoping then establishes the assessments' boundaries and emphasis. This step of the Castellón municipal solid waste research entailed assessing 24 distinct scenarios with varying collection structures, treatment technologies, and policy objectives (Bovea et al., 2010).

3.2. Baseline Study

A baseline study documents the environment's existing condition so that it can be compared to future consequences. For instance, particular data was collected on physical, chemical, biological, ecological, cultural, and social aspects using a Rapid Impact Assessment Matrix (RIAM) before evaluating several waste-disposal alternatives in industrial waste research conducted at Iran's Toos Industrial Estate. This baseline is essential for comprehending how landfills, recycling facilities, and energy recovery facilities may alter air quality, resource usage, or community well-being over time (Hoveidi et al., 2013).

3.3. Impact Prediction

A planned project's impact forecast calculates how the environment will change in comparison to the current state. You can do this with a variety of tools. For instance, a system dynamics model was developed to forecast emissions, energy consumption, and additional expenses from trash generated during building and demolition under recycling vs landfill scenarios (Marzouk & Azab, 2013). Recycling at a rate of around 25% in Nonthaburi, Thailand, lessens the damage that landfilling causes to ecosystems and natural resources. As per the studies, estimating impacts in waste-management EIAs frequently include modelling emissions, material flows, and the long-term benefits of avoided impacts (Menikpura et al., 2012).

3.4. Mitigation and Alternatives

Following impact predictions, the EIA process assesses feasible alternatives and finds mitigation strategies. A three-step hydrothermal–hydrometallurgical recycling method and incineration were contrasted in the LCD panel investigation by Yu et al. (2019). The new procedure may be a better option, as life cycle modeling revealed that it decreased adverse effects by more than 90% in some categories. The Toos Industrial Estate's RIAM score also indicated that resource recovery in conjunction with a sanitary landfill had more positive effects than either open dumping or incineration (Hoveidi et al., 2013).

3.5. Reporting and Monitoring

For decision-makers, an EIA report compiles the findings from screening, scoping, baseline research, impact projection, and mitigation. Monitoring is done to see if the anticipated effects and mitigation strategies indeed materialize as intended after approval. A few publications emphasize the importance of updating sensitivity assessments and life cycle inventories with actual operational data over time, even if few case studies provided comprehensive monitoring results (Bovea et al., 2010; Menikpura et al., 2012).

3.6. Role of Life Cycle Assessment (LCA) within EIA

Life Cycle Assessment (LCA) provides a "cradle-to-grave" perspective on environmental impacts, which is a significant addition to Environmental Impact Assessment (EIA). An LCA normally consists of four steps, in accordance with ISO 14040/44 standards: establishing the objective and scope, creating an input and output inventory, evaluating the effects, and interpreting the findings (Renou et al., 2007; Bovea et al., 2010). In garbage and recycling projects, LCA and EIA enable decision-makers to compare direct costs, such the electricity needed in recycling plants, with avoided burdens, like greenhouse gas reductions from material recovery (Bovea et al., 2010; Renou et al., 2007).

The research of municipal solid waste in Spain illustrated this by using life cycle assessment (LCA) in all 24 scenarios, from pre-collection to final disposal. According to Bovea et al. (2010), this method allowed for the testing of assumptions through sensitivity analysis and revealed important variables influencing environmental performance. Two distinct life cycle impact assessment techniques, Eco-indicator 99 and ReCiPe 2008, were used in the LCD panel recycling study to address topics such as human toxicity, terrestrial acidification, and metal resource depletion. This demonstrates how using multiple approaches can highlight trade-offs and improve findings (Yu et al., 2019). In order to increase the reliability of wastewater treatment results, Renou et al. (2007) also emphasized the importance of comparing various effect assessment techniques (Bovea et al., 2010; Renou et al., 2007). Further demonstrating the advantages of LCA-style thinking are examples of industrial waste. In order to score the physical, chemical, biological, social, and economic characteristics of different disposal strategies, Hoveidi et al. (2013) utilized the Rapid Impact Assessment Matrix. This multi-criteria matrix, while not a complete ISO-compliant LCA, provided a visible and semi-quantitative method of comparing choices, addressing the widespread complaint that standard EIA is inconsistent and has a clear rationale for decisions. Lastly, Ecosystem damage, abiotic resource damage, human health harm, and income-based community well-being were all specifically evaluated by the Nonthaburi study in Thailand. Menikpura et al. (2012) found that recycling approximately 25% of municipal solid waste helped counteract the adverse effects of landfilling the remaining 76%. This integrated approach provides a more comprehensive view of the effects of waste management systems, which is similar to the growing trend of combining environmental LCA with social LCA (sLCA) (Menikpura et al., 2012).

Environmental Impact Assessments (EIAs) for recycling and waste-management initiatives use a number of common techniques across various case studies. To ensure that important projects and their effects are appropriately identified, screening and scoping are employed initially (Bovea et al., 2010; Yu et al., 2019). Following that, baseline investigations set the reference conditions required for significant change forecasts (Hoveidi et al., 2013). Nowadays, dynamic modeling and life-cycle thinking are frequently used in impact prediction and mitigation to account for both the direct consequences of a project and the avoided impacts from recycling or recovery efforts (Marzouk & Azab, 2013; Menikpura et al., 2012). By revealing the outcomes and enabling managers to modify their course of action over time, reporting and monitoring round out the process. By evaluating cradle-to-grave impacts and avoided burdens using established techniques like ReCiPe, Eco-indicator 99, CML, or Rapid Impact Assessment Matrices, integrating Life Cycle Assessment (LCA) into EIA strengthens each level. When applied to waste management and recycling, this combined strategy supports the transition from a linear to a circular flow of materials, helps decision-makers anticipate trade-offs, and safeguards ecosystems and public health (Menikpura et al., 2012; Yu et al., 2019). The integration of Life Cycle Assessment (LCA) parameters within the EIA framework is represented in Table 1.

Table 1. Role of Life Cycle Assessment (LCA) in strengthening each stage of the Environmental Impact Assessment process.

EIA Stage	LCA Contribution	Key Parameters Considered
Screening and Scoping	Identifies relevant life-cycle boundaries and alternatives	Functional unit, system boundaries, scenario definition
Baseline Study	Establishes current material and energy flows	Waste composition, energy use, transport distances
Impact Prediction	Quantifies environmental impacts across life cycle	GHG emissions, acidification, eutrophication, toxicity, resource depletion
Alternatives Analysis	Compares treatment scenarios based on total impacts	Recycling vs landfill vs incineration performance
Mitigation Planning	Identifies improvement opportunities	Energy efficiency, emission reduction, material recovery rates
Monitoring & Reporting	Updates inventory with operational data	Real-time emissions, recovery rates, sensitivity analysis

The integration of Life Cycle Assessment within different stages of the Environmental Impact Assessment process is illustrated in Figure 3.

Figure 3. Integration of Life Cycle Assessment (LCA) within the Environmental Impact Assessment (EIA) framework for sustainable waste management.

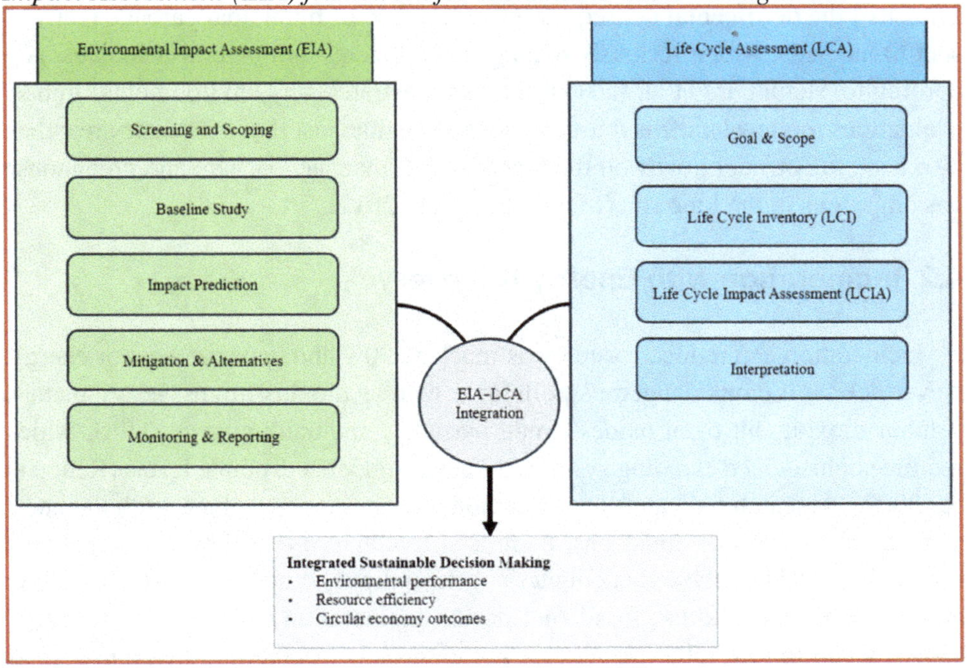

4. ENVIRONMENTAL IMPACTS OF COMMON WASTE MANAGEMENT PRACTICES

4.1. Landfilling

Around the world, landfilling remains the most popular method of garbage disposal. Today's landfills still damage the environment despite the adoption of protective liners, leachate collection systems, and gas recovery systems. According to Renou et al. (2007), if the leachate created when water flows through garbage is not properly handled, it may contain hazardous materials such as ammonia, heavy metals, hazardous compounds, and disease-causing organisms that could contaminate groundwater. When organic garbage decomposes without oxygen, methane, a potent greenhouse gas that is around 25 times stronger than carbon dioxide over a 100-year period, is released by landfills. This gas can still escape in huge quantities even with mechanisms in place to trap it. According to Spanish research, waste management systems that primarily rely on landfilling have the most negative effects on climate change and water pollution when compared to recycling or mixed approaches (Bovea et al., 2010). The 20%-25% recycling rate in Nonthaburi, Thailand, only marginally mitigated the detrimental consequences of the 76% of municipal garbage that was sent to landfills, which seriously damaged ecosystems and natural resources. According to Menikpura et al. (2012), this demonstrates why environmental impact evaluations for new landfills must consider more than just the immediate area; they also need to consider how well the landfills will manage leachate and greenhouse gas emissions in the long run (Menikpura et al., 2012).

4.2. Incineration with Energy Recovery

Incineration can reduce waste by as much as 90% and generate heat or energy. Nevertheless, it emits dangerous pollutants such as dioxins/furans, heavy metals, sulphur dioxide, nitrogen oxides, small particles, and acidic gasses, all of which require sophisticated cleaning systems to remain under acceptable levels (Renou et al., 2007). Research indicates that even though contemporary plants with the latest technology can reduce emissions, there is still a chance that they will negatively impact local air quality and encounter substantial public opposition (Forastiere et al., 2011). By substituting fossil fuel power, incineration combined with energy recovery can reduce total greenhouse gas emissions throughout the course of its whole life cycle. For instance, waste-to-energy facilities in Austria are thought to reduce CO_2 emissions by as much as 2.6% when compared to not recovering energy at all (Unegg et al., 2023). Marzouk and Azab (2013) found that energy recovery methods reduced overall external costs as compared to landfilling in Egypt's building and demolition waste study. These findings demonstrate why the benefits of averted impacts should be included in EIAs as well, as concentrating solely on stack emissions may understate the climate benefits of energy recovery (Marzouk & Azab, 2013; Unegg et al., 2023).

4.3. Recycling

Recycling frequently results in significant environmental advantages by reducing the need to extract new resources and save raw materials. Though the precise benefits vary depending on factors like contamination, collection systems, and how the recycled products replace virgin materials, life cycle studies consistently demonstrate that recycled materials generally produce less greenhouse gas emissions, use less energy, and deplete fewer resources than making materials from scratch (Bovea et al., 2010; Unegg et al., 2023). For example, a new discovery of about recycling nutrients from human waste reduing greenhouse gas emissions in comparison to advanced wastewater treatment, also raised the risk of acidification and water pollution. This indicates that recycling may require balancing various impacts (Spångberg et al., 2014).

Menikpura et al. (2012) discovered that recycling 24 percent of municipal waste decreased the harm that landfilling the remaining waste caused to ecosystems and natural resources in Thailand. Comparing bio-based or recycled polymer nanocomposites to virgin polymers, Carroccio et al. (2021) shown that the latter has less of an adverse effect on the environment, particularly when production is powered by renewable energy. These examples demonstrate why life cycle evaluations concerning various impacts should be included in EIAs of recycling systems in order to reveal hidden trade-offs and not only concentrate on climate consequences (Carroccio et al., 2021; Menikpura et al., 2012).

4.4. Composting and Anaerobic Digestion

Municipal waste's organic component is the focus of biological treatments including anaerobic digestion (AD) and composting. With the help of oxygen, composting breaks down organic matter, producing a soil conditioner but also generating ammonia, CO_2, and smells. A digestate that may be used in place of chemical fertilizers and biogas that can be converted to biomethane or used as energy are both produced by AD. Because they prevent methane emissions and recover valuable goods, life cycle studies demonstrate that both composting and AD have less of an impact on the climate than landfilling. Depending on how effectively the biogas is used, AD systems, for example, can cut greenhouse gas emissions by 30 to 80% when compared to landfills. EIAs must evaluate smells, bioaerosols, nutrient runoff, and market acceptability in addition to climate benefits since these processes require careful control to avoid stray emissions, leachate, and heavy-metal pollution. In a comprehensive environmental impact assessment (EIA), the primary waste management techniques exhibit clear advantages and disadvantages. Despite being cheap, landfilling has significant long-term concerns due to methane emissions and leachate (Bovea et al., 2010; Renou et al., 2007). Though it can lower greenhouse gas emissions by substituting fossil fuels, incineration with energy recovery raises issues about air pollution (Marzouk & Azab, 2013; Unegg et al., 2023). Recycling reduces emissions and saves resources, but it requires clean waste streams and stable markets to prevent shifting environmental consequences. For organic waste, composting and anaerobic digestion work well and stop methane from escaping, but they need to be carefully managed for bioaerosols, odors, and product quality. The inclusion of life-cycle data in EIAs facilitates the comparison of these options across many impacts by policymakers, allowing waste systems that safeguard ecosystems and human health while promoting the objectives of the circular economy (Carroccio et al., 2021; Menikpura et al., 2012). Biogas plants provide a viable means of treating organic municipal waste by diverting it off landfills, cutting down on greenhouse emissions, and producing renewable energy to meet the aims of the circular economy and sustainable waste management (Akhai, 2023). A comparative summary of environmental impacts is presented in Table 2.

Table 2. Comparative environmental performance of major waste treatment options based on life-cycle and EIA findings.

Waste Management Method	Environmental Benefits	Environmental Risks / Limitations	Key Impact Categories
Landfilling	Low operational cost; landfill gas recovery possible.	Methane emissions (≈ 25 × CO_2 global warming potential); groundwater contamination from leachate; long-term land use; resource loss.	Climate change, water pollution, and ecosystem damage.
Incineration with Energy Recovery	Reduces waste volume (up to 90%); generates electricity/heat; offsets fossil fuels.	Air pollutants (dioxins, nitrogen oxides, SO_2, particulates); toxic ash disposal; public opposition.	Air quality, human health, and climate trade-offs.
Material Recycling	Conserves raw materials; reduces energy use and GHG emissions; supports circular economy.	Energy use in processing; contamination issues; possible shifting impacts (e.g., acidification, transport emissions).	Resource depletion, climate change, toxicity, and eutrophication.
Composting	Converts organic waste into soil conditioner; reduces landfill methane.	Odor, ammonia emissions, bioaerosols, and nutrient runoff.	Air quality, eutrophication, and local nuisance.
Anaerobic Digestion	Produces biogas (renewable energy); reduces methane from landfills; digestate replaces fertilizers.	Potential leaks, odor, heavy metal contamination in digestate.	Climate benefits, nutrient management, and local environmental impacts.

A qualitative comparison of the relative environmental performance of different waste treatment methods, including greenhouse gas emissions, air pollution potential, and resource recovery benefits, are illustrated in Figure 4.

Figure 4. Relative comparison of environmental performance of major waste management methods based on key indicators including greenhouse gas emissions, air pollution potential, and resource recovery benefits.

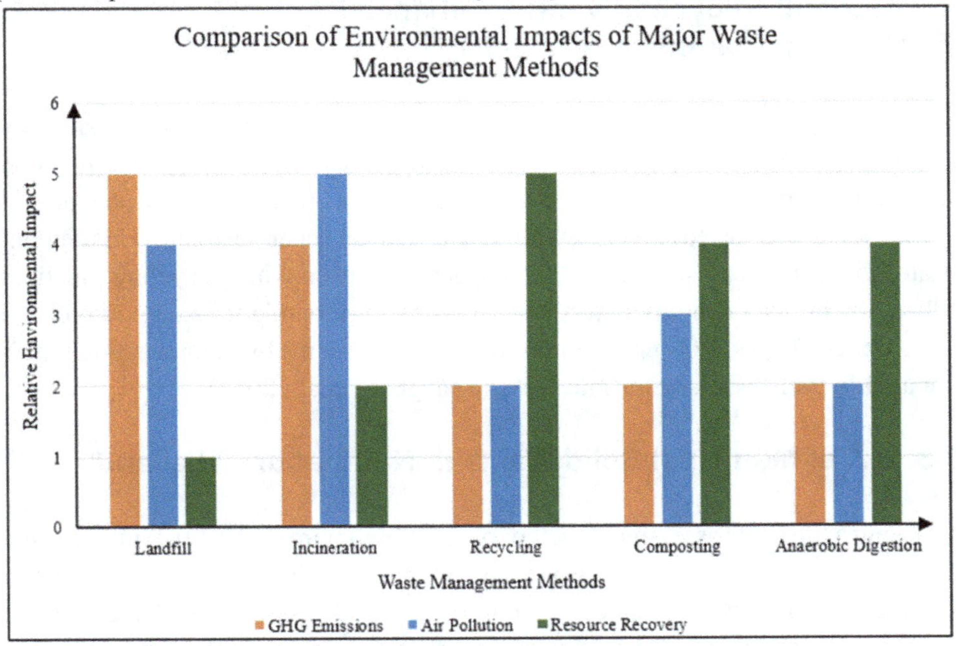

The values are presented on a qualitative scale from 1 (low) to 5 (high), derived from comparative findings reported in the literature reviewed in this study and intended to illustrate general trends rather than precise quantitative measurements.

5. CASE STUDIES

5.1. Municipal Solid Waste Scenarios in Spain

A comprehensive life cycle study of municipal waste management in Castellón, Spain, was carried out by Bovea et al. (2010). They tested 24 scenarios that incorporated landfilling, burning, mechanical-biological treatment, and selective collection. Recycling-focused strategies had the least impact on eutrophication, acidification, and global warming, while landfill-heavy strategies had the most because of leachate and methane. While incineration with energy recovery outperformed simple disposal, it required robust emission controls to control heavy metals and dioxins. In this instance, the use of life cycle assessment (LCA) in an EIA can assist in determining the optimal combination of waste-treatment choices rather than evaluating each technology independently (Bovea et al., 2010).

5.2. Nutrient Recycling Versus Enhanced Wastewater Treatment

Recycling nutrients from human waste reduced greenhouse gas emissions by replacing synthetic fertilizers, but it also raised the danger of water contamination and acidification due to nutrient losses, according to a 2014 study by (Spångberg et al., 2014) that contrasted recycling with advanced wastewater treatment. This demonstrates that issues can spread to other impact areas even when recycling practices adhere to circular economy objectives. Decision makers might better comprehend these trade-offs prior to implementing large-scale systems by incorporating multi-impact life cycle assessments into EIAs (Spångberg et al., 2014).

5.3. EIA of Municipal Solid Waste in Nonthaburi, Thailand

Using a life cycle assessment methodology, Menikpura et al. (2012) investigated the management of municipal garbage in Nonthaburi, where 20% to 25% of waste was recycled and the remainder was dumped in landfills. They evaluated human health, income-based well-being, resource depletion, ecosystem harm, and other environmental, economic, and social aspects. According to their findings, recycling improved the sustainability of the system as a whole by mitigating some of the adverse effects of landfilling. This example demonstrates how environmental performance and social and economic factors can be included in EIAs conducted in poor nations (Menikpura et al., 2012).

5.4. Industrial Waste Assessment in Iran

In Mashhad, Iran, Hoveidi et al. (2013) evaluated industrial waste management solutions in the Toos Industrial Estate using the Rapid Impact Assessment Matrix (RIAM). The best overall performance was discovered in sanitary landfills with resource recovery, after they graded physical, biological, and socioeconomic aspects for recycling, incineration, and sanitary landfilling. By providing decision makers with structured information to assess options and clarifying the process, RIAM's semi-quantitative approach demonstrated how EIA concepts may be applied at the industrial-estate level (Hoveidi et al., 2013).

5.5. High-Tech Waste Recycling in China

Yu et al. (2019) used the ReCiPe 2008 and Eco-indicator 99 techniques to conduct an environmental impact assessment of recycling trash LCD panels in China. In certain areas, such as human toxicity and metal resource depletion, they discovered that the innovative hydrothermal–hydrometallurgical recycling process reduced environmental consequences by more than 90% when compared to traditional burning. This example demonstrates how technical innovation can significantly enhance environmental outcomes for high-tech waste streams when assessed using a combination of EIA and LCA (Yu et al., 2019).

5.6. Integrating LCA and BIM for Construction Waste (Spain)

Researchers examined the environmental impacts of several construction waste management techniques during the design phase of a residential building project in Seville, Spain, using a method that integrates Life Cycle Assessment (LCA) and Building Information Modelling (BIM). Before construction started, the design team was able to comprehend the effects of recycling versus disposal thanks to this method. According to their investigation, almost all non-hazardous building debris might be recycled or utilized again, particularly steel, concrete, and plastics. Recycling was found to reduce greenhouse gas emissions by approximately 14.6 tons (about 1.4 times fewer emissions) and save approximately 148.5 GJ of energy (about 85 times more energy saved) when compared to disposal; floors and foundation slabs benefited the most from recycling. This example shows how early use of environmental assessment tools in the design phase can identify high-impact materials and direct waste and energy conservation techniques (Llatas et al., 2022).

5.7. Scenario-Based Assessment of Plastic Waste Management (Sweden)

To investigate several approaches to managing plastic trash in Sweden until 2030, Milios et al. (2018) developed a sustainability evaluation model. A business-as-usual method was used to examine three alternative tactics, taking into account social consequences (like jobs), economic considerations (costs and rewards), and environmental implications (like greenhouse gas emissions). Even though it was more expensive than less aggressive options, the most ambitious plan—meeting EU recycling standards while gradually lowering incineration—was the most beneficial to the environment, reducing emissions by about 545 kt CO_2 when compared to business as usual. This case demonstrates how scenario-based modeling can assist decision-makers in balancing environmental benefits against economic and social trade-offs, resulting in more realistic and well-informed recycling programs (Milios et al., 2018).

Table 3 gives a summary of key case studies on EIA applications in waste management.

Table 3. Representative global case studies demonstrating the application of EIA and LCA in waste management decision-making.

Study	Waste Type / System	Method Used	Key Findings
Castellón, Spain (Bovea et al., 2010)	Municipal Solid Waste	LCA across 24 scenarios	Recycling-focused systems showed lowest impacts; landfill-heavy systems had highest climate and water impacts.
Nonthaburi, Thailand (Menikpura et al., 2012)	Municipal waste	Life-cycle sustainability assessment	20–25% recycling reduced ecosystem damage and improved economic and social outcomes.
Toos Industrial Estate, Iran (Hoveidi et al., 2013)	Industrial waste	RIAM-based EIA	Sanitary landfill with resource recovery performed better than open dumping or incineration.
China (Yu et al., 2019)	Waste LCD panels	LCA (ReCiPe, Eco-indicator 99)	Advanced recycling reduced environmental impacts by >90% in several categories.
Sweden (Milios et al., 2018)	Plastic waste	Scenario-based sustainability assessment	High recycling strategy reduced ~545 kt CO_2 despite higher costs.
Austria (Unegg et al., 2023)	National waste streams	LCA comparison	Recycling and energy recovery reduced national CO_2 emissions significantly.
Spain (Llatas et al., 2022)	Construction waste	LCA-BIM integration	Recycling reduced emissions by ~14.6 tons CO_2 and saved significant energy

6. POLICY, LEGAL AND REGULATORY FRAMEWORKS

The international conventions provide the external limits in which nations handle waste and evaluate the effects of waste on the environment. The Basel Convention on the Control of Transboundary Movements of Hazardous Wastes and their Disposal (1989) commits the signatories to reduce the production of hazardous wastes and to make sure that they manage them in an environmentally friendly manner (United Nations, 2019). In the case of recycling and waste- management projects, it implies that any export of secondary materials, e-waste or residual fractions, should be included under prior informed consent, and by shipment tracking, which further influences the scopes of Environmental Impact Assessments (EIAs) prepared in the context of such facilities (Wilson et al 2006). Similarly, the Stockholm Convention on Persistent organic pollutants (2001) recommends that highly toxic chemicals like dioxins and furans, which are mostly waste products of incineration, open burning and certain recycling procedures, be removed or limited. A waste-to-energy plant today, EIA regularly incorporates dioxin-monitoring and dioxin abatement plans to prove to the Stockholm requirements. (Khoo, 2009).

On the national level, various countries have come up with waste-management project-specific EIA guidelines. The Waste Framework Directive and the EIA Directive of the European Union (Unegg et al., 2023), as well as India in its Ministry of Environment and Forests, have developed separate notification procedures of solid-waste, hazardous-waste and biomedical-waste facilities, requiring baseline collection of environmental and social data prior to approval. The Pollution Control Department of Thailand incorporates compliance with the EIA into its 3R approach (reduce, reuse and recycle); municipal governments like Nonthaburi have implemented the processes to assist in recycling and separation of waste conducted by communities (Menikpura et al., 2012).

An excellent example of the transformation of international commitments to local EIA practice is the Toos Industrial estate at Mashhad, Iran. Hoveidi et al. (2013) used Rapid Impact Assessment Matrix (RIAM) in order to compare open dumping, sanitary landfill, gasification, and incineration of industrial wastes. The analysis determined that the sanitary landfill using energy recovery had the least adverse effects on air, water, and social conditions, and this was more in line with the Basel principles, as well as the national standard of Iran regarding hazardous waste management. This is an example of the use of an internationally recognised assessment tool with local regulatory standards that can be used to enable decision makers to select the least harmful option (Hoveidi et al., 2013).

There is another example of Austria which created greenhouse-gas emission factors in different streams of recycling to meet the EU climate targets and waste requirements. Paper, glass, metals and plastic recycling data on life cycles were part of the permit application to the recycling plant, such that time saved in climate benefits could be directly balanced with the transport emissions, occupational hazards and local nuisance in the EIA process. This strategy demonstrates how the climate policy is becoming more and more aligned with the waste-management regulation (Unegg et al., 2023).

There are also the Extended Producer Responsibility (EPR) approaches which have been implemented in several countries which are in interaction with EIA. In Japan under the Home Appliance Recycling Law and in the EU under WEEE Directive, the take-back and recycling of electrical equipment and electronic equipment is financed by the producers. In China and India, new e-waste treatment plants are now being assessed based on the technical guidelines of the Basel Convention on

E-waste in terms of not only the control of the pollution but also the social state of the informal recyclers. These evaluations can prevent the relocation of susceptible employees and contribute to the idea of a fair transition to formal recycling systems (Wilson et al., 2006). The systems of sustainable waste management are becoming highly reliant on supporting industrial and innovation policies to promote green technologies, digital transformation, and adaptive infrastructure to improve environmental performance and long-term resilience of the system (Abbas et al., 2026).

Transparency and the involvement of the populace are influenced by policy frameworks, as well. In Europe the Aarhus Convention and in countries like India the Right-to-Information laws insist on disclosure of environmental analysis and monitoring information in the public. In the case of the Nonthaburi municipalities, frequent reporting of the recycling rates and monitoring of landfills created the trust and encouraged citizens to get involved in the waste sorting, which subsequently enhanced environmental performance reported in its life-cycle sustainability analysis. This is an example of how the legal provisions governing participation can directly contribute towards making waste policies effective (Menikpura et al., 2012). Lastly, case studies indicate that the multi-level governance which includes international treaties, national EIA regulations and local implementation has the strongest results. In Brazil and some of Latin America, cooperatives assisted by the World Bank and UN - habitat programmes integrated informal waste pickers into municipal recycling programmes, in part as the result of international occupational-health recommendations. These projects had EIAs that incorporated social indicators like income, child labour and community well-being, which went beyond the traditional biophysical perspective to fulfil new policy expectations (Aparcana & Salhofer, 2013).

7. ECONOMIC AND SOCIAL DIMENSIONS

7.1. Cost-Benefit / Cost-Effectiveness Analyses of Recycling Programmes

In their study, Menikpura et al. (2012) examined the management of municipal solid waste in Nonthaburi, Thailand, where roughly 25% of waste was recycled and 76% was landfilled. Using an evaluation of life-cycle sustainability that included life-cycle cost as an economic indicator, they contrasted this with other possibilities. Their findings demonstrated how recycling significantly increases economic sustainability since recycling saves money and resources, which more than offset the expenses and environmental damage associated with landfilling. Recycling has a net economic gain since it avoids landfill operations costs, uses less resources, does less environmental harm, and has fewer health hazards. However, these initial costs are offset by the savings on collecting, sorting, and processing (Menikpura et al., 2012). Several garbage treatment strategies were investigated in Spain by considering environmental and economic issues. They discovered that recycling scenarios lower pollution, health effects, and social costs by reducing environmental harm, even though they are frequently more expensive to operate and construct. This demonstrates that environmental impact assessments that miss economic trade-offs may provide a false impression of how viable recycling programs are (Bovea et al., 2010).

The environmental benefits of new technologies, such as bio-based or recycled polymer nanocomposites, are heavily influenced by economic considerations, including market demand, production costs, and processing charges. In their life-cycle study of polymer substitutes, Carroccio et al. (2021) discovered that the overall economic argument is significantly strengthened when recycled or renewable materials are priced competitively or when regulations and subsidies aid in cost reduction. As a result, recycling cost-benefit evaluations should account for both the direct financial expenses as well as the savings from avoiding negative effects on the environment, human health, and resource depletion (Carroccio et al., 2021). Recycling rate, between 20% to 25%, in Thailand has already significantly decreased the negative effects of landfilling on the environment, society, and economy. Further increasing the recycling rate would result in even greater cost reductions that might be avoided, with diminishing benefits only becoming apparent at a certain point. This demonstrates the potential for significant gains from early recycling improvements and emphasizes the necessity of using realistic assumptions on recycling levels, performance, and contamination rates in economic assessments in EIAs to prevent unduly optimistic projections (Menikpura et al., 2012).

7.2. Community Participation, Public Perception and Social Outcomes

In addition to financial considerations, social elements are critical to effective waste management. Costs and recycling outcomes can be impacted by community involvement. Menikpura et al. (2012) used income-related community wellbeing as a social indicator and discovered that recycling enhances local livelihoods in addition to helping the environment and economy by generating revenue from recyclables, jobs in collection and sorting, and a decrease in health issues brought on by pollution. These social advantages boost public support for recycling programs and aid in defending financial or legislative adjustments. Additionally, public opinion matters because it influences regulations, site selection, and compliance, as well as how people feel about recycling, landfills, and waste facilities. Despite the fact that many studies do not look closely at public views, the Thailand case demonstrates how setting legislative goals and incorporating stakeholders can increase community support for recycling. Locals are more inclined to participate in sorting and recycling operations when they think trash management is safer, cleaner, and more profitable (Menikpura et al., 2012).

In many areas, informal recyclers or waste pickers endure greater health hazards, unfavorable working conditions, and inconsistent revenues while receiving little protection or acknowledgment, which raises concerns about equity. Research employing social life-cycle assessments, such as the Thailand example and research on the manufacture of bio/recycled polymers, demonstrates that implementing policies like work formalization, safety enhancements, equitable compensation, and social well-being promotion can result in significantly better outcomes. The amount that households must pay and their willingness to pay is another important aspect. Although not all studies provide data on this, some indicate that families' willingness to pay more for improved collection or additional bins relies on how easy the system is and how much they think it will enhance cleanliness, lessen odors, and protect their health. For policymakers, these trade-offs are essential since costly programs may not succeed if the public does not participate or does not perceive the benefits (Menikpura et al., 2012).

8. TECHNOLOGICAL AND INFRASTRUCTURAL CHALLENGES

8.1. Sorting and Contamination Issues

High-quality recycling starts with efficient sorting since contaminants lower the value of recovered materials, raise processing costs, and potentially negate environmental gains if garbage needs to be disposed of or yields lesser-quality outputs. Bovea et al. (2010) demonstrated that the environmental benefits of recycling are highly dependent on the cleanliness of the incoming materials in their life-cycle assessment of municipal solid waste scenarios in Castellón, Spain. EIAs for recycling facilities should assess contamination levels and throughput capacity because the reductions in global warming and eutrophication potentials were significantly smaller when source separation was inadequate than in scenarios utilizing well-sorted streams (Bovea et al., 2010). Similar findings were made by Menikpura et al. (2012) in Nonthaburi, Thailand, who discovered that although a recycling rate between 20% to 25%, already counteracted many of the negative consequences of landfilling, more contamination or inefficient collection would swiftly undo these gains and raise the cost of recycling per unit. Therefore, in order to improve material purity, they suggested merging informal collectors and toughening home source separation (Menikpura et al., 2012).

Complex or high-tech waste streams also present sorting issues. When Yu et al. (2019) compared a new hydrothermal–hydrometallurgical process with conventional incineration for waste LCD panels, they discovered that while the new method reduced environmental burdens by over 90% in some categories, its effectiveness hinged on pre-separating panels and removing hazardous layers prior to treatment. The need of including such technical limitations in EIAs to generate accurate estimates of environmental performance is highlighted by the fact that high pollution decreased process efficiency and increased costs (Yu et al., 2019). Carroccio et al. (2021) assessed bio- and recycled polymer nanocomposites at the material level and observed that recycled polymers frequently contain mixed plastics or additives that degrade mechanical qualities, make processing more difficult, or raise emissions during remanufacturing. This indicates that in order to recycle plastics in a closed-loop manner, the market for secondary goods must be robust in addition to technological advancements such improved sorting, cleaning, and removal of legacy compounds. Recycling construction and industrial waste materials including ceramic tile waste and micro silica and using them in concrete manufacturing can enhance mechanical performance and lessen the use of natural resources and environmental footprint, which encourages circular construction (Dhiman et al., 2025). Accordingly, environmental impact assessments (EIAs) of new polymer recycling facilities should evaluate the effects on the environment as well as the quality-control procedures required to manage contamination safely and guarantee high-quality outputs (Carroccio et al., 2021).

8.2. Treatment of Hazardous Fractions:

One of the most difficult tasks is managing hazardous fractions that are concealed in waste streams. Certain plastics, batteries, electronics, and medical waste can all introduce heavy metals, brominated flame retardants, and persistent organic pollutants into regular urban waste. Compost, recyclables, and incinerator ash may get contaminated if these products are not properly separated. A nutrient-recycling study by Spångberg et al. (2014) shown that although human waste can be used in place of mineral fertilizers, stringent control is necessary to keep medicines and pathogens out of agricultural soils. This emphasizes the necessity of comprehensive sanitary controls and risk evaluations when recycling organic fractions (Spångberg et al., 2014).

Hazardous material handling is also influenced by international agreements. The Stockholm Convention's limitations on persistent organic pollutants, like furans and dioxins, have a direct impact on various recycling and incineration procedures.

Modern facilities use flue-gas cleaning, continuous emissions monitoring, and ash stabilization to meet tight criteria, but older European incinerators were suspected of posing greater health dangers to the populations around them. Because of this, dioxin modeling and mitigation strategies are frequently included in EIAs for such facilities. Because open burning and low-tech incineration still emit harmful pollutants in nations with laxer regulations, project permits must require strict technical criteria (Forastiere et al., 2011; Khoo, 2009).

Hazardous materials also make sophisticated recycling procedures more difficult. Yu et al. (2019) discovered that liquid crystals, mercury, and indium are among the desirable yet hazardous compounds found in LCD panels. Their new recycling process necessitated a three-step pre-treatment to eliminate these concerns prior to recovering glass and metals; without these precautions, the risks to the environment and workers would exceed the advantages. Thus, it is imperative that these pre-treatment actions be specifically evaluated and priced in EIAs (Yu et al., 2019). Regarding the plastics industry, research points out that recycled streams could contain legacy additives like phthalates or brominated flame retardants, which, if not appropriately eliminated, can remain in new products. As a result, there are downstream health and liability issues, which is why chemical hazard assessments must be included in polymer recycling plant EIAs in addition to energy usage and emissions metrics (Carroccio et al., 2021).

8.3. Infrastructure Gaps and Institutional Capacity:

Only when backed by appropriate infrastructure can new recycling technologies function effectively. There are currently insufficient transfer stations, sorting facilities, or hazardous-waste collection points in many emerging cities, which might compromise even well-thought-out recycling programs. As demonstrated by the Nonthaburi research, the city's municipal infrastructure and community-based collection had to be connected for the recycling system to be successful. It is impossible to fully realize the anticipated environmental benefits of recycling without enough storage, transportation, and treatment capacity (Menikpura et al., 2012). Like physical infrastructure, institutional capacity is essential. Regulators can examine various waste-management solutions with the aid of technologies like the Rapid Impact study Matrix, as the Toos Industrial Estate study showed. However, it also exposed significant data gaps on emissions and dangerous materials. Stronger monitoring and reporting procedures are necessary for EIAs to be successful and credible in these kinds of situations, as this research emphasizes (Hoveidi, 2013).

9. FUTURE TRENDS AND INNOVATIONS

9.1. Circular Economy Integration

Individual garbage projects are giving way to a more comprehensive circular economy (CE) strategy, which is a significant trend in recent research. Dry waste materials like rice straw, wheat straw, and sugarcane bagasse can be used as a construction material, which can enhance thermal and mechanical performance and minimize agricultural waste and carbon emissions to achieve the concept of the circle and sustainable management of resources (Walia et al., 2025). CE emphasizes reducing, reusing, remanufacturing, and high-quality recycling in order to prolong the life of materials and products. This reduces the demand for new resources and eases environmental pressures. Comparing landfilling versus recycling, Menikpura et al. (2012) demonstrated that even modest improvements in recycling significantly improved environmental, economic, and social outcomes in Thailand. Incorporating CE concepts into EIAs would require considering additional factors, such as product design prior to use and secondary markets following use, in addition to the impact of a single facility. According to growing CE policies in the EU and Asia, EIAs for new recycling plants should, for instance, include metrics like resource productivity, the amount of material that remains in a circular loop, or the possibility of cascading reuse (Bovea et al., 2010; Menikpura et al., 2012).

9.2. Extended Producer Responsibility (EPR)

A strategy known as Extended Producer Responsibility (EPR) incentivizes manufacturers to create products that are simpler to recycle or reuse by assuming some or all of the financial burden and accountability for managing things at the end of their life. Research on e-waste in Asia demonstrates that formalizing informal recyclers and implementing EPR programs can improve material recovery and reduce health hazards. For EIAs, this means that in addition to emissions from a facility, project planners may also need to take into account reverse supply chains, take-back systems, and the social effects on current informal workers. The EU's WEEE Directive and Japan's Home Appliance Recycling Law, for example, require manufacturers to cover the cost of product return, and environmental evaluations of new treatment facilities now cover the handling of hazardous compounds such as mercury or brominated flame retardants in order to comply with requirements. By include EPR requirements in EIAs, it is possible to make sure that new recycling facilities complement these policy objectives rather than work against them (Wilson et al., 2006; Habuer et al., 2014; Khoo, 2009).

9.3. Digital Tools for EIA

Using digital tools and big data to improve the accuracy, transparency, and efficiency of EIAs is a third emerging trend. EIAs for trash projects used to mostly rely on manual public hearings and fixed baseline investigations. Real-time sensors, remote sensing, GIS, and web dashboards are some of the technologies that are currently being used to map trash flows, monitor pollution, and engage communities. Menikpura et al. (2012) in Thailand employed a digital life-cycle sustainability assessment to explore various scenarios, while Austria developed greenhouse-gas emission factors for recycling streams using digital models and massive life-cycle information. By utilizing dynamic emission models, open-source LCA databases, and web-based portals to display findings and collect input, future EIAs might build on these techniques and make the process more flexible and interactive.

9.4. Integrating Innovations into EIA Practice

When EPR, digital technologies, and circular economy concepts are combined, EIAs require new measures and a wider focus. Regulations currently primarily consider the consequences on local air, water, and noise, but they frequently overlook broader effects like resource conservation, climate benefits, and social justice. Projects would be encouraged to recover more resources if CE indicators were added. EPR would also maintain them in line with producer duties and jobs in the unorganized sector, and digital technologies would enable ongoing monitoring and public participation rather than single reports. Studies on bio- and recycled polymer nanocomposites demonstrate that open data systems and supply chains are essential for achieving significant environmental benefits. This suggests that EIAs are no longer merely a legal requirement but are being actively used to create circular, sustainable waste systems (Carroccio et al., 2021; Roes et al., 2007).

9.5. Case Illustrations of Future Practice

Numerous actual situations demonstrate how these novel concepts can alter EIAs. A "circular EIA" model was created in Nonthaburi, Thailand, by combining typical EIA procedures with a life-cycle sustainability assessment of its recycling system. This created a decision tool that took into account social, economic, and environmental considerations. EPR regulations for electronics and packaging in Europe currently mandate that facilities use digital dashboards to communicate real-time recovery rates and contamination levels to national authorities. As advised by the Basel and Stockholm conventions, experimental initiatives in Brazil that incorporate informal waste pickers into municipal recycling cooperatives have improved environmental assessments by including social factors like income and occupational safety. The combination of CE, EPR, and digital tools can make EIAs more comprehensive and policy-relevant, as demonstrated by these instances (Bovea et al., 2010; Cointreau, 2006; Khoo, 2009; Menikpura et al., 2012).

Future EIAs for waste management and recycling are anticipated to shift from examining individual projects to more system-wide, data-driven, and collaborative evaluations. EIAs can assist firms and governments in anticipating trade-offs, safeguarding the environment and public health, and expediting the transition from linear to circular material use by combining concepts from the circular economy, extended producer responsibility, and technology. The research discussed here demonstrates the challenges as well as the potential of this shift. To achieve this, new regulations, improved data systems, practitioner and regulator training, and close collaboration with producers and communities are required (Carroccio et al., 2021; Menikpura et al., 2012).

10. CONCLUSION

The Environmental Impact Assessment (EIA) has evolved from a merely legal requirement to a crucial planning tool for creating environmentally friendly waste management and recycling systems. With careful application, EIA may steer projects away from "end-of-pipe" solutions and toward designs that keep trash at its source, recover valuable materials, and protect public health. Making use of life-cycle thinking at every stage enables decision-makers to comprehend not only the immediate implications of a facility but also its broader upstream and downstream ramifications, resulting in more intelligent and well-rounded choices. Additionally, the review demonstrates that resource recovery and recycling are not always advantageous. The quality of the supporting infrastructure, the way hazardous materials are managed, the way the informal and formal sectors collaborate, and the way materials are categorized all affect how successful they are. Recycling may no longer benefit the environment or may pose new hazards if these requirements are not met. Likewise, waste programs' economic and social outcomes depend on their costs, the market's desire for recycled goods, and the degree of community support and involvement.

The foundation for conducting Environmental Impact Assessments (EIAs) is made up of laws and regulations, including national standards, international agreements, and extended producer responsibility initiatives. Projects are more likely to aid rather than inadvertently undermine broader sustainability objectives when EIAs are in line with these regulations. At the same time, EIAs are becoming more transparent, flexible, and participatory thanks to new tools including digital data platforms, real-time monitoring, and public access to outcomes. A future in which

EIA is a genuine planning and design tool for creating circular economy systems rather than merely a stage in the approval process is indicated by all of these advancements. By integrating economic, social, and environmental data, connecting local decisions with global pledges, and utilizing contemporary instruments and measurements, EIAs can assist governments, corporations, and communities in developing waste-management systems that are more sustainable, equitable, and financially sound. For policymakers and practitioners going forward, this is a major problem and the primary lesson to be learned from the case studies in this chapter.

REFERENCES

Abbas, U., Akhai, S., Abbass, M., Abass, S., Chouksey, A., & Wadhwa, A. S. (2026). Sustainable Growth by Navigating Complexity Through Innovation and Adaptive Industrial Policies. In Tzouvelekas, M., Zarotiadis, G., & Varsakelis, N. (Eds.), *Industrial Policy, Innovation, and Complexity: Comparative Analysis* (pp. 29–82). Trends, and Prospects., DOI: 10.4018/979-8-3373-1767-0.ch002

Aggarwal, P., Rana, M., & Akhai, S. (2022). Briefings on e-waste hazard until COVID era in India. *Materials Today: Proceedings, 71*(2), 389–393. DOI: 10.1016/j.matpr.2022.09.507

Akhai, S. (2023). Biogas plants for sustainable municipal waste management. In Akhai, S., Pant, P., & Bandara, J. M. S. J. (Eds.), *Optimizing sustainable wastewater treatment in developing nations* (pp. 162–179). IGI Global., DOI: 10.4018/978-1-6684-8117-2.ch012

Aparcana, S., & Salhofer, S. (2013). Development of a social impact assessment methodology for recycling systems in low-income countries. *The International Journal of Life Cycle Assessment, 18*(5), 1106–1115. DOI: 10.1007/s11367-013-0546-8

Bovea, M., Ibáñez-Forés, V., Gallardo, A., & Colomer-Mendoza, F. (2010). Environmental assessment of alternative municipal solid waste management strategies: A Spanish case study. *Waste Management (New York, N.Y.), 30*(11), 2383–2395. DOI: 10.1016/j.wasman.2010.03.001 PMID: 20381331

Bremenkamp, I., & Gallagher, M. J. S. (2024). Life cycle assessment methods for investigating novel food packaging systems. *Environmental Science. Advances, 3*(10), 1337–1350. DOI: 10.1039/D3VA00380A

Carroccio, S. C., Scarfato, P., Bruno, E., Aprea, P., Dintcheva, N. T., & Filippone, G. (2021). Impact of nanoparticles on the environmental sustainability of polymer nanocomposites based on bioplastics or recycled plastics – A review of life-cycle assessment studies. *Journal of Cleaner Production, 335*, 130322. Advance online publication. DOI: 10.1016/j.jclepro.2021.130322

Cointreau, S. (2006). Occupational and environmental health issues of solid waste management. (Urban Papers, No. 2). The World Bank. https://www.worldbank.org/urban/

Dhiman, S., Seema, S., & Akhai, S. (2025). Performance evaluation of OPC 43 concrete incorporating micro silica and recycled ceramic tile waste as partial cement and coarse aggregate replacements. *Asian Journal of Civil Engineering, 27*(1), 299–317. DOI: 10.1007/s42107-025-01504-7

Forastiere, F., Badaloni, C., de Hoogh, K., von Kraus, M. K., Martuzzi, M., Mitis, F., Palkovicova, L., Porta, D., Preiss, P., Ranzi, A., Perucci, C. A., & Briggs, D. (2011). Health impact assessment of waste management facilities in three European countries. *Environmental Health : A Global Access Science Source, 10*(1), 53. Advance online publication. DOI: 10.1186/1476-069X-10-53 PMID: 21635784

Habuer, N., Nakatani, J., & Moriguchi, Y. (2014). Time-series product and substance flow analyses of end-of-life electrical and electronic equipment in China. *Waste Management (New York, N.Y.), 34*(2), 489–497. DOI: 10.1016/j.wasman.2013.11.004 PMID: 24332400

Hoveidi, H., Ahmadi Pari, M., Pazoki, M., & Koulaeian, T. (2013). Industrial waste management with application of RIAM environmental assessment: A case study on Toos Industrial State, Mashhad. *Iranica Journal of Energy and Environment, 4*(2), 142–149. DOI: 10.5829/idosi.ijee.2013.04.02.11

Khang, A., & Akhai, S. (2025). E-Waste and Lithium-ion Battery Recycling Insights for Sustainable Transportation. In Khang, A. (Ed.), *Driving Green Transportation System Through Artificial Intelligence and Automation* (pp. 203–230). Springer., DOI: 10.1007/978-3-031-72617-0_11

Khoo, H. H. (2009). Life cycle impact assessment of various waste conversion technologies. *Waste Management (New York, N.Y.), 29*(6), 1892–1900. DOI: 10.1016/j.wasman.2008.12.020 PMID: 19157835

Llatas, C., Quiñones, R., & Bizcocho, N. (2022). Environmental Impact Assessment of Construction Waste Recycling versus Disposal Scenarios Using an LCA-BIM Tool during the Design Stage. *Recycling, 7*(6), 82. Advance online publication. DOI: 10.3390/recycling7060082

Marzouk, M., & Azab, S. (2013). Environmental and economic impact assessment of construction and demolition waste disposal using system dynamics. *Resources, Conservation and Recycling, 82*, 41–49. DOI: 10.1016/j.resconrec.2013.10.015

Menikpura, S. N. M., Gheewala, S. H., Bonnet, S., & Chiemchaisri, C. (2012). Evaluation of the effect of recycling on sustainability of municipal solid waste management in Thailand. *Waste and Biomass Valorization, 4*(2), 237–257. DOI: 10.1007/s12649-012-9119-5

Milios, L., Davani, A. E., & Yu, Y. (2018). Sustainability Impact Assessment of increased plastic recycling and future pathways of plastic waste management in Sweden. *Recycling, 3*(3), 33. Advance online publication. DOI: 10.3390/recycling3030033

Renou, S., Thomas, J., Aoustin, E., & Pons, M. (2007). Influence of impact assessment methods in wastewater treatment LCA. *Journal of Cleaner Production, 16*(10), 1098–1105. DOI: 10.1016/j.jclepro.2007.06.003

Roes, A. L., Marsili, E., Nieuwlaar, E., & Patel, M. K. (2007). Environmental and cost assessment of a polypropylene nanocomposite. *Journal of Polymers and the Environment, 15*(3), 212–226. DOI: 10.1007/s10924-007-0064-5

Spångberg, J., Tidåker, P., & Jönsson, H. (2014). Environmental impact of recycling nutrients in human excreta to agriculture compared with enhanced wastewater treatment. *The Science of the Total Environment, 493*, 209–219. DOI: 10.1016/j.scitotenv.2014.05.123 PMID: 24946033

Unegg, M. C., Steininger, K. W., Ramsauer, C., & Rivera-Aguilar, M. (2023). Assessing the environmental impact of waste management: A comparative study of CO2 emissions with a focus on recycling and incineration. *Journal of Cleaner Production, 415*, 137745. Advance online publication. DOI: 10.1016/j.jclepro.2023.137745

Walia, J., Sharma, R., & Akhai, S. (2025). Stubble-reinforced materials for sustainable construction. In Akhai, S. (Ed.), *Mitigating unsustainable practices in construction and architecture* (pp. 99–118). IGI Global., DOI: 10.4018/979-8-3373-2555-2.ch004

Wilson, D. C., Velis, C., & Cheeseman, C. (2006). Role of informal sector recycling in waste management in developing countries. *Habitat International, 30*(4), 797–808. DOI: 10.1016/j.habitatint.2005.09.005

Yu, L., Moriguchi, Y., Nakatani, J., Zhang, Q., Li, F., He, W., & Li, G. (2019). Environmental Impact Assessment on the recycling of waste LCD panels. *ACS Sustainable Chemistry & Engineering, 7*(6), 6360–6368. DOI: 10.1021/acssuschemeng.9b00119

ADDITIONAL READING

Astrup, T. F., Tonini, D., Turconi, R., & Boldrin, A. (2014). Life cycle assessment of thermal Waste-to-Energy technologies: Review and recommendations. *Waste Management (New York, N.Y.), 37*, 104–115. DOI: 10.1016/j.wasman.2014.06.011 PMID: 25052337

Christensen, T. H. (2010). *Solid Waste Technology & Management*. https://doi.org/ DOI: 10.1002/9780470666883

Eriksson, O., Reich, M. C., Frostell, B., Björklund, A., Assefa, G., Sundqvist, J., Granath, J., Baky, A., & Thyselius, L. (2004). Municipal solid waste management from a systems perspective. *Journal of Cleaner Production, 13*(3), 241–252. DOI: 10.1016/j.jclepro.2004.02.018

Finnveden, G., Hauschild, M. Z., Ekvall, T., Guinée, J., Heijungs, R., Hellweg, S., Koehler, A., Pennington, D., & Suh, S. (2009). Recent developments in Life Cycle Assessment. *Journal of Environmental Management, 91*(1), 1–21. DOI: 10.1016/j.jenvman.2009.06.018 PMID: 19716647

Gavrilescu, M., Ghinea, C., Petraru, M., Simion, I. M., Sobariu, D., & Bressers, H. T. A. (2014). Life cycle assessment of waste management and recycled paper systems. *Environmental Engineering and Management Journal, 13*(8), 2073–2085. DOI: 10.30638/eemj.2014.230

Laurent, A., Bakas, I., Clavreul, J., Bernstad, A., Niero, M., Gentil, E., Hauschild, M. Z., & Christensen, T. H. (2013). Review of LCA studies of solid waste management systems – Part I: Lessons learned and perspectives. *Waste Management (New York, N.Y.), 34*(3), 573–588. DOI: 10.1016/j.wasman.2013.10.045 PMID: 24369845

Sandin, G., & Peters, G. M. (2018). Environmental impact of textile reuse and recycling – A review. *Journal of Cleaner Production, 184*, 353–365. DOI: 10.1016/j.jclepro.2018.02.266

Taheri, M., Gholamalifard, M., Ghazizade, M. J., & Rahimoghli, S. (2014). Environmental impact assessment of municipal solid waste disposal site in Tabriz, Iran using rapid impact assessment matrix. *Impact Assessment and Project Appraisal, 32*(2), 162–169. DOI: 10.1080/14615517.2014.896082

Tushar, Q., Sun, W., Zhang, G., Navaratnam, S., Hou, L., & Giustozzi, F. (2023). Evolution in impacts assessment for managing and recycling of waste: A scientometric analysis. *Journal of Cleaner Production, 430*, 139685. Advance online publication. DOI: 10.1016/j.jclepro.2023.139685

KEY TERMS AND DEFINITIONS

Carbon Footprint: A common method to measure the amount of greenhouse gases generated by a product, process, or activity expressed in units of CO 2 -equivalent.

Circular Economy: A model of economy that would ensure efficiency in resource utilization by reducing waste and encouraging reuse, recycling and regeneration of materials.

Environmental Burden: The obstructive environmental effects such as air pollution, greenhouse gas emissions, water contamination, and land degradation caused by waste management activities.

Environmental Impact Assessment (EIA): It is a step-by-step procedure applied to the assessment of the possible environmental impact of a project, policy or system prior to implementation.

Environmental Indicators: These are the mediums for quantitative measures (e.g., energy use, emissions, water consumption) in order to measure or assess environmental performance.

Extended Producer Responsibility (EPR): A policy that can be followed which holds manufacturers accountable in the entire lifecycle of their products including the management of post-consumer waste products.

Integrated Waste Management (IWM): A comprehensive approach combining multiple waste treatment methods (recycling, composting, energy recovery, and disposal) to achieve optimal environmental outcomes.

Landfill Leachate: Water filtered down waste in a landfill to form liquid that may contain toxic chemicals in the soil and groundwater.

Life Cycle Assessment (LCA): One of the ways to determine environmental impacts related to all steps of a product life, including raw material extraction, and disposal or recycling.

Municipal Solid Waste (MSW): Home, commercial and institutional generated waste such as organic, recyclable and non-recyclable, so basically waste from households, businesses, and shops is known as municipal solid waste (MSW).

Recycling: It is a well-known way and a method to gather, transform and turn waste materials into alternative products to decrease the excessive production as well as consumption of resources and environmental pollution.

Resource recovery: The extraction of useful materials or energy from waste streams to reduce the need for virgin resources.

Waste Management Hierarchy: A conceptual model that places emphasis on waste prevention, reduction, reuse, recycling, recovery and disposal in the decreasing order of environmental preference. It structures the various / different waste management and treatment options.

Waste-to-Energy (WTE): It includes incineration technologies that transform waste materials to useful energy like electricity, heat or fuel.

Chapter 2

The Role of English in Raising Global Awareness About E-Waste Management:

Leveraging the Power of a Global Language to Drive Environmental Action

Iqra Anwar
http://orcid.org/0009-0003-5832-995X
Government College University, Faisalabad, Pakistan

ABSTRACT

Electronic waste (e-waste) is the fastest-growing waste stream worldwide, posing severe environmental, economic, and health risks. Yet global awareness and coordinated policies remain limited, especially in low- and middle-income nations where informal recycling dominates. This chapter explores how English, as the world's lingua franca, plays a vital role in raising awareness of e-waste management and driving transnational policy change. Drawing on global data, case studies, and communication theory, it argues that English connects scientific research, treaties, NGO campaigns, and media coverage, enabling diffusion of best practices, policy transfer, and civil society mobilization. However, overreliance on English risks marginalizing non-anglophone populations and widening knowledge gaps. The chapter highlights opportunities and constraints, showing how successful campaigns

DOI: 10.4018/979-8-2600-1960-3.ch002

use a cascade model: English seed documents translated and adapted into local contexts. It proposes inclusive strategies—bilingual publishing, plain language, and participatory translation.

1. INTRODUCTION

Electronic waste (e-waste) has become one of the most topical environmental problems of the twenty-first century. Electronic wastes have been increasing at a rate that can only be described as alarming considering that global use of electronic devices is increasing rapidly as a result of fast technological progress, the lifecycle of products is shortening as well as consumer demands. The 2024 Global E-Waste Monitor shows that the total amount of e-waste produced in the world in 2022 reached almost 62 million metric tonnes, which is 82 percent higher than the one produced in 2010 (Baldé et al., 2024). It has been predicted that this amount will rise to over 80 million tonnes by 2030 making e-waste the most rapidly increasing domestic waste stream in the world.

The situation is a twofold one since e-waste has not only the precious components (gold, copper, elements of rare earths, and so on) but also dangerous components (lead, mercury, cadmium, brominated flame retardants). The losses caused by mismanagement amount to USD 91 billion per year of economic value and threaten the life and health of people and the environment severely (UNEP, 2023). The informal recycling activities in low- and middle-income countries (LMICs) expose the vulnerable population to the dangers of toxic pollutants, and most people are unaware of safe disposal methods.

It is against this backdrop that increasing awareness of e-waste in the global arena is the most crucial. But awareness does not only concern information release, it is about border framing, communication, and comprehension. It is here that language is a determining element. English language as the lingua franca of science, diplomacy, and commerce has no equal or superior reach and power. It carries most international treaties, scholarly publications and environmental campaigns. This capacity to bring global discourses together around e-waste and at the same time be translated into local languages renders the English language unparalleled in its power to act policy.

The chapter discusses the interface of language and environmental governance by looking at the use of English to create awareness about e-waste management to the whole world. It claims that English allows the world to coordinate, spread the best practices, and have a platform where searching campaigns are carried out. Simultaneously, excessive use of the English language threatens to marginalize non-anglophone communities and reinforce knowledge inequality. The chapter provides means to capitalise on more inclusively leveraging English to achieve environmental action and policy change based on trend analysis, case studies, and policy systems.

2. THE GLOBAL LANDSCAPE OF E-WASTE: TRENDS, RISKS, AND GAPS

2.1 Historical Growth and Projections

The historical growth and projections of the nation are presented in 2.1.

The production of e-waste has been reflective of the tendencies of digitization, urbanization and consumerism across the world. The increase of the e-waste by two times in the last decade, between 33 million tonnes in 2010 and 62 million tonnes in 2022, is indicative of the boom of the economy and the culture of disposal common to the modern societies (Forti et al., 2022). The greatest producer is Asia with 29 million tonnes in 2022, then the Americas (14 million tonnes), Europe (12 million tonnes), Africa (3 million tonnes), and Oceania (1 million tonnes).

Table 1. Regional Breakdown of E-Waste Generation and Recycling Rates (2022)

Region	E-Waste Generated (Mt)	Formal Recycling Rate (%)
Asia	29.0	18.2
Europe	12.0	42.8
Americas	14.0	24.7
Africa	3.0	0.9
Oceania	1.0	7.0

Note. Adapted from Baldé et al. (2024). Data compiled by the Global E-Waste Monitor.

The disparities across regions highlight the dual challenge of awareness and infrastructure. While Europe achieves relatively high recycling rates due to strong regulations and extended producer responsibility (EPR) schemes, Africa and parts of Asia remain heavily dependent on informal recycling.

Figure 1. Global E waste forecast

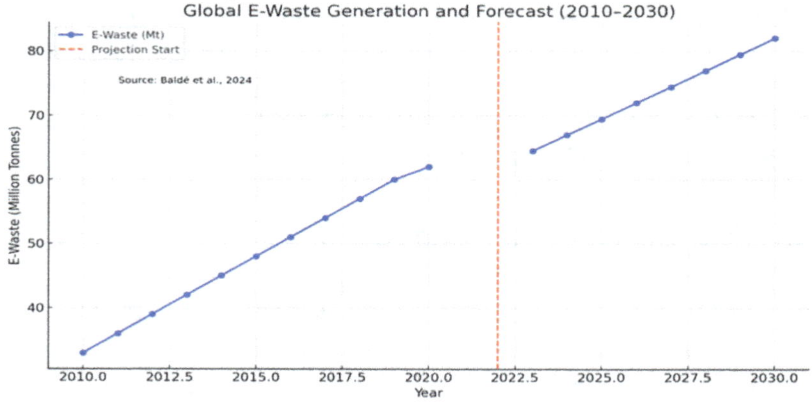

As shown in *Figure 1*, global e-waste generation has increased sharply from 33 million tonnes in 2010 to 62 million tonnes in 2022, with projections indicating more than 82 million tonnes by 2030 (Baldé et al., 2024).

2.2 Environmental and Public Health Impacts

E-waste also contains more than 1,000 various substances, most of which are hazardous. Lead in cathode ray tubes, mercury in fluorescent lamps, and cadmium in semiconductors are some examples of toxic materials that may leach in soil and groundwater when not properly disposed (WHO, 2022). Cables are also burned openly and this releases dioxins and furans, and workers are also exposed to serious chemical hazards when acid is being used to reclaim metals.

It is disproportionately impacted on children, women and informal laborers. According to the estimation of WHO (2022), there are over 18 million children and adolescents who engage in informal recycling of e-waste, resulting in high chances of respiratory diseases, cognitive impairment, and developmental disorders.

In addition, transboundary flows increase inequity. The e-waste is exported by high-income countries to the LMICs usually in the name of used electronic. Arrived there, devices which cannot be refurbished are discarded in open dumps or crude recycling yards with environmental costs being externalized to communities already in a vulnerable position (UNEP, 2023).

2.3 Policy and Governance Gaps

Although 81 countries have implemented e-waste laws, implementation and coverage is disproportionate. Not all (only 48) countries have collection rate goals, and less than 40 countries impose recycling regulations (Forti et al., 2022). In most instances, policies are on paper but not implemented because they are not well funded, there is no sufficient technical skills and poor monitoring measures.

The international legal instrument is the Basel Convention on the Control of Transboundary Movements of Hazardous Wastes and their Disposal of 1989. However, there is a patchy enforcement, and e-waste is exposed to cross-border flows under the category of donation or reuse (Clapp, 2021).

The slight but significant role is played by language. Most of the international conventions such as the Basel Convention are written in English before being translated into other UN languages. To smaller countries where translation ability is not high, this presents a problem in terms of interpretation, implementation, and compliance. This may be undermined by misunderstandings or ambiguities and result in the practices in different countries being different.

2.4 Economic Potential of Proper E-Waste Management

In addition to the environmental hazards, e-waste also indicates unexploited economic potential. The e-waste value of recoverable materials was estimated at USD 91 billion in 2022, consisting of 31 million tonnes of iron, 17 million tonnes of plastics and 2.5 million tonnes of copper (Baldé et al., 2024). Effective recycling would help diminish the use of the virgin mining, reduce the level of greenhouse gas emissions, and generate green jobs.

Nevertheless, this potential can only be actualized through concerted efforts across the globe with an emphasis on proper communication. The awareness campaigns must focus on the risks of e-waste and the potential economic advantages of recycling and the use of the circular economy strategies as well. Use of English as global language of business and trade plays a significant role in positioning e-waste as an economical opportunity and not a pure environmental liability.

Figure 2. Global Map of E-Waste Legislation (2023)

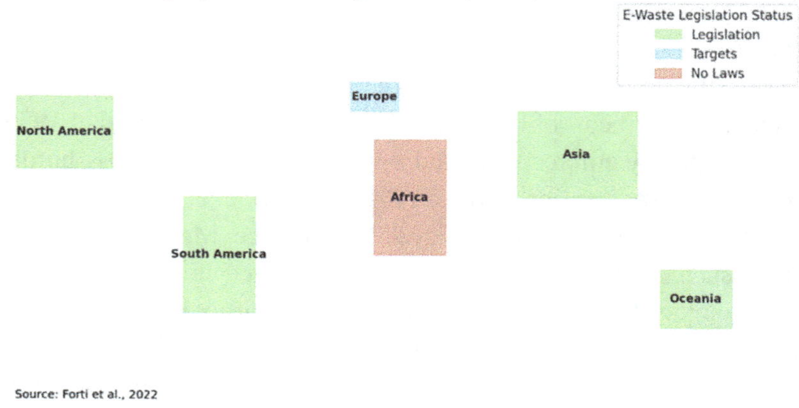

Figure *2* illustrates the uneven legislative landscape, where Europe has adopted collection targets, several regions maintain legislation without targets, and Africa remains largely without formal laws (Forti et al., 2022).

3. TRANSITION: FROM NUMBERS TO NARRATIVES

The global landscape of e-waste reveals a paradox: while the scale of the problem is well documented in English-language reports and statistics, awareness at the local level remains fragmented. Policymakers in developing countries may read UNEP or WHO reports in English, but informal recyclers on the ground often operate without access to such information in their own languages. Bridging this gap requires not only data but narratives that resonate globally and locally.

4. THE POWER OF LANGUAGE IN ENVIRONMENTAL AWARENESS AND POLICY

4.1 Communication Barriers in Environmental Issues

Environmental governance is inherently complex. It blends scientific, economic, legal, and social dimensions into a single policy arena. Communicating such complexity to diverse stakeholders is difficult even within one country, let alone globally. Three major barriers stand out:

1. **Technical Jargon.** Terms like *extended producer responsibility* (EPR), *circular economy*, or *transboundary flows* may be clear to specialists but remain opaque to the general public.
2. **Audience Diversity.** Policymakers, academics, business leaders, and local communities consume information differently. A technical report might influence regulators but fail to resonate with informal recyclers who lack formal education.
3. **Translation Gaps.** Concepts in English often lack exact equivalents in local languages. When translated literally, nuances may be lost or distorted, leading to misunderstanding and misapplication.

These barriers emphasize why the medium of communication is as important as the message. Without accessible language, awareness campaigns risk alienating the very groups they aim to mobilize.

4.2 English as a Global Lingua Franca

English has emerged as the dominant language of globalization. Approximately 1.5 billion people speak English at some level, and it is the official or working language of most international organizations (Crystal, 2019). In the environmental domain, English has several distinct advantages:

- **Academic Dominance.** Over 80% of peer-reviewed environmental science publications are in English, ensuring global dissemination of research findings (Montgomery, 2013).
- **Policy Instruments.** Treaties, conventions, and UN reports are first drafted in English, before translation into other official languages.
- **Media and Journalism.** International news outlets such as *BBC*, *Reuters*, and *The Guardian* report primarily in English, giving issues global visibility.
- **Network Effects.** English connects NGOs, donors, and researchers across regions, fostering partnerships that transcend borders.

This dominance has created what some scholars call the "Anglophone information cascade": knowledge and policy ideas spread rapidly through English, influencing national agendas and shaping local interventions.

4.3 Strengths of Using English in E-Waste Awareness

- **Standardization.** Using English ensures that key concepts (e.g., "hazardous waste," "formal recycling," "take-back schemes") are consistently defined across reports and treaties.

- **Visibility.** Global reports in English gain media traction, ensuring policymakers cannot ignore the issue.
- **Benchmarking.** Countries can compare themselves against global data sets compiled in English, such as the *Global E-Waste Monitor.*
- **Donor Alignment.** International funders require proposals in English, aligning local projects with global priorities.

4.4 Risks and Limitations of English Dominance

However, the hegemony of English also carries risks.

1. **Exclusion.** Communities with limited English literacy may be cut off from key knowledge. Informal workers in Ghana's Agbogbloshie e-waste site, for example, rarely have access to English reports about the hazards they face.
2. **Translation Errors.** Literal translations can lead to misinterpretation. A policy term like "extended producer responsibility" may be rendered in local languages as "producer's obligation," missing nuances about lifecycle management.
3. **Cultural Mismatch.** Campaigns framed in Western terms may not resonate in other contexts. For instance, appeals to "individual consumer responsibility" may conflict with collectivist cultural norms in parts of Asia and Africa.
4. **Knowledge Hegemony.** Overreliance on English can marginalize indigenous knowledge and local practices, reinforcing power imbalances between Global North and South.

Thus, English is best understood as a powerful amplifier — but one that must be complemented with inclusive, multilingual strategies.

5. CASE STUDIES: ENGLISH-LED GLOBAL CAMPAIGNS ON E-WASTE

5.1 The Basel Convention

The Basel Convention on the Control of Transboundary Movements of Hazardous Wastes (1989) is the cornerstone of international e-waste governance. Drafted in English, it set legally binding obligations to minimize hazardous waste generation and control cross-border shipments.

Over time, the Basel Secretariat has issued *technical guidelines on e-waste management* (2019, 2022 editions), published first in English. These guidelines became templates for national laws in countries such as Nigeria, India, and Brazil.

Policymakers in these countries adapted English guidelines into local legislation, but ambiguities in translation sometimes produced enforcement challenges.

This illustrates the dual role of English: it provides a common foundation but requires localization to be effective.

5.2 UNEP and the International E-Waste Management Network

The United Nations Environment Programme (UNEP) established the International E-Waste Management Network (IEMN) in 2011. Its working language is English, enabling experts from over 30 countries to exchange practices, draft policy frameworks, and build capacity.

Through IEMN, best practices such as *extended producer responsibility* were shared from Europe to Asian and African nations. Reports were distributed in English but later adapted into French, Spanish, and local languages for on-the-ground implementation.

UNEP's training modules for policymakers also rely heavily on English terminology, which ensures technical precision but poses accessibility issues for participants from non-English-speaking backgrounds.

5.3 Academic and NGO Collaborations

Academic journals like *Waste Management* and *Resources, Conservation & Recycling* publish predominantly in English. Research on informal recycling in India or Ghana, once published in English, gains visibility in the global discourse, attracting funding and influencing donor strategies.

NGOs like the Global E-Waste Statistics Partnership publish the *Global E-Waste Monitor* in English, making it the gold standard reference. Its tables and graphs are widely cited in policy briefs, conference presentations, and media coverage worldwide.

Table 2. Prominent English-Language Institutions Driving E-Waste Awareness

Institution / Report	Primary Language	Global Impact
Basel Convention Secretariat	English	Treaty, technical guidelines, policy frameworks
UNEP International E-Waste Network	English	Knowledge exchange across 30+ countries
Global E-Waste Monitor (GESP)	English	Benchmark data, global media visibility
Academic Journals (*Waste Management*)	English	Dissemination of scientific evidence
International NGOs (BAN, GAIA)	English	Advocacy campaigns, funding mobilization

5.4 Media and Public Awareness Campaigns

English-language media outlets amplify e-waste issues beyond specialist circles. For example:

- The *BBC* documentary *The World's E-Waste Crisis* (2019) reached millions of viewers globally.
- *The Guardian* ran a series of articles in 2020 linking e-waste to climate change, spurring debate in policy circles.
- Social media campaigns with English hashtags (#Ewaste, #CircularEconomy) trended globally, later adapted into multilingual variants.

This visibility matters because media pressure often pushes governments to act. English thus acts as a "megaphone" for environmental advocacy.

5.5 Regional Adaptations: The Case of India and Nigeria

India adopted its first *E-Waste (Management and Handling) Rules* in 2011, heavily influenced by English-language guidelines from the Basel Convention and EU directives. While the rules were drafted in English, implementation required translation into Hindi and other regional languages for industry workshops and public campaigns.

Similarly, Nigeria's National Environmental Standards and Regulations Enforcement Agency (NESREA) partnered with UNEP to develop e-waste legislation in English. However, public awareness campaigns in Lagos were later translated into Yoruba and Pidgin English to reach local audiences.

These cases demonstrate the pipeline effect: English provides the initial legal and technical template, but localized communication ensures grassroots adoption.

5.6 Lessons from the Case Studies

The case studies highlight three key insights:

1. **English as a Policy Originator.** Global treaties, guidelines, and academic research originate in English, shaping national agendas.
2. **Translation as a Necessity.** Without translation, English-only documents remain confined to elites. Translation and adaptation make them actionable at the community level.
3. **Media Amplification.** English media ensures visibility but must be complemented by local storytelling.

6. TRANSITION: TOWARD MECHANISMS OF POLICY TRANSFER

The analysis so far shows that English plays a critical role in setting the agenda, shaping global narratives, and mobilizing action. Yet the mechanisms by which English facilitates actual policy transfer and behavior change remain underexplored.

The next section (Part III) will examine how English functions as an awareness and policy transfer mechanism, the behavioral economics of communication, and the challenges of ensuring that English-based global narratives lead to local environmental improvements.

6.1 Information Diffusion and Network Effects

One of English's greatest strengths is its ability to act as a carrier of information across borders. When global e-waste statistics, policy briefs, or academic studies are published in English, they are rapidly cited, reprinted, and shared across a variety of institutions. This creates what economists call network effects: the value of information increases as more actors can access and share it within a common language.

For instance, the *Global E-Waste Monitor* published in English becomes the reference document for governments, NGOs, and donors alike. Its data are repeatedly cited in national strategy documents, academic papers, and advocacy campaigns. Because English dominates both academic and policy spheres, knowledge spreads faster and more uniformly than if multiple languages were used without translation.

6.2 Policy Transfer and Benchmarking

English serves as the common denominator for policy transfer—the process by which governments borrow ideas, policies, and practices from each other.

- **Templates.** English-language guidelines from the European Union and the Basel Convention provide templates for national legislation.
- **Benchmarking.** Countries can measure their progress against international targets set in English reports. For example, the European Commission's documentation of its 65% e-waste collection target is frequently cited by policymakers in Asia and Africa as a benchmark.
- **Legitimacy.** English references confer legitimacy. When a national government cites an English-language UN report, it signals alignment with global norms.

Thus, English enables countries to "borrow legitimacy" by linking their domestic policies to internationally recognized English-language standards.

6.3 Funding, Grants, and International Coordination

Another mechanism is access to funding. Almost all international donor agencies—including the World Bank, the Global Environment Facility (GEF), and the European Union—require project proposals, budgets, and evaluation reports in English.

This creates a feedback loop:

1. Local NGOs prepare English-language proposals citing English-language reports.
2. Donors fund projects framed within global discourses.
3. Project reports in English are published, reinforcing the cycle.

By making English the language of funding, donors unintentionally reinforce its centrality in awareness-building. NGOs that cannot operate in English are often excluded, highlighting the importance of multilingual capacity-building.

6.4 Civil Society Mobilization

Civil society campaigns frequently begin in English before spreading globally. Hashtags such as #Ewaste, #CircularEconomy, and #RightToRepair first gained traction on Twitter (now X) in English, before being localized. English provides the initial "signal" that resonates across borders, then local adaptations ensure inclusivity.

Transnational advocacy networks (Keck & Sikkink, 1998) often rely on English for coordination. Groups like the Basel Action Network (BAN) issue reports in English exposing illegal e-waste shipments, which then trigger media coverage and government investigations.

6.5 Behavioral Economics of Language

Communication effectiveness is not only structural but also psychological. English as a global medium can influence behavior through framing effects:

- **Loss vs Gain Framing.** English reports often emphasize economic value lost in e-waste (USD 91 billion annually). This "loss frame" motivates policymakers by highlighting opportunity costs.
- **Norm Framing.** Global English publications create the perception that "responsible e-waste management is the global norm," encouraging governments to conform.

- **FOMO (Fear of Missing Out).** Countries may act to avoid reputational loss if English reports rank them poorly in recycling rates.

By framing narratives in English, institutions shape not just awareness but the incentives and behaviors of policymakers.

7. CHALLENGES: FROM ENGLISH MESSAGING TO LOCAL IMPACT

While English enables global communication, its dominance also creates significant challenges in ensuring equitable and effective environmental governance.

7.1 Translation, Local Relevance, and Cultural Fit

The first challenge lies in translation. English terminology in environmental governance often lacks direct equivalents in local languages. For example:

- "Circular economy" may be rendered literally but fail to convey cultural connotations of reuse and repair in societies where informal recycling already exists.
- "Extended Producer Responsibility" (EPR) has been mistranslated in some contexts as simply "producer's duty," omitting the lifecycle responsibility for post-consumer waste.

These translation gaps can distort meaning, leading to partial or incorrect implementation of global policies.

Moreover, cultural framing matters. An English-language campaign that emphasizes "individual responsibility" for recycling may resonate in Western contexts but clash with collectivist cultures where community norms, not individual action, drive behavior.

7.2 Illiteracy, Digital Divide, and Access

Another limitation is access. English literacy is uneven across the world. In many LMICs, policymakers and urban elites may operate in English, but the broader population does not. Informal recyclers, who are most exposed to e-waste hazards, often lack any access to English-language resources.

This is compounded by the digital divide. Many English-language awareness campaigns are disseminated online, through PDFs, websites, or social media. Yet millions of people in Africa and South Asia lack reliable internet connectivity, leaving them excluded from English-based awareness efforts.

7.3 Linguistic Dominance and Knowledge Hegemony

Critics argue that English dominance perpetuates knowledge hegemony. Environmental knowledge framed in English often reflects Global North perspectives, sidelining indigenous practices.

For example, communities in India have long practiced repair and reuse traditions, which align with circular economy principles. However, when global reports in English present the circular economy as a “new paradigm,” they erase local histories and create the perception that innovation originates only in Anglophone contexts.

This risks alienating local actors and diminishing the legitimacy of global campaigns.

7.4 Overload, Fatigue, and Credibility

English-based communication also faces the risk of information overload. Policymakers and NGOs are inundated with English-language reports, guidelines, and toolkits. Without effective prioritization, this abundance leads to fatigue, diminishing the impact of campaigns.

Credibility is another issue. If English-language campaigns are perceived as external impositions rather than locally grounded, they may be resisted. For instance, some governments in the Global South dismiss English-language NGO reports as biased or agenda-driven, especially if they conflict with national economic priorities.

7.5 Inequality in Participation

Finally, English-centric governance structures create unequal participation. In international conferences, delegates with strong English skills dominate discussions, while those less fluent may remain silent. This skews outcomes toward Anglophone perspectives.

In academic publishing, researchers from non-English-speaking countries face higher barriers to entry, limiting the diversity of knowledge in global e-waste discourse (Tardy, 2021).

Table 3. Challenges of English Dominance in E-Waste Communication

Challenge	Manifestation Example	Consequence
Translation Gaps	Misinterpretation of "EPR"	Policy errors
Cultural Mismatch	Individual vs collectivist framing	Weak resonance
Literacy/Access	Informal workers excluded	Awareness gaps
Knowledge Hegemony	Circular economy framed as "Western innovation"	Marginalization
Overload and Fatigue	Dozens of English toolkits	Decreased impact
Inequality in Participation	English dominance at conferences	Skewed outcomes

7.6 Balancing Global English with Local Multilingualism

The core challenge, then, is balance. English is indispensable for global coordination, but its effectiveness depends on how well it is integrated with local languages, contexts, and practices. Awareness in English must cascade into multilingual forms: translations, visual campaigns, oral communication, and culturally resonant narratives.

8. TRANSITION: TOWARD SOLUTIONS AND STRATEGIES

The preceding sections highlight a paradox: English is both the enabler and barrier of global awareness on e-waste. Its global reach standardizes discourse and mobilizes resources, but its exclusivity risks alienating those who most need awareness.

The final part of this chapter (Part IV) will propose strategies to leverage English more effectively while ensuring inclusivity, describe a hypothetical multinational campaign, and suggest metrics for evaluating impact.

To maximize the role of English in raising global awareness about e-waste while minimizing exclusion, a set of integrated strategies is required. These strategies balance global communication power with local cultural resonance.

8.1 Bilingual and Multilingual Publishing

Reports and campaigns should be launched in English for global visibility, but immediately paired with translations into major world and regional languages (e.g., Arabic, Mandarin, French, Spanish, Hindi, Swahili). Crucially, translations must be contextualized, not literal, ensuring key terms are meaningful in local idioms.

Example: Instead of translating *extended producer responsibility* word-for-word, campaigns should adapt it as "manufacturer's duty from cradle to grave" in languages where lifecycle concepts are better understood.

8.2 Plain English and Visual Communication

Global e-waste publications are often heavy with jargon. Shifting toward plain English ensures wider accessibility even among non-native speakers. Plain English emphasizes short sentences, active voice, and minimal technical terminology.

Visual storytelling also matters: infographics, flowcharts, and animations in English can be easily localized into other languages. Visuals reduce reliance on linguistic proficiency and can communicate across literacy divides.

8.3 Local Partnerships and Co-Creation

International organizations should partner with local NGOs, universities, and community leaders to co-create campaigns. English-based materials can serve as the backbone, but local partners should adapt narratives into culturally resonant forms — songs, theater, radio dramas, or storytelling traditions.

This approach ensures bottom-up legitimacy while maintaining global alignment.

8.4 Digital Platforms with Built-In Translation

Advances in AI translation make it feasible to develop multilingual platforms. A global e-waste portal could host English reports but automatically generate high-quality translations in multiple languages, including less widely spoken ones. Feedback loops would allow local actors to annotate or improve translations, creating a participatory knowledge commons.

8.5 Training and Capacity Building

Workshops and online courses can be designed in English but paired with interpretation or subtitling. Training government officials, journalists, and community leaders in interpreting English-language policy documents empowers them to act as "knowledge bridges" to their communities.

8.6 Media Engagement and Storytelling

English media should not only publish reports but tell human-centered stories about e-waste workers, communities, and innovators. Narratives in English can attract global attention, funding, and solidarity, while inspiring local media to reframe the same stories in regional languages.

8.7 Two-Way Communication

Finally, global English discourse should not be a one-way pipeline. Mechanisms must exist for feedback from local actors. For instance, informal recyclers' experiences in Ghana or India should be documented in local languages, then translated upward into English to enrich global reports. This ensures mutual learning and avoids the perception of English dominance as "top-down."

9. HYPOTHETICAL MULTINATIONAL CAMPAIGN

To illustrate how English could be leveraged effectively, consider a Global E-Waste Awareness Initiative (GEWAI) coordinated by UNEP, NGOs, and academic partners.

9.1 Phase 1: English Global Report

GEWAI launches with an English-language *State of the World's E-Waste 2030* report. Written in plain English and rich in visuals, it highlights economic losses (USD 100 billion), environmental impacts, and best practices.

9.2 Phase 2: Multilingual Adaptation

Within three months, the report is translated into 20 major languages, including Hindi, Urdu, Hausa, and Swahili. Local case studies are added to ensure contextual resonance.

9.3 Phase 3: National Workshops

Governments receive English templates for policy development but are encouraged to co-host workshops with local NGOs. At these workshops, English terms are debated, translated, and adapted into local policy vocabularies.

9.4 Phase 4: Community Campaigns

Local NGOs roll out campaigns in local languages using posters, radio shows, and mobile apps. English hashtags (#Ewaste2030) are paired with local hashtags to ensure both global visibility and community connection.

9.5 Phase 5: Feedback and Iteration

Local actors share impact stories back to the global platform. These are translated into English and added to global databases, closing the communication loop.

10. MEASURING IMPACT: METRICS, EVALUATION, AND FEEDBACK

To ensure accountability, campaigns must define measurable outcomes.

10.1 Awareness Metrics

- Pre- and post-campaign surveys of awareness in target populations.
- Tracking English hashtag trends vs local language hashtags on social media.
- Downloads and translations of English-language toolkits.

10.2 Policy Outcomes

- Number of new or revised national e-waste laws citing English-language reports.
- Inclusion of English-based benchmarks (e.g., Global E-Waste Monitor statistics) in government policy documents.

10.3 Practical Outcomes

- Increases in e-waste formally collected and recycled.
- Reduction in informal burning or dumping practices.
- Growth in circular economy enterprises (repair shops, recycling plants).

10.4 Linguistic Analytics

- Number of translations produced and distributed.
- Diversity of languages represented in feedback loops.
- Citation analysis of English vs translated documents.

Table 4. Example Metrics for Evaluating Global E-Waste Campaigns

Category	Metric Example	Indicator of Success
Awareness	% increase in public awareness surveys	>30% increase
Policy	# of new laws referencing English reports	20+ countries
Practical Impact	% rise in formal recycling	+15% globally
Linguistic Reach	# of translations distributed	25+ languages

11. CONCLUSION AND FUTURE OUTLOOK

English plays a paradoxical role in the global struggle against e-waste. On one hand, it is the unifying language that enables data standardization, global visibility, and international coordination. Reports, treaties, and campaigns in English provide the backbone for policy diffusion and mobilization. On the other hand, English risks reinforcing inequalities, excluding non-anglophone communities, and perpetuating knowledge hierarchies.

To harness its benefits without deepening inequities, English must function as both a global amplifier and a local enabler. The strategies outlined—bilingual publishing, plain English communication, local partnerships, digital translation platforms, and feedback loops—point to a more inclusive model.

Looking forward, AI-assisted multilingualism offers promise. English can remain the core medium of global communication, but real-time translation technologies will increasingly allow local voices to feed upward. Imagine a global dashboard where community reports in Hausa, Tamil, or Quechua are instantly translated into English, influencing global policy debates.

E-waste is both a material and linguistic challenge. It requires managing toxic flows of discarded electronics and addressing the communicative flows of information and awareness. By leveraging English strategically — not as a gatekeeper but as a bridge — the global community can accelerate the transition toward sustainable e-waste management, protect vulnerable populations, and unlock economic value.

Ultimately, the goal is not to make the world "speak English," but to use English as a platform for mutual translation, inclusive dialogue, and collective action. In doing so, English becomes more than a language of global power; it becomes a tool of global responsibility.

REFERENCES

Baldé, C. P., Forti, V., Gray, V., Kuehr, R., & Stegmann, P. (2024). *The global e-waste monitor 2024.* United Nations University, United Nations Institute for Training and Research, International Telecommunication Union, & International Solid Waste Association.

Clapp, J. (2021). *Waste trading among rich nations: Building a new theory of environmental regulation.* Routledge.

Crystal, D. (2019). *English as a global language* (3rd ed.). Cambridge University Press.

Forti, V., Baldé, C. P., & Kuehr, R. (2022). *E-waste statistics: Global, regional, and national trends.* United Nations University.

Keck, M. E., & Sikkink, K. (1998). *Activists beyond borders: Advocacy networks in international politics.* Cornell University Press.

Montgomery, S. L. (2013). *Does science need a global language? English and the future of research.* University of Chicago Press. DOI: 10.7208/chicago/9780226010045.001.0001

Tardy, C. (2021). Academic publishing and English dominance: Challenges for scholars in multilingual contexts. *Journal of English for Academic Purposes, 53,* 101034. DOI: 10.1016/j.jeap.2021.101034

United Nations Environment Programme (UNEP). (2023). *E-waste management and the circular economy.* UNEP-IETC.

World Health Organization (WHO). (2022). *Children and digital dumpsites: E-waste exposure and child health.* WHO.

Chapter 3
Optimizing Waste Management in Smart Cities:
Tracing and Tracking Solutions

Neeru Sidana
http://orcid.org/0000-0003-1087-4214
Amity University, Noida, India

Disha Tiwari
Amity University, Noida, India

Gaurangii Hari
Amity University, Noida, India

ABSTRACT

In the densely populated urban hubs of India, the sight of brimming garbage containers and the pervasive stench of waste have become a routine aspect of daily life for many. This predicament is not limited to India alone, but is a global issue, with the projected production of municipal solid waste expected to reach 3.8 billion tonnes by 2050. Accumulations of uncollected waste can give rise to the proliferation of disease-carrying pests and severe air and water contamination, which have a severe impact on public health. Furthermore, inadequate disposal methods, such as the open incineration of waste, make a significant contribution to air pollution, releasing harmful toxins and greenhouse gases into the atmosphere. The rapid pace of urbanization exacerbates this issue, overwhelming traditional waste disposal methods and creating severe environmental and health risks. As cities expand and populations increase, the strain on conventional waste disposal systems intensifies.

DOI: 10.4018/979-8-2600-1960-3.ch003

INTRODUCTION

Today's cities across the globe in the twenty first century are in the process of transformation due to new technologies, increase in population density and the quest for sustainable urbanization. It has brought about the concept of "Smart Cities," education portraying a vision where enhanced technological features are used to enhance the quality, sustainability as well as effectiveness of the urban construction (Ojadi et al., 2025)

That is why the concept of "smart cities" appeared within the context of the modern phase of urbanization and technological advancement, with the aim to finally revolutionize the functioning as well as communication with and within the urban landscapes. Embracing everything from powering the smart grid to providing green and proactive public services, smart cities represent what is often referred to as technology and urban planning in contemporary cities correlated to effectively assist in meeting the multifaceted needs of fast-growing metropolitan areas (Gracias et al., 2023).

However, if it is expedited to the ultimate, smart cities are not just a collection of the novel technologies or progressive architecture; they represent an idea of the city and its development targeted for improving the quality of their populations' lives as well as their effectiveness, and sustainability (Subbanarasimha et al., 2025).

In numerous such trailblazing endeavours, one must cite India's Smart City Mission which is an ambitious project to revolutionize cities across the country. The preceding schemes formed the basis of the Smart Cities Mission and helped in the transition of managing and developing the Indian cities. They highlighted the intensity that the government of India has realized about tackling the multifaceted issues of urbanization and for supporting sustainable and inclusive development of cities and towns across India (Javed et al., 2021).

The phenomenon of urbanisation is rapidly phenomenon which is engulfing the Indian landscape with the cites growing at a phenomenal pace. Such growth comes with several problems mostly in the field of infrastructure and sustainability. Another challenge is mostly waste disposal., which is the major factor for hygiene and health in the urban areas as well as sustainability (Thakur et al., 2022).

A number of indices raise in connection with the growth of cities and the population such as the amount of build-up waste. The current models for waste disposal., riddled with numerous problems related to effectiveness and sustainability, are unable to cope with the challenges produced by constantly growing metropolitan areas. This is why in smart cities there is a necessity to use innovative technologies in waste management for several reasons. First of all, the matter of proper waste disposal is of great importance in preserving the health of a population. Lack of proper waste disposal results in the spread of diseases, infestation of pests and pollution of the environment (Faizan M., 2021). Secondly, sustainable waste management leads to recycling of wastes in addition to reducing the frequency of visits to the landfills and hence the carbon footprint on wastes disposal. At last, waste management when properly done can lead to considerable savings among the municipalities majorly because of lower costs in manpower and transportation provided by efficient automation and uncomplicated collection networks (Yadav et al., 2020).

Looking at the waste management aspect, the Smart Cities Mission picture now has the vision to change this scenario by adopting Intelligent waste management techniques that help in making use of technology intervention data analysis. Proper waste disposal is an essential factor that concerns urban sustainability since it entails the health of residents, the quality of citizens' environment, and the ability or suitability of cities for habitation. The realization of the mission has led to the use of several technology-based strategies and activities in the various fields (Yuvaraja, H., 2020).

Taking into consideration all the above-mentioned aspects, it can be concluded that the further development of modern society has encouraged some breakthroughs in waste management, but at the same time, contemporary world is still experiencing certain difficulties and obstacles in the path of a global and efficient waste management process. Difficulties due to initial costs of implementing complex processes form one of the hurdles. Another major challenge that has been identified is technological integration, which means that the different systems and gadgets in an organization should interconnect. There are also issues of security and privacy of data since smart system includes the collection of large amounts of data. Besides the support from the Government, public acceptance and participation are also important for these initiatives; however, rallying up a community engagement is not always easy (Yuvaraja, H., 2020).

In this context, this paper analyses the smart waste management techniques which have been adopted in a case study of the Surat to demonstrate how it has been conspicuous in incorporating smart techniques to solve its waste management issues.

However, this paper estimates that by 2024, there has been growth in the implementation of the Smart City Mission. This Waste management in these cities is going through several changes among them; Smart bins and sensors – collection trucks have been fitted with GPS, WtE technologies, big data & AI, The citizens' engagement platforms. Inherent to these efforts is the utilization of what may be referred to as ''technological solutions' understood as tools that assist in promoting improved efficiency as well as sustainability (Saha, A., 2020).

There are several approaches to implement waste management and overcome above challenges in developing smart cities as follows; Public Private Partnership, Pilot Projects and phased implementation, Community visibility and awareness, Policy and Regulatory framework, and Monitoring and Evaluation.

Smart City Mission in India is a revolutionary concept sewn with a vision to remodel the Urban structure and promote Sustainable, Inclusive and Integrated society. The incorporation of enhanced waste targeting strategies is crucial in the development of smart cities. The reach of its mission lies in the premise that by embracing technology and innovation coupled with collaboration, the vision has the capability to bear solutions to the urban's most critical issues, enhancing the lives of millions of people (Singh et al., 2022).

Due to the growing need for the implementation of Smart City Mission, more efforts should be made to overcome existing problems and take advantage of identified opportunities. Thus, as the population of dwellers in cities grow, there is need to enhance proper waste management that is efficient, productive and sustainable to environment. The use of technology and big data in smart cities ensures that the management of wastes from becoming major barriers of sustainability can be transformed into best shots at innovation.

REVIEW OF LITERATURE

Since waste management is still considered a immanent problem for rapidly developing cities, smart city strategizes as a constructive approach to the matter. 'Smart cities' means the active application of technology and big data tools in managing the life of cities, what refers to, in particular, more efficient waste disposal. The implication of technology in smart cities also aims at the effective, efficient and sustainable management of waste as a form of waste minimisation mechanism.

Actual ambitious concept in India is The Smart City Mission initiated at 2015 that encapsulates Indian long-term goals to transform urban environments through integrated approaches and deploy smart products. Adjacent to this goal., waste management is one of the strategies that seek to promote sustainable developments and improve the people's quality life within cities. Primary literature evaluates and investigates different aspects related to this mission such as technological issues, policy aspects and sociological issues in proper waste management affecting the general environment among others.

In the view of this, Kumar et al. (2017) has argued that waste management in India has its inhibitions with regard to sustainable and efficient management of wastes. The authors acknowledge some challenges that surround waste management with special consideration to the Indian context particularly in relation to infrastructure, policies and the general people attitude of waste disposal.

Moreover, Sabeet (2017) over the success of sustainable communal solid waste management in India focusing on the environmental and public health consequences of waste management failures. These conclusions explain the necessity for new and efficient ideas in waste management in the context of India.

In the paper by Rana &. Z. S. Rana & Kumar, &Singh, (2018), the challenges associated with development of smart cities with reference to the Indian situation, with involvement of multiple interrelated aspects of urban development including waste management is described. The authors identified some challenges which include; inadequate infrastructure; and Technology barriers that may hinder the shift towards smart cities. This insight should help one appreciate the fact that waste management should be considered as part of the larger strategies of developing smart cities.

Datta et al. (2018) expressed a detailed evaluation of the existing concerns that specify the biomedical waste management in India and its main focus, which is the appropriate disposal of healthcare waste. In the same context, the authors point out the risks which a society might likely be exposed to through poor biomedical wastes disposal thus pointing to a significant component of waste management that should be effectively addressed within the smart city framework. Chauhan et al. (2020) have extended the debate about circular economy principles and Industry 4. They identified citation, collection and disposal of 0 technologies in the area of healthcare waste disposal., however, due to the smart city development potential of the field, new solutions could exist.

Smart garbage management system is introduced by Jajoo et al. (2018) and expressively contributes to the goal of smart city. As for the key points of focus for the authors, these include the application of technology and data management in waste collection and disposal. This insight has the implication of inspiring the use of available technological innovation for improvement of waste management in smart cities.

According to Mani and Singh earlier works in 2016, the policy directions for the development of sustainable municipal solid waste management in India are detailed, with policies considered as required tools in tackling the issues that arise from waste management. The authors also emphasize that the government and other relevant authorities should incorporate social., economic and environments factors into the waste management regulations; the mentioned may be a possible direction of research and policy in smart cities.

Bharadwaj et al., (2016) give a conceptual model of an IoT based solid waste management system for smart city services. The authors also put an emphasis on how the application of IoT technology would be useful when it comes to trash management and its corresponding tracking and real-time services they can provide in connection to innovational approaches to the smart city's waste management process.

Concerning the smart city discussions, the advancements in the Internet of Things concepts are even more relevant in presenting progressive solutions to the conventional issues of waste. In some papers, the integration of smart sensors in waste bins, for example, is discussed on how IoT allows the monitoring of waste collection flexibility and the identification of the most efficient pick-up routes that minimize operating expenses and the environmental impact (M et al., 2020).

On the other hand, handling of industrial waste in addition to the municipal waste increases the challenge in the smart city waste management systems. Investigation also focuses on a successive employment of waste treatment and recycling facilities applying to the industrial belts to protect marine environment as well as support the circular economy. It is in this light that the accomplishment and implementation strategies of the efficient industrial waste management systems are argued to remain as a crucial element in the general framework of the smart sustainable city.

Also, academic conferences and journals concern the role of sustainable solid waste management practices in smart cities. Topics that are debated at such interfaces include ways and manner in which waste segregation, popularization of recycling, and waste reduction campaigns can be incorporated into smart cities (Open Access International Journal of Science & Engineering, 2016).

Finally, while analysing the barriers to waste management in smart cities, the literature also follows the decomposition of size and capacity planning of renewable energy sources linked with the waste management system.

The incorporation of technological advancement and application of policies and sustainable practices appears as the major theme in solving the hurdles associated with waste management. Further research opportunities could be focused on modern technologies using Internet of Things and data analysis to improve trash collection and recycling processes in smart cities.

Furthermore, more studies are required on the application of waste management in the overall framework of smart city development with the focus on the interaction of infrastructure, technology, and the environment. Altogether, one can observe that there is a comprehensive evaluation of waste management in smart cities that is available now to researchers and policymakers, thus laying the groundwork for further advances as well as measures in this sphere.

THE SMART CITIES MISSION

The Indian government introduced the Smart Cities Mission (S.C.M.), an innovative program for urban redevelopment and retrofitting, in 2015. Its main goal was to spur the creation of smart cities, which use innovation and technology to improve government, infrastructure, services, and general urban living standards. The genesis of this mission lies in recognizing the rapid pace of urbanization and the associated challenges such as inadequate infrastructure, environmental degradation, traffic congestion, and inadequate public services (Ministry of Housing and Urban Affairs, 2024).

According to the Ministry of Housing and Urban Affairs (MoHUA), smart cities in common parlance are understood to be cities that use appropriate technologies to improve the quality of lives of their citizens. The following six guiding principles—community at the center; more from less; cooperative and competitive federalism; integration, innovation, sustainability; technology as a tool, not an end in itself; and convergence—are the foundation of smart cities. These robust guiding principles of the mission will assist Smart Cities in addressing three core issues namely Liveability, Economic-ability and Sustainability.

The priority areas of work are the implementation of citizen-oriented initiatives related to the further development of each of the cities; The selection of 100 cities was carried out within three years from 2015 to 2018. Every city is different, the projects, goals and subjects of interest are defined by the local parties. This approach reserves the choice to the cities and different cities have different agendas; hence, the concerns may be the areas such as health, mobility, housing, water, and the remaining and more. Mohd Housing and Urban Affairs India (2024).

Concerning the selection process, there was a cooperative and competitive federalism whereby the ICM used an inclusive challenge method in two stages. Thus, the selected cities developed the Smart City Proposals (SCPs) that define the objectives and strategies in detail. They have further stressed that such problems are highly likely to increase in the future as indicated by Nasir and his team in their 2019 study. The mission has been extended to June, 2024 to capture more cities in order to foster efficient and effective urban development in accordance with government's vision for the same. This extension appreciates the fact that urbanization comes with complex problems and that all the problems must be dealt with simultaneously. According to recent studies, it can be concluded that the problem is urgent (Prasad et al., 2023).

The Smart Cities objectives have to are achieved through partnerships linking the governments, the organizations, the private sector, the civil society and the citizens. Smart Cities Mission is an umbrella of infrastructural., technological., and governance interventions accompanied by the development strategy. Smart infrastructure, smart mobility, smart energy, smart water, smart city envelopes, effective ICT infrastructure, non-physical infrastructure, and smart social infrastructure are the core concentration areas. It focuses on the use of technology within the various regions of cities for efficient and effective use of resources, and better delivery of services with an aim at enhancing the liveability of cities today (Mane S. R, 2021).

By the end of 2018 smart city projects have started being adopted in many sectors in numerous cities. Such projects were also as varied as intelligent traffic management projects, integrated public transport projects & last mile connectivity projects. Also, smart governing systems were introduced, and smart meters for utilities were placed, intelligent transportation systems were evolved, and IoT-based infrastructure for waste management was implemented. Internet cafes were established, and basic sectors such as health and education became technological (Singh et al., 2022). They also aimed at sealing more smart energy meters, encouraging people to utilize renewable energy and adopting green building codes for energy sustainability.

Additionally, there was a better disposition in the adoption of smart governance solutions such as e-governance portals, citizens' service delivery centers, and improvement of digital structures for better service delivery and people engagement. There were also advertisements to enhance the urban physical structures like smart water management system, Sewage treatment plants and SWM solutions in many Cities. Besides, continuous efforts were being made to improve social infrastructure through augmenting intelligence solutions in healthcare facilities, schools, entertainment areas and other community facilities (Prasad et al., 2023).

Thus, the Smart Cities Mission in India is a visionary concept aimed at changing the paradigms of urbanization and developing sustainable cities. Thus, with the help of the presented mission, in combination with technological., creative approaches and collaborative management, people's lives in urban settlements will improve, the challenges of urbanization will be solved. Therefore, it is necessary to keep improving the status of Mission by observing objectives such as inclusiveness, more sustainable, innovation, and citizen's oriented to maximize the potential of future urban risk of India.

NEED FOR WASTE MANAGEMENT IN SMART CITIES

Municipal waste is generated in proximity to human habitations and is shaped by the actions of individuals worldwide, encompassing their daily consumer choices and household waste management strategies. The way individuals procure, utilize, and discard materials dictates the level of energy and raw resources consumed, as well as the volume of waste produced. (UNEP & ISWA, 2024).

The generation of municipal solid waste is projected to surge from 2.1 billion tonnes in 2023 to 3.8 billion tonnes by 2050. In 2020, the direct global expenditure on waste management was estimated at USD 252 billion. When factoring in the concealed expenses associated with pollution, public health issues, and climate change resulting from inadequate waste disposal practices, the cost escalates to USD 361 billion. Without immediate action on waste management, the global annual cost could nearly double to an astounding USD 640.3 billion by 2050. (UNEP & ISWA, 2024)

Smart cities rely on interconnected infrastructure and data-driven optimization, necessitating a robust waste management system for effective operation. Improper waste disposal throws a wrench into this delicate system affecting lifestyle, healthcare, environment, recycling, disposal., and various industries. (Aazam et al., 2016)

The need for a proper waste management system can be translated in the form of the following reasons:

- Poor waste management leads to environmental pollution, contaminating air, water, and soil undermining the Sustainable Development Goals (S.D.G.s) specifically–
    - S.D.G. 6: Clean Water and Sanitation
    - S.D.G. 13: Climate Action
    - S.D.G. 14: Life Below Water
    - S.D.G. 15: Life on Land

In addition, it hinders the achievement of environmental sustainability goals in smart cities, which depend on clean and healthy ecosystems.

- Uncollected waste gives rise to unpleasant odours, attracting pests that can contaminate water sources, undermining efforts to establish a clean and healthy urban environment – a fundamental objective of any smart city.
- Improper waste management also translates to increased costs, fuel consumption, operational costs, and increased maintenance costs and disruptions to the overall city infrastructure.
- Inadequate waste collection systems impose substantial financial burdens on cities. The expenses incurred in managing inefficient waste collection systems divert budgetary resources that could otherwise be allocated to essential municipal services, thereby affecting the overall financial sustainability of the city. (UNEP & ISWA. (2024). Global Waste Management Outlook 2024 - Beyond An Age of Waste: Turning rubbish into a resource.)
- The absence of a well-established waste sorting and recycling infrastructure hinders the realization of a circular economy, where resources are effectively conserved and reused. Instead, linear disposal practices persist, leading to detrimental environmental and economic outcomes. (MoHUA, NIUA, & Smart City, 2024)
- Inefficient waste collection routes not only impede the effective management of waste but also contribute to traffic congestion in urban areas. The increased number of vehicles traversing city streets due to inefficient waste collection further exacerbates the existing traffic problems, impacting the overall efficiency of transportation systems.

WAYS OF WASTE MANAGEMENT

The population of India has increasingly adopted the urban lifestyle in the last decade and following this there has been a proportional increase in waste production. Current untidy and uncoordinated structures that are in place to facilitate the management of waste are still being stretched beyond their capacities to suit the needs of the increasing population of developing countries today posing great threats to the public health hygiene and also the image of our environment as we continue to witness immeasurable cuts in our pocket. Thus, the problem of proper waste management surfaces as an imperative for sustainability and enhanced quality of life of the target smart cities in India.

While the rate at which urbanization is increasing in Indian cities, the quantity of the rubbish being generated is rising at an even higher rate. The municipal solid garbage of India is believed by the World Bank to be over 62 million tonnes every year, which is on a constant rise in the coming years. This increase in waste generation is from rapid urban development, alteration in consumption pattern, industrialization and source, and this puts pressure on the urban municipalities (Yadav et al., 2020).

Solid urban waste which is if disposed inappropriately poses great health and environmental hazards to the population of our cities. Sanitary dumps exposure of chemicals in the open, and burning of waste exposes dangerous chemicals to the air, soil, and water polluting the air, and degrading the soil. These pollutants not only result in the deterioration of environment quality but also other health complications that make the residents more susceptible to respiratory diseases, waterborne diseases, and other diseases (Faizan, M., 2021).

Besides, environmental and health issues that are connected with waste flow, the ill-considered waste management supposes considerable economic losses for cities and their inhabitants. The haulage [collection, transportation, and disposal] costs in particular impact the physical treasury through direct costs and ties up lots of public resources offering non-essential services. In addition, the social costs such as pollution of the environment, cost of public health and lost productivity because of poor waste management make the economic losses even bigger (Yadav et al., 2020).

Hence, the requirement for integrating smart waste management systems in India can be deemed evident based on the indicated dimension and issues of waste management in the country. Current systems of waste management still involve the use of labor for collecting the waste and poor infrastructure to contain the volumes of waste produced. Smart waste management holds the potential that can help the cities to manage these difficulties with the help of tools and technologies (Cheela et al., 2021). Modern cities are adapting better strategies in an effort to properly collect wastes, many are using technology, data tools and citizens to make urban environment cleaner and eco-friendly.

Smart waste collection is one of the revolutionary strategies of defined smart waste management in Indian cities. These systems determine proper waste pick-up schedules and itineraries with the help of modern technologies such as sensors and GPS navigation. Smart bins designed with sensors are able perceive the status of waste in real-time and alert waste agencies when it is full and is due for collection, thus eliminating instances of full bins by the time the agency arrives to collect the waste. The use of advanced vehicles, which are traceable on GPS, helps to optimize the area and save costs due to efficient fuel consumption and routes planning (Sati & Sudhakar, 2018).

Encouraging source segregation is another key area that is also undertaken as part of smart waste management in Indian cities. Compaigns that make residents of cities put their waste in two categories that of being recycled and that which cannot be recycled help cut the amount of waste that is disposed off as waste and put in the landfills. Cities are coming up with unique strategies that can promote waste separation; for instance, the provision of several bins for various kinds of wastes; public sensitisation, and incentives for households that undergo recycling. Also, more urban authorities are putting capital into recycling infrastructure and equipment to handle recyclable products locally, hence cutting on transport costs and pollution (Sedláková, S., 2016).

In Indian smart cities, big data and AI are mandatory for efficient operations of the waste management processes. Cities can examine the information about the trends in the generation of trash, routes of collection, and means of disposal and determine what needs to be changed. It can forecast coming waste generation rates to improve those cities efficiency and plans for their further infrastructural needs. Moreover, art of data science is used to track the efficiency of interventions aimed at managing waste, assess the dynamics of sustainable development indicators, and consider the potential for further development (Cheela et al., 2021).

Citizens' involvement and participation are required for this smart trash management necessary for effective functioning in Indian cities. What is more, through residents' engagement in decision-making processes and the call for citizens' active participation in waste management, the cities can enhance the latter's sense of responsibility (Thakur et al., 2022). To involve the communities in smart cities some of the measures are launching of mobile applications and website for reporting the waste issues, conducting cleaning drives and awareness programmes and associating with local organizations and enterprises. Cities can, therefore, make urban spaces cleaner, healthier and more livable for all by engaging the residents in the management of waste (Gorkova et al., 2021).

These are some of the innovations in handling wastes that are making Indian smart cities better places to live in through improved environmental cleanliness. Technology and data analytics, With the help of community engagement, cities are innovating the old traditional method of waste collection, processing, and disposal (Cheela et al., 2021). Starting from intelligent waste management solutions to Waste to Energy technologies these innovations provide eco-friendly solutions to the issues of urbanization and waste management.

ISSUES ENCOUNTERED IN MANAGING SOLID WASTE IN SMART CITIES

Managing solid waste in urban areas poses numerous challenges that hinder waste management systems' effectiveness and efficiency, a few of which are highlighted in the following lines:

- Large-scale urbanization and simultaneous population boom has resulted in an increase of waste, pressuring the already exhausted infrastructural capacity.
- Waste management facilities have a significant influence on the process of waste disposal. Inadequate resources, poorly equipped waste containers, and greater distances to disposal sites can all contribute to an increased likelihood of waste being improperly discarded in open areas and along roadsides. (Aazam et al., 2016). Poor waste segregation at the source complicates recycling and disposal., leading to increased environmental pollution.
- Local governments often face limitations that hinder their capacity to invest in technologies and infrastructure required for effective waste management. Widespread lack of public awareness and involvement in proper waste disposal practices, worsening the situation. (Aazam et al., 2016). Inefficiencies in waste collection and transportation, such as delays, vehicle breakdowns, and route mismanagement, exacerbate these challenges.
- Inadequate training and lack of support for waste management personnel significantly contribute to operational inefficiencies. Insufficient guidance and resources may impede personnel from efficiently fulfilling their responsibilities, thereby obstructing the smooth execution of the waste management process as a whole.
- The persistence of illegal dumping and littering poses a challenge to formal waste management efforts, necessitating specific interventions to combat these issues. (Aazam et al., 2016)
- The deployment and upkeep of intelligent waste management systems entail significant expenses, requiring the establishment of viable financial frameworks and effective cost recuperation strategies to ensure enduring viability. (UNEP & ISWA, 2024).

- The presence of poorly equipped waste containers and long distances to disposal sites contributes to improper waste disposal in open areas and along roadsides, further complicating waste management efforts.
- The lack of interoperability among smart waste management systems developed by different companies can impede cities' ability to effectively integrate systems and share data. (UNEP & ISWA, 2024).
- The incorporation of the informal waste collection sector into the formal waste management systems in smart cities presents a complicated challenge that necessitates meticulous planning and coordination. This issue involves finding a balance between the efficient disposal of waste and the livelihoods of informal waste pickers, while also ensuring environmental sustainability and social equity within the context of smart urban development.
- In various countries and regions, there exist substantial hurdles in terms of waste data and its availability. A critical problem is the lack of consistency in measurement and reporting, coupled with the underdeveloped monitoring systems in numerous countries. Consequently, accurate estimates are unavailable for basic indicators like the total volume of waste collected and the percentage of waste deposited in managed landfills. (UNEP & ISWA, 2024).

Addressing these issues necessitates comprehensive strategies involving technological integration, public education, strong regulatory frameworks, and sufficient funding to establish a resilient and efficient solid waste management system.

THE CASE OF SURAT

This case derives its inputs from the Smart cities and Academia towards Action and Research (S.A.A.R) Report published by the National Institute of Urban Affairs (N.I.U.A.) in 2023.

Surat's Solid Waste Management (S.W.M) system serves as a prime example of the criticality of a well-coordinated and efficient waste management strategy. At the heart of the city's system lies the Integrated Command and Control Centre (I.C.C.C.), which shoulders a crucial role in amalgamating the diverse facets of the waste management infrastructure. Substantial investments have been made by the Surat Municipal Corporation to build up a comprehensive infrastructure of garbage collection vehicles, disposal equipment and cradle-shaped dustbins and disposal mechanisms. The SWM system in Surat boasts a crucial strength in its capability to meticulously track waste collection routes in real-time. This feat is primarily achieved through the utilization of vehicle-tracking systems and mobile applications, which play a pivotal role in optimizing waste collection efficiency by continuously providing valuable data. In cases of breakdowns, spare vehicles are promptly deployed to ensure that collection operations proceed as planned. Moreover, the system integrates cutting-edge technologies such as RFID tags, weighbridges, CCTV cameras, and biometric authentication at transfer stations to streamline waste segregation and measurement processes. By incorporating automation into Surat's waste management services, the precision and effectiveness of the entire system are significantly enhanced. The strategic implementation of automated processes has enabled the urban metropolis to elevate the accuracy of its services, reduce grievances, and reinforce the accountability of its contractors.(Bhardwaj et al., 2024) This was made possible through the implementation of Service Level Agreements (S.L.A.s) and Key Performance Indicators (K.P.I.s) based on data. The use of automation has undoubtedly contributed to Surat's management excellence, leading to its recognition as one of India's cleanest cities in the 2023 Swachh Survekshan Awards, where it secured the first place alongside Indore. The success of Surat's waste management strategy highlights the significant role it plays in promoting urban health and sustainability. By efficiently managing waste through a well-coordinated system, Surat has been able to create a better, cleaner as well as healthier environment for its residents. Surat's achievement in securing the top spot in the Swachh Survekshan Awards is a testament to the effectiveness of its waste management strategy and serves as an inspiration for other cities to prioritize and invest in similar initiatives. (MoHUA, NIUA, & Smart City, 2024).

RECOMMENDATIONS

Thus, successful waste management is presented as a fundamental aspect of sustainable city development given that India is rapidly urbanizing at present. Smart City Mission has a significant place for waste management and there are certain recommendations that will assist to proceed with proper waste management in Indian smart cities sustainably (Thakur et al., 2022).

This is one of the basic recommendations to be followed for efficient trash management to the smart city in India that is to build smart waste collection mechanisms (kumar, 2024) . This entails coming up with smart bins that use sensors to help in monitoring the level of trash at any given time, using GPS to determine the best route for collection trucks and having the right waste disposal systems in place. Collection infrastructures can be upgraded to smarter technologies; the various cities can operate efficiently, cutting on expenses while offering better services (Cheela et al., 2021).

Another major recommendation that was given was that waste collection should be Source Separation and Recycling with the residents (Addas et al., 2024). To foster people's awareness of smart cities, authorities could conduct periodic advertisements on the radio, TV or through the media to encourage people to sort their waste, offer certain privileges to those who sort their waste properly, or create convenient spots for sorting recyclable waste in their cities. For this reason, an integrated management of recyclable waste means that cities can contain pollution, protect resources and practices the principles of circular economy within the context described by Gorkova and others, 2021.

Waste-to-energy (WtE) technologies offer the provision of renewable power and a proper means of disposing of waste products derived from organic materials. Indian smart cities should incorporate WtE technologies through practicing the investment in the formulation of WtE plants and biogas plants. In absorbing organic waste whereby, the cities pump it into generation of biogas or electricity, which will help in managing the flow of fossil-based products and at the same time managing carbon dioxide emission and, in the process, generate more income streams (Paulchamy et al., 2018).

Data analytics and artificial intelligence also play a great role in smart cities where they can be quite useful in the procedures relating to trash. Municipalities should relies on gathering SDG data and should invest into gathering media, big data analysis technologies, and AI to forecast the tendency of waste generation and estimate demand for the resources. Thus, data can be used to make the right choices as well as bring a positive change in the functioning of a city, as well as quality of services that it provides.

Smart cities need waste management that can be funded, backed, and established by operational PPPs. According to the law, local governments and their municipalities should join forces with organizations of the private business and with non-profit organizations and universities to develop and introduce the most effective practices relating to waste management. This may build on the technology management and the ability towards better and sustainable waste management as concluded in the case of Nivas, A. H. (2019) where PPPs may push the improvement of the technology transfer, knowledge dissemination and the PPP capacities.

Therefore, in my view, waste management can only be effectively implemented if there are good legal frameworks backed by rock-hard polices that have to be complied with to the latter. (Ahmed et al., 2024). Smart cities of India need enhancement and effective implementation of the existing laws concerning disposal of wastes, segregation and recycling. Local government should make sure that it is implemented through, awareness, penalties in case of non- compliance and check-up frequently. As such, it calls for city authorities and other state agents to enforce existing rules and regulations in order to influence improved mechanisms of handling trash. Trash management systems can be assessed in the course of implementation and it is significant to realize useful information from the systems' feed backs

(Nesmachnow et al, 2025) . They should also develop polices and legislation that call on citizens, institutions and companies to integrate the opportunities of smart waste management into existing practices. Apart from that, through pilot projects, the cities will apply smart strategies of waste management with some limitations and then generalize these approaches as long as such approaches are to be escalated to the absolute scale; thus, the overall risks and costs of innovations in policies can be controlled easily (Gorkova et al., 2020).

Waste management in smart cities requires the involvement of people in the community for the initiatives to succeed. Residents, community groups and civil society organizations should be involved in offering feedback on waste management matters and being taken through sensitization processes on the need to support waste reduction processes in municipalities. Hence, empowering citizens and engaging them into projects, cities can develop stronger and coherent urban environment (Gorkova & Bulychev, 2021; fang et al., 2023)

The principles as well as practice of Waste Management shall always remain a cornerstone in achieving Sustainable development goals in the new smart city mission for India and developing urban centres in the country. When the mentioned recommendations will be put into practice, Indian smart cities can improve waste management, lobby for sustainable environmental practice and (%Gathered %) generally improve the quality of life of the inhabitants. It is equally important to acknowledge waste management as one of the crucial areas of focus in the further evolution of cities and their infrastructures. Consequently, creating cross-sector partnerships and partnering with the communities, Indian cities can secure a clean, green and prosperous future for the upcoming generations (Thakur et al., 2022).

CONCLUSION

Smart Cities revolve around emerging technologies that can transform urban life and support environmental sustainability over a long period of time. The waste management system is an essential part of the health and welfare infrastructure within cities, so it needs to be promptly upgraded.

With urban populations on the rise, the volume of waste generated is also increasing, straining traditional waste collection methods. Therefore, it is imperative to embrace innovative waste management solutions to tackle the growing waste levels. Smart waste management technologies, such as monitoring bin fill levels in real-time and optimizing collection routes, offer data-driven approaches to address the complexities of modern waste management. The potential for enhancing efficiency and reducing environmental impact lies within these proposed solutions. Nevertheless, achieving successful execution necessitates synchronized endeavors and detailed strategizing. It is imperative to make significant investments in sensor technology, communication infrastructure, and data management systems, while also prioritizing data security and community involvement. By overcoming obstacles in implementation and promoting cooperation among various stakeholders, urban areas can advance towards a greener and more sustainable future that benefits all inhabitants.

To conclude, the utilization of tracing and tracking technologies offers a hopeful prospect for transforming waste management procedures in intelligent urban areas. This refined approach guarantees enhanced sanitation, reduced ecological contamination, and the potential for creating a circular economy in waste management. Through prioritizing data-informed decisions and engaging with the public, this collaborative approach is essential for unlocking the full advantages of smart waste management for a healthier planet and a more sustainable urban environment.

REFERENCES

Aazam, M., St-Hilaire, M., Lung, C. H., & Lambadaris, I. (2016, October). Cloud-based smart waste management for smart cities. In 2016 IEEE 21st international workshop on computer aided modelling and design of communication links and networks (CAMAD) (pp. 188-193). IEEE. DOI: 10.1109/CAMAD.2016.7790356

Access, O. (2016).. . *International Journal of Science and Engineering*.

Addas, A., Khan, M. N., & Naseer, F. (2024). Waste management 2.0 leveraging internet of things for an efficient and eco-friendly smart city solution. *PLoS One, 19*(7), e0307608. DOI: 10.1371/journal.pone.0307608 PMID: 39083562

Ahmed, K., Dubey, M. K., Kumar, A., & Dubey, S. (2024). Artificial intelligence and IoT driven system architecture for municipality waste management in smart cities: A review. *Measurement. Sensors, 36*, 101395. DOI: 10.1016/j.measen.2024.101395

Bhardwaj, V., Anooja, A., Vermani, L. S., Sunita, , & Dhaliwal, B. K. (2024). Smart cities and the IoT: An in-depth analysis of global research trends and future directions. *Discover Internet of Things, 4*(1), 19. DOI: 10.1007/s43926-024-00076-3

Chauhan, A., Jakhar, S. K., & Chauhan, C. (2020). The interplay of circular economy with industry 4.0 enabled smart city drivers of healthcare waste disposal. *Journal of Cleaner Production, 279*, 123854–123854. DOI: 10.1016/j.jclepro.2020.123854 PMID: 32863607

. Cheela, V.R., Ranjan, V.P., Goel, S., John, M., & Dubey, B.K. (2021). Pathways to sustainable waste management in Indian Smart Cities. Journal of Urban Management.

Datta, P., Mohi, G., & Chander, J. (2018). Biomedical waste management in India: Critical appraisal. *Journal of Laboratory Physicians, 10*(1), 6–14. DOI: 10.4103/JLP.JLP_89_17 PMID: 29403196

Faizan, M. (2021). Solid Waste Management in India under Covid 19 Pandemic: Challenges and Solutions. *International Journal of Waste Resources, 11*, 1–4.

Fang, B., Yu, J., Chen, Z., Osman, A. I., Farghali, M., Ihara, I., Hamza, E. H., Rooney, D. W., & Yap, P.-S. (2023). Artificial intelligence for waste management in smart cities: A review. *Environmental Chemistry Letters, 21*(4), 1959–1989. DOI: 10.1007/s10311-023-01604-3 PMID: 37362015

. Gorkova, N.V., & Bulychev, S.N. (2021). Modern ways of solving the problem of waste management in different countries.

Gracias, J. S., Parnell, G. S., Specking, E., Pohl, E. A., & Buchanan, R. K. (2023). *Smart Cities—A Structured Literature Review*. Smart Cities. DOI: 10.3390/smartcities6040080

Javed, A. R., Zikria, Y. B., Rehman, S. U., Shahzad, F., & Jalil, Z. (2021). *Future Smart Cities: Requirements*. Emerging Technologies, Applications, Challenges, and Future Aspects.

Kumar, A. (2024). A novel framework for waste management in smart city transformation with industry 4.0 technologies. *Research in globalization, 9*, 100234.

Kumar, S., Smith, S. R., Fowler, G., Velis, C., Kumar, S. J., Arya, S., Rena, , Kumar, R., & Cheeseman, C. (2017). Challenges and opportunities associated with waste management in India. *Royal Society Open Science*, *4*(3), 4. DOI: 10.1098/rsos.160764 PMID: 28405362

Mane, S.R. (2021). A Systematic Overview of India's Smart City Mission.

Mani, S., & Singh, S. (2016). Sustainable Municipal Solid Waste Management in India: A Policy Agenda☆. *Procedia Environmental Sciences, 35*, 150–157. DOI: 10.1016/j.proenv.2016.07.064

Ministry of Housing and Urban Affairs. (2024). Smart Cities Mission. *Evaluation*.

MoHUA, NIUA, & Smart City. (2024). Smart Cities and Academia towards [+ Case Studies of Innovative Projects of Smart Cities Mission.]. *Action Research*, •••, 75.

Nasir, M., Salman, M., Ahmad, F., & Siddiquee, T. A. (2019). Critical Assessment of Smart City Mission - A Research of Gwalior Smart City Proposals. *International Journal of Innovative Technology and Exploring Engineering*.

Nesmachnow, S., Rossit, D., & Moreno-Bernal, P. (2025). A Literature Review of Recent Advances on Innovative Computational Tools for Waste Management in Smart Cities. *Urban Science (Basel, Switzerland)*, *9*(1), 16. DOI: 10.3390/urbansci9010016

Nivas, A.H. (2019). A Critique on Need of Private Sector Participation to Enhance Solid Waste Management Practices in India.

Ojadi, J. O., Owulade, O. A., Odionu, C. S., & Onukwulu, E. C. (2025). AI-Driven Optimization of Water Usage and Waste Management in Smart Cities for Environmental Sustainability. *Engineering and Technology Journal.*, *10*(3), 4284–4306. DOI: 10.47191/etj/v10i03.36

Paulchamy, D., Alwar, E.B., Anbarasu, K., Hemalatha, R., Lavanya, R., & Manasa, K.M. (2018). IOT Based Waste Management in Smart City.

Prasad, D., Alizadeh, T., & Dowling, R. (2023). Smart city planning and the challenges of informality in India. *Dialogues in Human Geography.*

Rana, N. P., Luthra, S., Mangla, S. K., Islam, R., Roderick, S., & Dwivedi, Y. K. (2018). Barriers to the Development of Smart Cities in Indian Context. *Information Systems Frontiers*, *21*(3), 503–525. DOI: 10.1007/s10796-018-9873-4

Sabeet, . (2017). Sustainable Solid Waste Management in India. International Journal of Scientific Research in Science and Technology, 3, 1317-1324 .

Saha, A. (2020). The Socio-political Relevance of the Indian Smart City Mission:A Critical Analysis.

Sati, A., & Sudhakar, K. (2018). Smart Dustbins: An Approach for Sustainable Solid Waste Management in Smart Cities.

Sedláková, S. (2016). Separation of Waste - One of the Ways to Reduce the Cost of Waste Management Municipality.

Singh, T., Solanki, A., Sharma, S. K., Nayyar, A., & Paul, A. (2022). A Decade Review on Smart Cities: Paradigms, Challenges and Opportunities. *IEEE Access : Practical Innovations, Open Solutions*, *10*, 68319–68364. DOI: 10.1109/ACCESS.2022.3184710

Singh, U., & Upadhyay, S. P. (2022). Fractured smart cities: Missing links in India's smart city mission. *Environment and Planning. B, Urban Analytics and City Science*, *50*(7), 1790–1805. DOI: 10.1177/23998083221144321

Subbanarasimha, R. P., Venumuddala, V. R., Prakash, A., & Chaudhuri, B. (2025). Embracing Complexity in Smart City Services: A Case of Waste Management. *Journal of Innovation & Knowledge*, *10*(2), 100661. DOI: 10.1016/j.jik.2025.100661

Thakur, V., Parida, D., & Raj, V. (2022). Sustainable municipal solid waste management (MSWM) in the smart cities in Indian context. *International Journal of Productivity and Performance Management.*

UNEP & ISWA. (2024). Global Waste Management Outlook 2024 - Beyond An Age of Waste: Turning rubbish into a resource.

Yadav, S. K., Rajagopal, K., Priya, A. K., & Sharma, G. D. (2020). Smart Waste Management and Energy Extraction from Waste in Indian Smart Cities –. *RE:view*.

Yuvaraja, H. (2020). Challenges in development and implementation of Indian smart city mission. International Journal of Advance Research. *Ideas and Innovations in Technology*, *6*, 534–540.

Chapter 4
AI-Driven Cognitive Robotics for Autonomous Multi-Material Sorting and High-Value Recovery in Circular Supply Chains

Angadi Kavya Sri
Woxsen University, Hyderabad, India

Abhishek Allapur
Woxsen University, Hyderabad, India

Sujit Singh
http://orcid.org/0000-0002-3643-5183
Woxsen University, Hyderabad, India

Ale Udaya Sri
Woxsen University, Hyderabad, India

Afzalur Rahman
http://orcid.org/0000-0002-1561-561X
Woxsen University, Hyderabad, India

ABSTRACT

The chapter addresses the dis-ruptive power of AI-based cognitive robotics to revamp autonomous multi-material sorting and high-value recovery, in the case of the e-waste industry. Based on the analysis of cur-rent academic literature (2019-2025), this

DOI: 10.4018/979-8-2600-1960-3.ch004

paper will generalize the emergence of artificial intelligence (AI), machine learning (ML), deep learning (DL), computer vision (CV), rein-forcement learning (RL), and enhanced robotic manipulation to outline an innovative con-cept of intelligent waste management. The methodologies include the wide range of robotic architecture, advanced AI routines to recognize and classify materials, and improved sorting routines that can classify complex waste streams more precisely and efficiently than ever before. In the chapter, a conceptual design of an AI-based cognitive robotic system that in-corporates sensory gain, cognitive automation, increased physical abilities, sensory-motor convergence, and human-machine cooperation to develop the highest value of material re-covery is suggested.

1. INTRODUCTION

The unstoppable increasing population of urban areas and the changing nature of the consumption has driven the generation of wastes to unprecedented levels presenting enormous threats to environmental sustainability, human health, and security of resources. The electronic waste (e-waste) is one of the most urgent issues in the diversity of waste streams because of the rapid increase in its spread, the diversity of materials, and the availability of toxic components (Likotiko et al., 2023). The quantity of municipal solid waste (MSW) produced by all countries on the planet annually is frightening, as it is estimated to grow to 3.4 billion tonnes by 2050 and the amount of e-waste is already a part of this growing crisis. More concerningly, a small part of e-waste namely 17 percent is already gathered and recycled which highlights a severe ineffectiveness of the contemporary waste management methods. This lack is not only leading to greater pollutions in the environment and health risks but also is an enormous waste of precious resources of a finite nature that can always be reintroduced into the economy.

As a reaction to this world crisis, the principle of circular supply chains has been introduced as a critical paradigm shift and proposes a more systemic reaction to the conventional linear framework of take-make-dispose. The purpose of a circular economy is to ensure that the resources are used in the most efficient and comprehensive way by reducing waste, maximizing the value of the material and reusing the resource through continuous reuse instead of disposing it after one use. The key to the success of the circular supply chains is to get high-value recovery out of complex multi-material waste, particularly e-waste, which will lower the requirement of virgin resources, save energy, and lower the greenhouse gas emissions due to the new material production. (Vukicevic et al., 2025)

A solution to these issues is provided through the fast development of artificial intelligence (AI) and robotics, which provide a radically new way to solve them.

The fundamental paradigm shift in the manner in which waste is collected, sorted and recycled is due to the fact that these digital technologies represent a means through which systems can make smart decisions on the basis of information and therefore there is increased accuracy and efficiency in the overall waste management lifecycle. Intelligent sorting systems with the use of AI, advanced robotics, and intelligent automation can optimize the rates of recycling greatly by eliminating the necessity of manual labor, recovering more materials, and providing higher levels of transparency and sustainability in the lifecycle of the material. The combination of artificial intelligence and robots allows them to learn, evolve, and have more and more elaborate interactions with the physical environment, beyond automation to cognitive processes. This combination is essential to operate with the highly contaminated and heterogeneous e-waste, in which distinguishing the various types of materials, regularly down to a micro-scale, is an important part of high-value recovery. (Lakhouit, 2025)

The chapter is designed to offer an in-depth discussion of AI-based cognitive robotics to autonomous multi-material sorting and high-value recovery in circular supply chains, specifically when it comes to e-waste management. The primary objectives are:

1. To put the e-waste crisis in the world in perspective and highlight the necessity of the circular supply chain solutions.
2. To assess the conceptual premises of the circular economy, the issues with e-waste recycling and the present level of robotic and AI technology in waste management.
3. To develop a theoretical model that blends artificial intelligence robots, cognitive computing, machine learning, deep learning, reinforcement learning, and sensor fusion, and connects these ideas to the concepts of sustainability and Industry 4.0.
4. To describe the method and techniques used to create and adopt AI-based robotic systems to identify and assign multi-materials during recognition and sorting, including applications and industrial examples.
5. To present a new model of an AI-based cognitive robotic system specifically, high-value recovery in e-waste supply chains would be proposed.
6. To find the technical, economic, environmental and ethical issues and constraints which arise with these advanced technologies.
7. To outline the future research and development, such as future technologies like neuromorphic chips, swarm robotics and blockchain-enabled supply chains.

The conceptual framework of this chapter is the theoretical foundations, technological progress, practical applications, and what lies ahead of cognitive robotics as far as multi-material sorting and high value recovery of e-waste is concerned. It is based on recent peer-reviewed articles, published in 2019-2025 to assure currency and scholarly quality.

This chapter offers a profound overview of the existing knowledge, connecting the piecemeal research in the field of AI, robotics, waste management, and related literature about the circular economy. Because it suggests an extensive system of AI-based cognitive robotics applied to e-waste, the proposal provides a conceptual map of the path that academic researchers and industry players should follow to achieve greater efficiency, sustainability, and cost-effectiveness of e-waste recycling. Moreover, by overcoming the obstacles and providing the perspectives on the future, this chapter will encourage further innovation and interdisciplinary cooperation, therefore, leading to the creation of cleaner and more resilient urban areas and a faster shift towards a truly circular economy.

2. LITERATURE REVIEW

Ancient linear model of economic relations, which is known as a take-make-dispose, resulted in a world-wide waste crisis, unparalleled by the level of depletion of resources and the destruction of the environment. The circular economy (CE) has responded by becoming a paradigm shift and proposing an economic system that seeks to help resources take advantage of the maximum value they can by ensuring that products, components, and materials remain in circulation as long as possible, and the loop of resource flows is therefore closed. This model, compared to the traditional recycling that usually leads to downcycling, focuses on high-value recovery and recovery of the highest possible utility and value of the products and materials. Several major concepts of the circular economy are the design elimination of waste and pollution, circulating products and materials at the greatest value, and regenerating natural systems. (Farhin Faiz et al., 2024)

The need of CE is supported by the fact that global use of primary materials is projected to increase twofold between 89 gigatonnes used in the year 2017 and 167 gigatonnes used in the year 2060. This requires a shift to a more circular and resource efficient economy so as ensure there are no material supply shortages in the future. The circular economy through the combination of AI and IoT can ensure transparency and sustainability in the life cycle of recyclable materials and that they are not redirected to landfills as they are collected, processed, and recycled. Such traceability is critical for boosting consumer confidence and ensuring that recycling targets are met. (Lakhouit, 2025).

2.1 E-Waste Challenges

Electronic waste (e-waste), which consists of the disposed electrical or electronic equipment, is one of the most rapidly developing and the most complicated streams of waste worldwide. The rising amount of e-waste due to the fast technological progress and the consumption of the goods by people is a huge threat to the environment and health. E-waste has many different materials, valuable metals (e.g., gold, silver, copper, platinum) and hazardous materials (e.g., lead, mercury, cadmium, brominated flame retardants). Mishandling or mismanagement of e-waste may result in discharge of such toxic elements into the environment resulting in respiratory diseases, aggravating health conditions in the local populations and soil and water pollution. (Lakhouit, 2025)

In 2020, the world e-waste monitor stated that approximately 17 percent of electronic waste is collected and recycled. This is a critical problem of sustainability and resource preservation due to this low rate of recycling. Moreover, e-waste materials usually cause problems when it comes to identification and classification because they have various shapes, contain a mixture of substances, and are often in a damaged condition. This heterogeneity renders conventional sorting solutions ineffective and expensive, constraining the recovery of valuable materials and continuing with linear consumption framework.

2.2 Recycling Status

The existing recycling processes, particularly of mixed municipal solid waste (MSW) and e-waste, typically use large numbers of people and mechanical separators to carry out the tasks. The conventional sorting sites are mostly relying on human hands to sort the garbage by hand by visual observation. Such operation is not only expensive and time-consuming but also prone to errors where large proportions of recyclable materials are either missed or polluted (Lakhouit, 2025) 5. The process of manual sorting subject's workers to unhealthy materials and unhealthy environments, which are very dangerous to their health.

The precision and productivity of sorting, even in the case of mechanical systems, is low especially when handling multi-material items or fine fractions. Indeed, as an example, although there are sorting methods in place that serve small sized waste, coarse waste is still sorted by use of old sorting methods, and this has led to a lot of losses in the form of recyclable waste. Europe has a relatively low recycling rate of 31% of plastics, which mostly refers to rigid containers, which demonstrates the larger problem of recovering the material. The quality and marketability of the recovered products in the recycling streams is compromised by the presence of contamination mainly caused by incorrect sorting, increasing the costs involved in the reprocessing process and making them less economically friendly. This situation has led to the need to switch to more sophisticated and automated sorting technologies to enhance the purity of materials and their recovery.

Introduction of robotics into waste management facilities, especially material recovery facilities (MRFs) is transforming the efficiency and accurateness of the material recovery. Robotic recovery systems are aimed at automating the process of detection, sorting and separation of various forms of waste materials with incredible accuracy, which eliminates the need to engage human labour and reduces the health risk involved. Such systems use machine learning algorithms and advanced sensors that attach to robotic arms in identifying different wastes such as different types of plastics, metals, glass, and paper, which are usually difficult to discern using conventional systems (Lakhouit, 2025).

Robotic recovery has got many advantages:

- Improved efficiency and speed: Robots are able to process larger amounts of waste within a shorter duration, resulting in an increase in throughput in MRFs.
- Increased accuracy and decreased contamination: AI robots with massive datasets trained can detect minor differences in materials and achieve precise sorting and cleaner streams of the materials, which are more useful in reprocessing.
- Greater worker safety: Automating sorting operations, human working with hazardous materials is reduced, and workers do not come into contact with unsanitary conditions and possible injuries.
- Response to changing waste streams: AI-based systems can be trained and changed to identify and sort emergent materials or packaging patterns without needing to reconfigure them manually. (Farhin Faiz et al., 2024)

Economic viability: The better the quality of the recycled products is, the more valuable they would be in the market, which will produce positive revenues and make more people be interested in investing in the recycling facility.

The first instances are Zen Robotics in 2011 who presented robotic waste sorting based on algorithm AI, machine learning, and computer vision to sort and find recycling materials in conveyor belts. The systems of the company can sort different types of waste, constructions, demolition waste, industrial waste, metals, and plastics, the systems are Heavy Picker and Fast-Picker.

2.3 Cognitive AI

Cognitive AI are AI systems that emulate human cognitive capabilities including learning, reasoning, problem solving, perception and interaction. Cognitive AI specifically within the waste management process allows the systems to make smart decisions based on information, resulting in increased accuracy and efficiency in the different processes. This is not just automation, though robots can learn to think about their surroundings and change their behavior. (Ganesan, 2023)

2.3.1 The concept of machine learning (ML)

This is a subset of AI that enables systems to learn through data and optimize their work over the time without having to be programmed explicitly. The predictable waste generation patterns, schedule optimization and detection of maximum waste volumes are predictions that can be made using ML models and used to respond proactively to complex datasets. When it comes to sorting, the ML algorithms increase the accuracy by doing repetitive tasks and being exposed to various waste materials.

2.3.2 Deep Learning (DL)

This is a more sophisticated implementation of ML, which involves the implementation of artificial neural networks (ANNs) with more than one or two layers and imitates human brain learning processes. DL has been applied especially in high-order pattern recognition, as is the case with detecting slight distinctions of the waste materials in image data. (Lakhouit, 2025)

2.3.3 Computer Vision (CV)

A branch of AI, which allows computers to see and perceive what is in the real world. CV is used to examine the shape, color, size, and texture of objects in a conveyor system in real-time, identifying what they are made of, meaning that they can be sorted with a high degree of efficiency.

2.3.4 Reinforcement Learning (RL)

This field of ML deals with the question of how smart agents should act in a world to maximize the concept of cumulative reward. The RL is especially adapted to the training of robotic arms to control their tasks of precise manipulation, learning optimal gripping forces and movement trajectories with the help of trial and error. (Tiedemann et al., 2021).

Although originally created to model the food supply chain, the Cognitive Automation of Food (COGAF) framework can be used to model the concept of cognitive robotics, dividing the research into the sensory improvements, cognitive automation, physical automation, sensory-motor fusion, and collaborative automation. These streams can directly be applied to the creation of cognitive robots to sort waste (Sarc et al., 2019).

2.4 Current Sorting Technologies

The waste sorting technology is also developing, shifting towards the primitive manual and mechanical approaches and advanced AI-enabled systems.

Manual Sorting is the sorting of waste has been very dependent on human workers. The separation of waste by workers is done manually depending on what the eye sees. This is slow, costly, ineffective, and subject workers to great health hazards of dangerous materials. (Lakhouit, 2025).

Mechanical Sorting involves technologies as screens, balers, shredders, eddy current separators and air classifiers. These systems are useful in bulk separation, yet generally do not have the accuracy to separate similar materials or sort complex and multi-layered objects, resulting in contamination and low-value recovery. (Sarc et al., 2019)

Sensor-Based Sorting is a type of mechanical sorting which involves more advanced sensors to detect the material properties. These include:

Near-Infrared (NIR) Spectroscopy is a commonly used method in the identification of various materials in plastics and paper with respect to their molecular structure.

Laser-Induced Breakdown Spectroscopy (LIBS) is a high-energy pulse of a laser is used to ablate a small sample of the material, and the resulting light is used to analyze the composition of the element, which is applicable to both metals and complex substances. (Sarc et al., 2019)

X-Ray Fluorescence (XRF) is used to determine elemental composition, which is useful in detecting heavy metals and brominated flame retardants in e-waste plastics.

X-ray Transmission (XRT) is also applied in separating materials according to density, e.g. glass and ceramics or various alloys of metals.

Electromagnetic Sensors (EMS) for Metallic material.

Thermal Imaging able to sort viable recyclables of a municipal solid waste or electronics-based metallic/non-metallic fractions.

AI-based Automated sorting used sensor-based sorting by the introduction of AI, especially computer vision and machine learning. The pick and place operations of robotic arms fitted with such AI systems are very accurate and fast.

YOLOv5 is a deep learning model popularly known as YOLOv5, which is a You Only Look Once (YOLO) model and is a high accuracy (80%), fast (28 FPS) and energy-efficient (waste classification and segregation) model, and can be used in low-power and real-time applications on edge devices, such as Jetson Nano and Raspberry Pi (Prasath, 2025). It can sort out various kinds of waste, separate plastics, metals, glasses, paper and even various kinds of plastics.

oSegment Anything Model (SAM) family of deep learning models is under evaluation on the possibility to remove highly varying objects in robotic waste sorting. It does not require specific waste detection and segmentation algorithms, which reduces the adaptation time and costs of waste types (Vukicevic et al., 2025).

Convolutional Neural Networks (CNNs) are broadly used in image classification in waste management, and high accuracy has been attained in identifying plastics, metals, paper, and glass.

Specialized Robotic Systems such as ZenRobotics and AMP Robotics have implemented industrial-scale AI-controlled sorting robots (e.g., Heavy Picker, Fast-Picker, Cortex, MAX-AI AQC) which use sensor, computer vision, and machine learning combinations to recognize and sort a variety of materials, including mixed waste streams and construction/demolition waste which are challenging. These systems are capable of being conditioned to accept new fractions and provide future adaptability (Sarc et al., 2019).

The disadvantages of the current sorting technologies especially in case of e-waste require the creation of highly flexible, adjustable, and smart systems that could be able to work with multi-material objects with different conditions and characteristics. The combination of innovative sensor technologies with intelligent AI offers a good avenue to realize high-value recovery in circular supply chains.

3. THEORETICAL FRAMEWORK

A sound theoretical framework that incorporates a number of hybridized advanced computational paradigms underpins the development of AI-based cognitive robotics to autonomously sort natural resources of different materials with high value recovery in circular supply chains. This framework is a conceptualization of intelligent robotic systems as agents that can perceive, learn, reason and act in complex and dynamic waste management environments and thus change them into active members of the circle economy and the principles of Industry 4.0.

3.1 Conceptual Models

3.1.1 Artificial Intelligence Robotics and Cognitive Computing

AI robotics is a field of artificial intelligence that utilizes robotic systems to intelligently, similar to human beings, learn, adapt and make decisions. The vision of cognitive robotics is an ability to supply robots with cognitive functions to enable them to perceive the surrounding world, make inferences on causation and consequences, and acquire context-sensitive symbols on high-level planning. This is not just automation, but cognitive robots can exercise self-control and adjust themselves to different environments and problems and therefore can make a connection between apparently unrelated spheres of use. They deal with their surroundings independently and work independently with their abilities without the ongoing human intervention. As an example, a robot system can learn to drive a machine by relating visible control inputs by human operators with the resultant machine behaviour, in essence guessing the relationship between a command and the response of a machine. This Machine Brain method entails a number of AI based software functions of managing manipulators and grippers, identifying and sorting things and materials (Tiedemann et al., 2021).

3.1.2 Machine Learning (ML) and Deep Learning (DL)

Machine Learning is one of the fundamental elements of cognitive AI, which offers the algorithms and systems that enhance their predictive and decision-making ability as they learn and acquire the new experience. ML models can be very important in waste management to predict the trends of waste generation, a sequence of collection routes, and also improve the accuracy in sorting of wastes. These algorithms are able to process large amounts of data to detect trends as well as predict what lies ahead in the waste production to assist in the development of specific diversion programs (Farhin Faiz et al., 2024). Deep Learning, a branch of ML, involves the use of artificial neural networks (ANNs) to compute complex data and detect complex trends and imitates the human brain learning mechanisms. Image recognition and classification are particularly strong tasks that can be performed by the DL models, enabling robotic systems to recognize and separate different kinds of plastics, metals, glass, and other recyclable products with very high accuracy, even being able to distinguish between different grades or between black plastics. As an example, CNN websites are effectively used to identify recyclable substances with a high rate of accuracy. Segment Anything Model (SAM) line of DL architectures is being applied to versatile waste sorting, with the aim of extracting and classifying waste objects, requiring less effort to label the dataset with detailed labels (Vukicevic et al., 2025).

3.1.3 Reinforcement Learning (RL)

Reinforcement Learning is a behavioral model in which the agent is encouraged or discouraged to accomplish a goal by interacting with the environment and being rewarded or punished in response to its actions. The trial-and-error learning approach is best suited to teaching robotic manipulators how to perform some delicate and precise tasks, such as grabbing deformable items (e.g. in food waste, which also has parallels with some e-waste parts) or operating in a sorting environment with a lot of complexities. An ANN can be trained with RL to control a machine, so that the robot learns to predict the response of a given control command to the machine, and it can use that to move to an exact position in the target position. The vision-based robotic manipulation is also being investigated using scalable deep reinforcement learning (Zocco et al., 2025).

3.1.4 Sensor Fusion

Sensor fusion is the integration of multi-sensor data in order to obtain a better, more complete and robust view of the environment in comparison to a single sensor. Multi-material sorting multi-material sorting uses different sensors (e.g., optical cameras, 3D laser cameras, metal sensors, and NIR, LIBS, XRF and thermal images) to measure various types of data about waste objects. The sensor fusion algorithms combine this heterogeneous data to give a complete profile of material which allows the robot to make more intelligent sorting decisions. An example of this is when RGB image data is combined with other spectral data such as object classification can be done, but also accurate material classification, particularly with difficult objects such as broken or unique form during construction and demolition waste, which is similar to the heterogeneity of e-waste. The aim of future studies is to increase the fine-tuning of the sorting accuracy with the increase of ML algorithms and sensor technology (Lakhouit, 2025).

4. CONNECTION WITH SUSTAINABILITY AND INDUSTRY 4.0

4.1 Sustainability

AI-driven cognitive robotics have a theoretical framework that is intrinsically associated with the notion of environmental sustainability. These systems decrease the amount of waste that goes to the landfill, recovery of valuable materials is grown, and resources reused, greatly decreasing the need of virgin materials and saving energy. An example of recycling is recycling aluminium, which conserves 95 percent of energy used to make aluminium using raw materials and recycling plastic which conserves the emission and resources used to manufacture new plastic. Predictive analytics based on AI can facilitate better consumption habits and curb unnecessary production of waste by offering consumption habits. In addition, automation of hazardous sorting operations enhances the health and safety of workers, which helps to make the community environmentally sustainable. The general aim is to reduce the environmental footprints in urban areas and promote a circular economy (Lakhouit, 2025).

4.2 Industry 4.0

The concept is also fundamentally embedded into the concepts of Industry 4.0 and is a convergence of digital technologies, cyber-physical systems, and intelligent automation. The waste management sector is also being elaborated on Industry 4.0 approaches, including digital networking and robotics, in order to advance it as a circular economy. Industry 4.0 is characterized by the arrival of AI and IoT, which has made it possible to create smart waste disposal solutions that help to analysed, predict, and optimize information in real-time (Lakhouit, 2025). Typically, smart waste factories are designed as digitally connected networks of recycling and recovery processes of the highest quality, whereby physical objects, machines, and control systems can communicate and interact independently (Sarc et al., 2019).

- IoT sensors and AI algorithms dynamically process real-time data, optimizing decisions in route planning, predictive maintenance and resource allocation.
- AI-based robotics can be used to support adaptive and scalable waste management systems that can adapt to the new waste content and market requirements.
- Digital communication and Machine-to-Machines (M2M) enables machines to optimally adjust to one another and provide stakeholders such as local governments and recycling organizations with data on operational matters and enable partnerships through collaboration.
- Digital twin technologies applied, especially with neuromorphic computing, have potential to streamline the operations of sustainable supply chains by tracking, preempting, and determining the sustainability metrics in real-time. (Tathavadekar & Mahankale, n.d.)

The application of these conceptual models to the setting of Industry 4.0, AI-driven cognitive robotics offers a comprehensive and efficient way of turning waste management into a highly efficient, sustainable, and cost-effective industry that will push the world toward a legitimate circle of the economy.

5. METHODOLOGIES/APPROACHES

The successful implementation of AI-driven cognitive robotics for autonomous multi-material sorting and high-value recovery in circular supply chains necessitates sophisticated methodologies and approaches spanning robotic design, AI algorithms, material recognition, and integrated sorting techniques. These methodologies leverage cutting-edge technologies to overcome the inherent complexities of heterogeneous waste streams, particularly e-waste.

5.1 Robotic Architectures

Effective sorting of multi-material is dependent on the design and choice of robotic architectures. Waste management involves industrial robots because they are precise, have a load capacity, and are fast (Sarc et al., 2019).

Several forms of robots are used:

The industrial robotic: arms are usually multi-degree of freedom fixed-base manipulators, which have the ability to execute pick-and-place tasks with ease. They include ABB delta robots (FlexPicker) employed in such systems as MAX-AI AQC and AMP Cortex to sort lighter materials at high speed. The Heavy Picker by ZenRobotics is a construction and demolition waste picker, which has powerful gripper arms that can pick objects up to 30 kg. (Sarc et al., 2019).

Cartesian robots: are driven on three linear axes (X, Y, Z) and are particularly precise and may be utilized in applications where there is a requirement of a linear motion, e.g. the ZenRobotics Fast-Picker.

Mobile Robots: Mobile robots can be used in collection (e.g. DustBot project, autonomous street cleaning and household waste collection) but can also be deployed in sorting facilities (internal logistics) or on specific tasks.

Anthropomorphic gripping: robots typically cannot meet the flexibility, flexibility and dexterity necessary to handle soft, fragile and deformable objects of a wide range of shapes and sizes, as some waste streams involve soft or deformable objects. The development of the soft grippers is enabling manipulations of delicate tasks to be performed, which is similar to that of the human hand. In the case of sorting wastes; pneumatic grippers are specialized grippers that are quick and easy to pick and place or more complicated manipulations such as vacuum grippers and parallel grippers. (Zocco et al., 2025)

ROS- Based Frameworks: The Robot Operating System (ROS) is a de-facto standard in robotics, which makes it easier to develop and test cognitive architectures to control robotic agents. The ROS-related frameworks allow the combination of option learning, symbolic reasoning, in the case of various fields, and the facilitation of the complicated interactions, such as the work with objects, navigation, and experience learning. (Beraldo et al., 2024).

5.2 Artificial Intelligence Algorithms

These robots have a set of sophisticated AI algorithms that make their cognitive abilities strong:

Computer Vision (CV): CV systems form the eyes of the robotic sorter, which allows real-time detection and classification of waste materials. They interpret visual stimuli like shape, color, size and texture.

Image Recognition with Deep Learning (DL): CNNs are widely applied to recognize and categorize waste and identify it with high precision. Neural networks such as YOLOv5 provide the best possible compromise between performance, performance, and energy consideration, and as a result, they can be used to implement the sorting facilities in real time. These systems are capable of scanning conveyor belts in real time and placing items to classify in near real time.

Segment Anything Model (SAM): The SAM family of deep learning models have recently been evaluated based on its capability to capture highly variable waste objects in raw images, avoiding the need to localize them during the localization stage and eliminating the need to perform significant amounts of manual annotation. This greatly reduces the time and costs of repackaging solutions to new wastes (Vukicevic et al., 2025).

Machine Learning (ML): ML algorithms are important to constant enhancement and flexibility. They enable robotic systems to be taught based on the information and improve their functionality with time, adjusting to new or changing materials in the waste stream. The ML models also have the ability to predict future trends of wastes using past information, which would enable the MRFs to plan ahead of the expected changes in the composition of the wastes and ensure that the recovery remains high. (Lakhouit, 2025)

Reinforcement Learning (RL): RL is used to impart a robot with complex manipulation skills. It is through the interactions with the environment and reward-based feedback on successful sorting behavior (e.g. correctly pick-and-place) that robots can be taught to adopt optimal gripping strategies, movement behaviors and decision-making policies. This method enables the robots to perfect their movements and adjust to unpredictability of waste materials whereby a perfect and delicate handling is needed. The fundamental concept of AI-based automation is to train an artificial neural net (ANN) to control a machine by allowing it to learn how observable inputs to control manipulate specific machine behavior. (Tiedemann et al., 2021)

5.3 Material Recognition

Effective multi-material sorting is based on accuracy of recognizing material. The risks of various waste characteristics, contamination, and complicated material structures, in particular, of e-waste are countered through the use of advanced methods.

Deep Learning Models the CNNs and other types of DL models are trained on large amounts of waste images to detect patterns and features that belong to different types of materials, such as different types of plastics, metals, glass, paper, and even subtle visual features that indicate certain types of plastics. These are models that keep on getting better and better in terms of accuracy. (Farhin Faiz et al., 2024)

Multi-spectral Imaging and Spectroscopy: In addition to the conventional RGB cameras, systems have multi-spectral sensors that allow the detection of properties that cannot be noticed by the human eye.

Near-Infrared (NIR) Spectroscopy: Determines a variety of plastic materials and paper.

Laser-Induced Breakdown Spectroscopy (LIBS): Identifies the elemental composition of metallic and complicated materials.

X-Ray Fluorescence (XRF): It identifies heavy metals and particular elements, which are important in e-waste.

Thermal Imaging: Has the capability of sorting metal and non-metallic debris of e-waste and valuable recycled material of MSW.

Sensor fusion A sensor fusion is an improvement that integrates these diverse sensors to offer a more detailed and dependable material identification, thus allowing one to distinguish between materials that are similar in appearance but made of different materials (Lakhouit, 2025).

Dataset Development: DL models primarily depend on high quality datasets that are varied. Researchers create certain datasets according to various applications, including floating waste, municipal waste, e-waste (e.g., aluminum, copper, brass), and smart bin waste. The quick adaptation strategies towards new types of waste like the use of SAM to extract objects, followed by an object classifier fine-tuning is essential to the applicability of the industrial application. (Vukicevic et al., 2025)

5.4 Sorting Methods

Upon the recognition of materials, robots sort them using specialized sorting methods:

Automated Pick-and-Place: Robotic arms are able to pick and place items identified in a moving conveyor belt into specified bins in specific material streams. This reduces the human factor in the process and optimizes the sorting process.

Real-time Control: Self-learning machines can control the speed of conveyor belts in real time, resulting in continuous flow of material and lessening of the bottlenecks. With the help of AI algorithms, it is possible to analyze object characteristics in real-time and dynamically change the actions of robots. (Farhin Faiz et al., 2024)

Multi-tasking Robots: Superior robot systems, with AI incorporated into them, are able to carry out the work of sorting two or more types of fractions together, classifying and sorting several fractions at once. This is important in the management of mixed waste streams that are complex.

Quality Assurance Systems: Robots are also used at the end of sorting lines and used to check the flows of output in order to control the quality of processed fractions particularly in plastics to be used in high-value products or solid recovered fuels. (Sarc et al., 2019)

6. PILOT PROJECTS AND INDUSTRIAL CASES.

A number of industrial use cases show how AI-based cognitive robotics can help in managing waste:

6.1 AMP Robotics (Cortex System)

It is an AI-based computer vision and machine learning (in particular Deep Learning) that can identify and sort different types of paper and plastic packaging waste (Farhin Faiz et al., 2024) The Cortex system or Clarke, with an ABB delta robot (FlexPicker) can cycle 3600 picks per hour, much higher than the old system and greatly enhances performance and less contamination. The technology of AMP Robotics takes advantage of computer vision and machine learning to separate using the highest level of precision and increase the recycling rates and the contribution to the circular economy (Farhin Faiz et al., 2024).

6.2 ZenRobotics

A Finnish company, ZenRobotics created its Heavy Picker during 2009, which is used to sort construction and demolition wastes, industrial wastes, metals, wood, and hard plastics and even color-sorted bags (Al et al., 2020). It is an AI-controlled system, that is, the robotic arms are controlled to pick up to 2000 items per hour of up to 30 kg with sensor information gathered by metal sensors, 3D laser cameras, and display cameras. The Fast-Picker works with smaller products (to 1 kg) with 4000 picks per hour. They are flexible to changing waste streams as their systems can be continuously trained on a new sample fraction. (Sarc et al., 2019)

6.3 BHS (MAX-AI AQC)

It is a quality control (AQC) robot created in cooperation with Sadako and NRT that seeks to replace the human operator with Deep Learning and VIS-sensors. It is able to pick as many as six different fractions simultaneously attaining 3900 picks an hour. It is commissioned in plants such as Athens Services Material Recovery Facility in California where it is applied in fully automated PET sorting. (Sarc et al., 2019)

6.4 Apple Daisy

It is not described in its own AI or robotic architecture in the given sources, but the iPhone recycling robot created by Apple is referred to as Daisy (Sarc et al., 2019) 2. These systems are also intended to dismantle and recover materials of complex electronic equipment with the highest degree of accuracy, which in practice is the practical implementation of high-value recovery.

6.5 TrashBot

TrashBot is an AI-powered robot trash can that was created by Clean Robotics and has sensors and machine learning capabilities, allowing it to identify the nature of the object thrown into it with an accuracy of 90 percent, extract any fluid, and place it into the appropriate tank. It fits well in large and dense spaces such as shopping malls, stadiums and airports (Al et al., 2020). Specifically, one can discuss the German SmartRecycling Project (2020-2021): this project aimed at creating a technical solution to sorting coarse and bulky wastes by applying AI-based robots. One of the strategies was to automate standard off-the-shelf hydraulic and electro-hydraulic machines through AI techniques such as Reinforcement Learning, and such that legacy machines could be used in recycling plants without much alteration. (Tiedemann et al., 2021)

6.6 Danish Institute of Technology (DTI) AAWSBE1 Project

Researchers created a robotic system based on AI to sort hazardous waste, namely, battery recovery of the Waste Electrical and Electronic Equipment (WEEE). In this project, signals of various cameras and Deep Learning technology were employed to identify devices with batteries and single batteries in material flows. (Sarc et al., 2019). These various projects and commercial implementations indicate that AI-based cognitive robotics in multi-material sorting are now mature and being innovated upon, and thus can make substantial contributions in the recycling endeavor and in the targets of a circular economy economy in different waste streams, such as the difficult e-waste sector.

7. PROPOSED FRAMEWORK: AI-BASED COGNITIVE ROBOTIC SYSTEM OF HIGH-VALUE RECOVERY.

In order to deal with the multifactor of autonomous multi-material sorting and high-value recovery in e-waste circular supply chains, this chapter suggests an integrated AI-Driven Cognitive Robotic System (AI-CR System). This system combines the innovative abilities of cognitive AI, robotics and sensory technologies and applies them together in a networked, adaptable, and learning based architecture to produce as much material recovery and purity, the least environmental impact and the most efficient operations. The proposed model is based on the currently established ideaș of cognitive automation, including the COGAF model, and consists of the synergistic approach to AI and IoT to manage wastes in an adaptive and scalable manner. (Sharma, 2020)

7.1 System Design

AI-CR System is envisioned as a scalable and modular architecture, a real-time data processing architecture, an energy efficient architecture, designed to support the adaptive decision-making and green technology adoption in complex urban supply networks. The main philosophy of it consists in lifelong learning and change, which allows the company to remain versatile and adapt to the changing waste streams and market demands.

7.2 Key Design Principles

- Modularity: The system consists of self-sufficient interchangeable modules containing every element, the system can be configured flexibly, upgraded easily, and repaired. This guarantees flexibility to the various sizes of operation (small batch to large facilities) and material compositions. (Vukicevic et al., 2025)
- Scalability: The architecture can be expanded (augmented) not only in terms of robotic units, sensor arrays, and computational power capacity but also as the volume of waste grows, as well as the size of the granular sorting demands.
- The emphasis on intelligent decision making, self-learning and self-optimization without constant human intervention.
- Real-time Processing: The need to acquire and analyze data in real time is critical to dynamic sorting processes on conveyor belts, which is necessary to identify and take immediate action on materials. (Sharma, 2020)
- Energy Consciousness: Architectural design factors are conscious of low-power computing and optimal robotic control so as to minimize the environmental impact of the system.
- Human-Robot Collaboration: Although the system is very autonomous, there are human interfaces, such as human supervision, human intervention, human training, and human maintenance, as it has realized the importance of humans in more complex environments.

The AI-CR System of e-waste recovery would be depicted by a high-level diagram that would represent some interconnected modules:

Figure 1. E-Waste Items on Conveyor

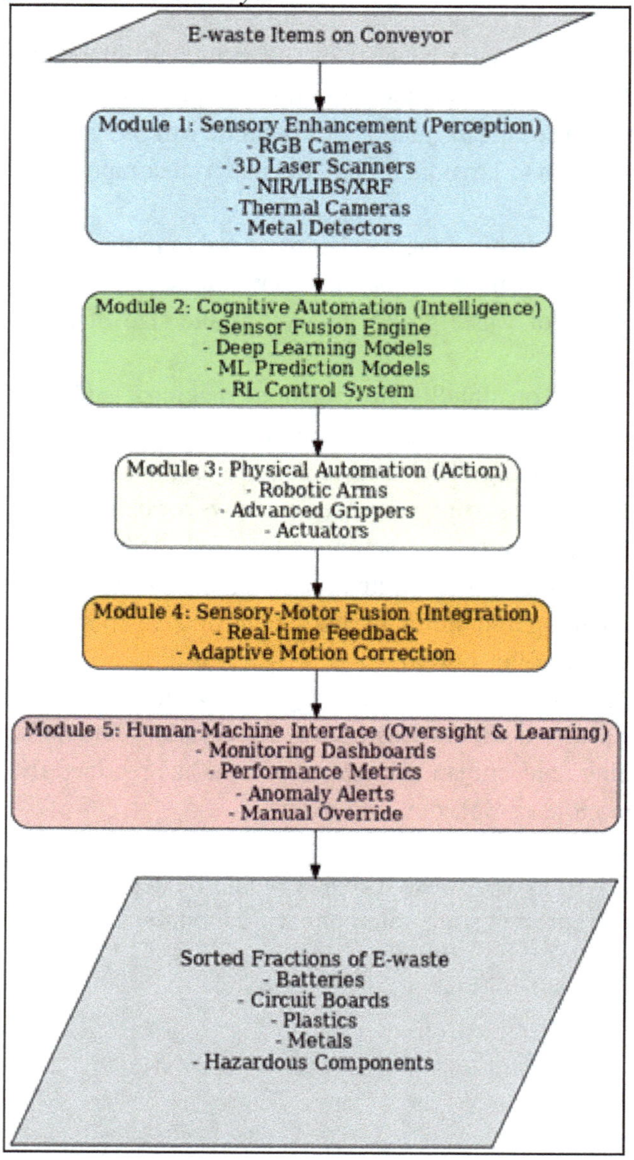

Module 1: This module involves the enhancement of rate of perception. Sensory Enhancement Module (Perception)

- fitted with a high number of sensors to characterize materials in a multifaceted manner. This comprises high-resolution RGB visual feature cameras, 3D laser scanners, volumetric data and 3D shape analysis and high-resolution chemical and elemental composition spectroscopic sensors (NIR, LIBS, XRF). Thermal imagers have the ability to differentiate between metallic and non-metallic e-waste fractions. (Vukicevic et al., 2025)

Module 2: This module is another element of cognitive automation. Cognitive Automation Module (Intelligence)

- Sensor Fusion Engine: This is the main component that combines and matches the data of all sensors into a single, strong, and precise image of each e-waste piece.
- Deep Learning Classification Algorithms: Based on the state of the art DL models (e.g., YOLOv5, SAM using MobileNetV2 classifiers) to detect and classify specific e-waste materials (e.g., various types of plastic, ferrous/non-ferrous metals, types of circuit boards, batteries) with high accuracy. These models are constantly improved and changed with the exposure to new data. (Farhin Faiz et al., 2024)
- Machine Learning Prediction Models: Data analysis based on the past and present data and predictions of how the material can be moved, where sorting priorities based on high-value recovery can be given, and where some contaminants may be found. (Lakhouit, 2025)
- Reinforcement Learning Control System: Trains and optimizes robot tasks, such as gripping force, path optimisation, and pick sequence optimisation, so as to maximise damage-free and efficient handling and sorting of fragile e-waste parts. (Beraldo et al., 2024)

Module 3: Increased Physical Automation Module (Action)

- Robotic Arms: Multiple degrees of freedom, have been selected due to their speed, reach and payload capacity to work with e-waste (e.g. ABB FlexPicker with lighter weight objects, ZenRobotics Heavy Picker with heavier components). (Sarc et al., 2019)
- Advanced Grippers: A collection of specialized grippers, such as a vacuum suction cup on flat objects, multi-finger parallel grippers on variety of shapes, and soft grippers of delicate parts (such as circuit boards or screens). The system automatically chooses the right gripper depending on the material and property that is detected. (Sharma et al., 2020)
- Actuation System: This system is used to guarantee accurate and fast motions of robotic arms and grippers so that they can perform high throughput sorting.

Module 4: Sensory-Motor Fusion Module

- This module is the close connection and feedback between the perception (sensory) and action (motor) systems. Cameras with NVIDIA Jetson Nano microcomputers as part of robot flanges provide real-time visual feedback to constantly correct and inform robot actions, providing adaptive and precise manipulation (Zocco et al., 2025)! This enables the robot to modify its behaviour with regard to the current situation in the environment and object.

Module 5: A team of automation

- Enables easy communication between humans and robots, especially in the event of those activities that might need human oversight, difficult problem-solving, or training the AI system. This incorporates monitoring and diagnostic reporting and manual intervention with intuitive Human-Machine Interfaces (HMI). The main point of this module is that human beings help to optimize AI algorithms and deal with exceptions to maintain a steady flow of operations and ensure a continuous learning process of the whole system. (Sharma et al., 2020)

8. SUPPLY CHAIN INTEGRATION

ASI-CR System is aimed to become a significant link in an even greater circular supply chain in the realm of e-waste, improving the level of transparency, traceability, and resource recoveries:

- Traceability and Accountability: This system will be able to establish unalterable records of e-waste materials during collection, processing, and recycling by linking it with blockchain technology. This makes sure that no high-value materials are diverted by the recycling steam and also gives the stakeholders verifiable information on the flows of the materials, hence building a sense of trust and adherence to regulations. (Lakhouit, 2025)
- Data Sharing and Insights: The AI and IoT systems produce the operational data on waste decomposition, volume, and recovery rates, which can be shared with local governments, manufacturers, and recycling companies. This is an evidence-based method that can guide policy formulation, resource distribution and awareness efforts to achieve more effective and targeted interventions. It also allows manufacturers to create products that can be easily

recycled (Design for Disassembly) and incorporate a greater amount of recycled materials. (Lakhouit, 2025)

The AI incentive-based solutions may include incentive programs based on data that rewards consumers and businesses engaged in appropriate e-waste sorting and disposal patterns (perhaps with the help of blockchain smart contracts).

- Quality Recyclates Demand: The AI-CR System will elevate the market value of the secondary raw materials by creating higher quality, purer recycled materials, which will encourage a greater uptake of recycled materials in manufacturing and lessen the necessity of obtaining virgin material resources. This goes directly to the circular economy as it enhances the economic sustainability of recycling.
- Predictive Maintenance: The AI-based analytics can be optimized to streamline the maintenance schedule of the robotic system itself and other machines within the recycling facility, which will minimize downtime and operational expenses. (Sarc et al., 2019)

To conclude, AI-CR System is an intelligent, holistic, and adaptive method of managing e-waste. It prioritizes the change to a circular economy with the emphasis on multi-material sorting and high-value recovery that relies on innovative AI and robotics to establish a more efficient, sustainable, and transparent supply chain of essential resources.

9. APPLICATIONS/CASE STUDIES

One of the areas where AI-assisted cognitive robotics is being used in waste management is efficient multi-material sorting and high-value recovery. First is the Cortex system created by AMP Robotics that sorts plastics, paper, and packaging with high speed (up to 60 picks/min) with the help of AI-guided computer eyes and an ABB FlexPick robot. This increases recycling, pollution reduction, and the circular economy, implemented in four continents. ZenRobotics is a company based in Finland and provides Fast-Pickers and Heavy Picker systems, which combine AI, computer vision, and various sensors. They use the ZenRobotics Brain to learn new materials, and are capable of processing difficult waste streams with ease through a process of continuous learning.

1The AI version of robotics is transforming the waste management sector into a highly automated industry. Heavy Picker developed by ZenRobotics is a system with maximum loading of 30 kg at 2000 picks/hours, the Fast-icker with 1 kg at 4000 picks/hours. The systems enhance hazardous environment safety and recycling. The MAX-AI AQC of BHS is an optical quality control system that applies VIS-sensors and deep learning to sort six fractions with up to 3900 picks/hour, which is highly pure. Apple Daisy robot is a specialized dismantler of iPhones, which recycles the rare earths, precious metals, and plastics to e-waste sustainability. The TrashBot by Clean Robotics uses AI and sensors during the disposal process with 90% effectiveness, sorting waste into the right containers in the streets (Lakhouit, 2025). These new technologies in combination point to the benefits AI-based robotics has on efficiency, safety, and high-value recovery that advance the circular economy with scalable waste management systems.

10. CHALLENGES/LIMITATIONS

Robotics based on artificial intelligence can transform e-waste sorting and recovery, yet a number of technical, economic, environmental, and ethical issues restrict its massive usage. Technical difficulties involve the problems with the quality of data, because various and damaged e-waste makes the reliable training of AI models difficult. CNNs and other deep learning models find it difficult to generalize to vastly diverse waste and need to be retrained at a high cost. Dust, moisture as well as lighting changes interfere with sensors making them less accurate. Robotic dexterity is also restricted especially with fragile or irregularly shaped objects. Moreover, the process of connecting new robotic systems to the already existing waste management facilities is not simple and the realization of real-time processing needs a lot of computational power, which causes latency problems.

High initial investment in hardware and software and recurrent energy, maintenance, and highly skilled operation place an economic strain on the economy. It may be challenging to justify the cost of return on investment, particularly because depending on the prices of recycled materials, they may not necessarily offset the cost of running the business. To ensure economic feasibility, it is important to ensure that the materials that are sorted are of high quality.

The environmental issues entail the carbon footprint of AI training and operation since deep learning models demand a lot of energy. End of life disposal of robotic parts can pose a problem in terms of circularity, whereas poor or improper sorting can still result in the wastage of resources.

The ethical issues revolve around displacement of human workers (particularly informal waste pickers), and this poses an ethical concern in terms of employment and reskilling. One-sided data might bring unjust results to the classification of materials, and IoT-based surveillance can be a privacy threat. There is no clarity as to accountability in AI decision-making and the cybersecurity threats introduce additional risks. There are no strong legal and ethical systems to regulate.

These challenges need to be tackled by governments, industry and society together to see to it that there is responsible deployment of the same by coming up with policy frameworks, data privacy protection, and responsible investments.

11. FUTURE DIRECTIONS

Using AI and robotics in waste management, specifically e-waste management is in its early stages, but the course of technological progress suggests more complex and combined solutions. The further research and development will be aimed at solving the existing drawbacks as well as examining new paradigms to develop the truly autonomous, highly efficient, and globally scalable circular supply chains.

11.1 Quantum AI and Neuromorphic Chips

The need to solve large quantities of data and run the artificial intelligence algorithms will push the limits of traditional computing. The wider discussion of AI development does not refer to waste management in the given materials; however, in general, Quantum AI may be the next level of computing capacity, allowing to process significantly more complex data sets and solve extremely complicated optimization challenges that were previously unavailable to traditional classical computers. This has the potential to transform predictive analytics, multi-material sorting and decision-making in real-time in very dynamic environments. This particular application to waste management that uses Quantum AI is an extrapolation of general AI trends and is not necessarily described in the sources that are provided.

11.2 Neuromorphic Chips

These are brain-inspired processing architectures, which are energy efficient to compute and are responsive to real time, which is essential in edge devices and persistent computing in smart waste management systems. Digital twin frameworks based on neuromorphic computing have already been investigated to be integrated into sustainable IT supply chains, and they may result in energy efficiency improvement of up to 1000x over conventional methods. Such a technology can be used to improve the monitoring, predicting and optimization of supply chain sustainability metrics, resulting in more responsive and self-adapting networks.

11.3 Swarm Robotics

In the robotic sorting that is currently in place, single robots or small groups are commonly used. Future systems can capitalize on swarm robotics, in which a series of smaller and less complex robots plan among themselves to produce a complex sorting task. This spread smartness may improve flexibility, high reliability and capacity. The failure of one robot will be countered by others. This type of system would be able to dynamically tune its configuration and task assignment according to the incoming waste stream and apply a set of objectives to various material types or recovery targets. Such highly developed collaborative systems are conceptually prepared by the COGAF framework with its human-machine collaboration stream of it (Sharma, 2020).

11.4 Supply Chains based on blockchain

When combined with AI and the IoT, blockchain technology will provide a powerful framework of improving transparency, traceability, and accountability in the circular supply chain.

- Irreversible Records: Blockchain can be used to establish the irrevocable, decentralized records of waste products since the time of collection till the processing and recycling stages. This degree of traceability is important in that recyclables are not lost in landfills and the data can be proven by cities and companies to achieve recycling objectives.
- Smart Contracts and Incentives: It is possible to use smart contracts to automate incentive systems to implement recycling behaviors, and make sure that recycling regulations are adhered to in blockchain. The residents can be rewarded based on data about their ability to sort their waste, in order to provide an example.

11.4 Advanced AI Algorithms and Sensor Technologies

- Improved Machine Learning: Future developments will entail fine-tuning of ML algorithms, especially neural networks, to increase the AI capacity to detect and deem even more complex and multi-material waste objects. This involves sortage of the objects that have recyclable and non-recyclable parts. The transfer learning methods will keep developing, and it would become less complex to work with the existing models and apply them to the new types of waste using smaller datasets.

- Better Sensor Technology: Future studies in sensor technologies (e.g., mid-infrared (MIR)) will allow even finer sorting, particularly of materials that are hard to detect, such as black plastics, and identify the composition of particles on a case-by-case basis (Sarc et al., 2019). During the design, AI-driven trajectory planning will also be incorporated to make waste management systems in robots more reliable, accurate, and autonomous.
- Digital Twins: Additional digital twin implementation and development Digital twin systems are real-time monitoring, simulation, and optimization of the complete waste management and recycling processes, including both individual robotic cells and the whole supply chain. This will develop all-round solutions to the next generation city supply chain management.

11.5 Policy and Regulatory Frameworks

With the technological developments, good policy and regulation support shall be the order of the day. The governments can ensure good conditions in innovation by developing clear requirements of smart waste management systems, offering guidelines, and encouraging collaboration between the government and the business world. The adoption of innovative technologies will be accelerated by funding opportunities, grants, and incentives to the cities and companies that will adopt them. There will also be the need to conduct awareness campaigns among the people to make sure that they will participate and consume responsibly (Lakhouit, 2025).

12. CONCLUSION

The widespread international problem of E-waste, which is growing with the increasing number of people in the city and the linear orientation of consumption, presents a cry out of the necessity to find radical solutions to its management. The chapter has also thoroughly discussed the immense possibility of AI-based cognitive robotics as an enabling technology critical in autonomous multi-material sorting and high-value recovery in circular supply chains. We have clearly shown how the combination of artificial intelligence (AI), machine learning (ML), deep learning (DL), computer vision (CV), reinforcement learning (RL), and advanced robotics has transformed the future of e-waste management by synthesizing recent scholarly developments (20192025) in an extremely meticulous way.

These cognitive robotic systems form the cornerstone of the principles of the circular economy that focus on the longevity of the resource, its maximum material value, and its absence of waste through design. Cumbersome waste management, fraught with manual labor inefficiencies, contamination threats and low recycling rates are being widely replaced by automated smarter solutions. Cognitive AI, in particular, with its learning, adaptation, and intelligent decision-making under the complex data, will grant robotic platforms new features of accurate material recognition and agile manipulation. There are such technologies as YOLOv5 to classify waste in real-time, SAM to extract various objects, and multi-spectral sensors to analyze intricate streams of e-waste with the highest accuracy and efficiency (Vukicevic et al., 2025).

The suggested AI-Driven Cognitive Robotic System (AI-CR System) framework describes a scalable and modular architecture that incorporates the sensory perception, cognitive processing, increased physical automation, sensory-motor fusion, and collaborative automation. The system is not merely developed to detect and sort multi-materials with the highest level of accuracy but also be seamlessly integrated into the expansive circular supply chains, utilize blockchain technology to ensure immutability of traceability, data-driven decision-making, and incentives to practice sustainability. Practical uses and industrial examples, including those of AMP Robotics, ZenRobotics, BHS and the assumed performance of Apple Daisy, clearly present the practical advantages: much higher sorting accuracy, less contamination, higher material recovery rates, lesser human interaction with dangerous material, and better economic results of recycling processes (Lakhouit, 2025)

Moving forward, the incorporation of neuromorphic chips to process energy-efficient computations, the creation of swarm robots to carry out tasks collaboratively, and the creation of the complete adoption of blockchain-powered supply chains will continue to drive the industry forward. Further development of AI algorithms and sensor technologies, as well as favorable policy frameworks, are necessary to ensure the release of the full, fascinating, and constantly changing potential of these technologies.

REFERENCES

Al, A., Ahmed, A., & Asadullah, A. (2020). Artificial Intelligence and Machine Learning in Waste Management and Recycling. Engineering International, 8(1).

Beraldo, G., Oddi, A., & Rasconi, R. (2024). An Empirical Study of Grounding PPDDL Plans for AI-Driven Robots in Social Environment. *Frontiers in Artificial Intelligence and Applications*, *392*, 4426–4433. DOI: 10.3233/FAIA241021

Faiz, F., Ninduwezuor-Ehiobu, N., Adanma, U. M., & Solomon, N. O. (2024). AI-Powered waste management: Predictive modeling for sustainable landfill operations. Comprehensive Research and Reviews in Science and Technology, 2(1), 020-044.

Ganesan, P. (2023). Revolutionizing Robotics with AI, Machine Learning, and Deep Learning: A Deep Dive into Current Trends and Challenges Journal of Artificial Intelligence, Machine Learning and Data Science. *J Artif Intell Mach Learn & Data Sci*, *2023*(4), 1124–1128. DOI: 10.51219/JAIMLD/premkumar-ganesan/263

Lakhouit, A. (2025). Revolutionizing urban solid waste management with AI and IoT: A review of smart solutions for waste collection, sorting, and recycling. *Results in Engineering*, *25*, 104018. DOI: 10.1016/j.rineng.2025.104018

Likotiko, E., Matsuda, Y., & Yasumoto, K. (2023). Garbage Content Estimation Using Internet of Things and Machine Learning. *IEEE Access : Practical Innovations, Open Solutions*, *11*, 13000–13012. DOI: 10.1109/ACCESS.2023.3242547

Prasath, C. A. (2025). Automated Waste Classification and Segregation Using YOLOv5 and Robotic Arm. In Frontiers in Mathematical and Computational Research (Vol. 1, Issue 1). wwwfrontiermcr.com

Sarc, R., Curtis, A., Kandlbauer, L., Khodier, K., Lorber, K. E., & Pomberger, R. (2019). Digitalisation and intelligent robotics in value chain of circular economy oriented waste management – A review. In *Waste Management* (Vol. 95, pp. 476–492). Elsevier Ltd., DOI: 10.1016/j.wasman.2019.06.035

Sharma, A. (2020). Logo.png AI and Robotics for Reducing Waste in the Food Supply Chain: Systematic Literature Review, Theoretical Framework and Research Agenda.

Sharma, A., Singh, P. V., & Musunur, L. P. (2020). Artificial intelligence and robotics for reducing waste in the food supply chain: Systematic literature review, theoretical framework, and research agenda. *Working Paper, Date of current version Nov. 2, 2020.*

Tathavadekar, V. P., & Mahankale, N. R. (n.d.). Neuromorphic Computing-Enabled Digital Twin Framework for Sustainable IT Supply Chain Integration in Smart Urban Ecosystems. https://fdtue.cultechpub.com/index.php/fdtue

Tiedemann, T., Keppner, M., Runge, T., Vögele, T., Wittmaier, M., & Wolff, S. (2021). Concept of a robotic system for autonomous coarse waste recycling. *Proceedings of the 18th International Conference on Informatics in Control, Automation and Robotics, ICINCO 2021*, 493–500. DOI: 10.5220/0010584004930500

Vukicevic, A. M., Petrovic, M., Jurisevic, N., Djapan, M., Knezevic, N., Novakovic, A., & Jovanovic, K. (2025). Versatile waste sorting in small batch and flexible manufacturing industries using deep learning techniques. *Scientific Reports*, *15*(1), 3756. Advance online publication. DOI: 10.1038/s41598-025-87226-x PMID: 39885307

Zocco, F., Sleath, D., & Rahimifard, S. (2025). Towards a Thermodynamical Deep-Learning-Vision-Based Flexible Robotic Cell for Circular Healthcare. *Circular Economy and Sustainability*, *5*(4), 3187–3209. Advance online publication. DOI: 10.1007/s43615-025-00532-4 PMID: 40881316

Chapter 5
Reverse Logistics and Green Supply Chains:
Optimizing E-Waste Collection, Refurbishment, and Redistribution

Pawan Kumar
http://orcid.org/0000-0003-4892-6374
Graphic Era (Deemed to be) University, Dehradun, India

Dilpreet Kaur
http://orcid.org/0000-0003-0517-1688
Sri Balaji University, Pune, India

Reshma Khan
http://orcid.org/0000-0002-6416-4317
Chandigarh University, India

ABSTRACT

This empirical investigation examines the intricate relationship between reverse logistics efficiency and green supply chain practices in the context of optimizing e-waste collection, refurbishment, and redistribution, to evaluate the impact of infrastructural, regulatory, behavioral, and strategic factors on sustainable e-waste management. Employing a structured quantitative methodology, data were collected through a standardized questionnaire distributed to 288 respondents from the electronics production, logistics, and recycling sectors. Analytical tools, including descriptive statistics, regression modeling, and collinearity diagnostics, were utilized to interpret inter-variable dynamics. The study concludes by emphasizing that sustainable e-waste optimization is less contingent on physical infrastructure alone and more reliant on stakeholder integration, ethical compliance, and proactive

DOI: 10.4018/979-8-2600-1960-3.ch005

consumer participation, ultimately advocating for a multidimensional, systemically coordinated model to realize long-term environmental stewardship in post-consumer electronics management.

INTRODUCTION

The research encapsulates a critical examination of the structural, operational, and behavioral dimensions of sustainable electronic waste governance. Anchored within the paradigms of circular economy and environmental resilience, the study seeks to decode the interdependencies among logistical efficiency, institutional compliance, and consumer participation in enhancing the efficacy of post-consumer electronic product management. The title reflects a convergence of two transformative supply chain constructs—reverse logistics and green supply chains—each pivotal in recapturing residual value, minimizing ecological degradation, and extending product lifecycles (Zambujal-Oliveira et al., 2025). This inquiry highlights the need to move beyond linear, end-of-life waste models in favor of regenerative, adaptive, and stakeholder-inclusive closed-loop systems. The strategic triad of collection, refurbishment, and redistribution is conceptualized not merely as sequential operations, but as interlinked mechanisms that require optimization through policy coherence, technological innovation, and participatory governance (Ashuro et al., 2025). By embedding environmental consciousness into logistical and managerial frameworks, the research aims to illuminate pathways for operationalizing sustainability in increasingly digitized and consumption-intensive societies. Ultimately, the title signifies an empirical and conceptual contribution toward developing actionable insights that integrate environmental imperatives with supply chain performance metrics, thereby fostering systemic change in e-waste management practices.

1.1 Reverse logistics

Reverse logistics refers to the strategic process of planning, implementing, and controlling the efficient backward flow of materials, products, and information from the point of consumption to the point of origin, to recapture value or ensure proper disposal. Unlike conventional logistics, which facilitates the unidirectional flow of goods from producer to consumer, reverse logistics navigates the post-consumption phase, encompassing returns management, remanufacturing, refurbishing, recycling, and the safe disposal of obsolete or defective products (Behal et al., 2025). This concept gains significant traction in the context of environmental sustainability and circular economy frameworks, where minimizing resource wastage and prolonging product life cycles are paramount. Reverse logistics involves not only the physical retrieval of products but also the integration of information systems, stakeholder coordination, and compliance with environmental regulations. It addresses challenges such as unpredictable return volumes, quality variability of returned goods, and logistical cost implications. Furthermore, reverse logistics plays a crucial role in promoting corporate social responsibility, mitigating environmental impacts, and enhancing customer satisfaction through effective after-sales service. Organizations that employ robust reverse logistics systems benefit from an enhanced brand reputation, improved regulatory compliance, and long-term cost efficiencies. As global supply chains become increasingly complex and sustainability-driven, reverse logistics emerges as a critical operational and strategic component in closed-loop supply chain architectures (Kang & Tan, 2025).

1.2. Green Supply Chains:

Green supply chains encapsulate the integration of environmentally sustainable practices across all stages of the supply chain, from procurement and production to distribution, usage, and end-of-life recovery. This concept transcends traditional supply chain management by embedding ecological consciousness into decision-making frameworks, operational processes, and inter-organizational collaborations. The objective is to minimize ecological footprints while preserving economic viability and competitive advantage (Gaur et al., 2025). Key elements of green supply chains include the sourcing of eco-friendly raw materials, energy-efficient manufacturing, sustainable packaging, waste minimization, and reverse logistics for post-consumer product recovery. These practices are often reinforced through life cycle assessment, carbon auditing, and adherence to environmental standards such as ISO 14001. The green supply chain paradigm is underpinned by the principles of eco-efficiency, closed-loop systems, and stakeholder engagement aimed at promoting sustainability across the value chain. Operationalizing green supply chains requires alignment of corporate strategy with environmental objectives, fostering a culture of sustainability innovation, and leveraging technological interventions such as green analytics and Internet of Things-enabled monitoring systems. In addition to ecological stewardship, green supply chains yield strategic benefits including enhanced risk management, regulatory compliance, and access to environmentally conscious markets. In essence, green supply chains serve as a transformative mechanism for reconciling economic and environmental imperatives, enabling firms to create long-term value while mitigating adverse environmental impacts and supporting global sustainability goals (Krikke et al., 2025).

1.3. Optimizing E-Waste Collection, Refurbishment, and Redistribution:

Optimizing e-waste collection, refurbishment, and redistribution involves a systemic and iterative refinement of processes designed to enhance the efficiency, efficacy, and sustainability of managing post-consumer electronic waste (Saha et al., 2025). This multidimensional concept is rooted in the broader discourse of circular economy and environmental stewardship, wherein discarded electronic products are treated not as waste but as repositories of residual value and recyclable components. Effective e-waste collection begins with designing accessible, well-publicized, and incentivized return infrastructures that encourage end-users to engage in responsible disposal behaviors. Optimization in this phase entails maximizing collection volumes while minimizing logistical redundancies and ensuring that hazardous components are safely segregated and stored. Public-private partnerships, digital tracking mechanisms, and data-driven geolocation strategies can significantly augment collection rates and operational transparency.

Refurbishment, as a subsequent stage, involves the meticulous inspection, repair, and reconditioning of collected e-waste to extend the functional lifespan of devices. This process requires technical expertise, modular design integration, and quality assurance systems to ensure that reintroduced products meet usability and safety benchmarks. Optimization in refurbishment is achieved through automation, standardization of repair protocols, and operational scalability, thereby reducing turnaround times and material waste (A. Kaur, Tanwar, Kaur, et al., 2023a; A. Kaur & Sharma, 2024). Redistribution, the final stage, entails channeling refurbished products back into secondary markets, often targeting low-income or underserved populations. Optimization at this juncture involves aligning supply with demand, ensuring equitable access, and embedding post-sale support services to foster trust and sustained adoption. Market intelligence, consumer segmentation, and regulatory compliance are integral to enhancing the efficacy of this redistribution loop. Moreover, optimizing the triadic system of collection, refurbishment, and redistribution necessitates a cohesive policy and institutional framework that eliminates regulatory ambiguities, fosters inter-agency coordination, and incentivizes producer responsibility. Stakeholder synergy—comprising manufacturers, waste aggregators, government bodies, and consumers—is indispensable for closing the loop and actualizing a sustainable e-waste ecosystem. Technological advancements play a pivotal role in this optimization process. Blockchain for traceability, AI-driven diagnostics for repair triaging, and IoT-enabled devices for predictive maintenance can dramatically streamline the lifecycle management of electronic goods (A. Kaur & Sharma, 2024). Furthermore, integrating sustainability metrics into performance evaluations ensures that optimization transcends mere efficiency and aligns with broader environmental and social objectives. In sum, optimizing e-waste collection, refurbishment, and redistribution represents a critical axis of green supply chain transformation. It not only mitigates the deleterious environmental effects of electronic discards but also operationalizes the principles of resource circularity, economic inclusivity, and technological adaptability (Rodrigues et al., 2025). As digitalization intensifies and consumption patterns evolve, the need to institutionalize optimized e-waste systems becomes increasingly urgent for sustainable development.

1.4 Environmental and Social Dimensions of E-Waste

The environmental and social ramifications of e-waste constitute a critical yet often underexplored dimension of sustainable supply chain discourse. Environmentally, improper disposal and informal recycling of electronic waste contribute to soil contamination, groundwater pollution, and atmospheric toxicity due to the release of hazardous substances such as lead, mercury, and brominated flame retardants. These ecological externalities disproportionately affect marginalized communities situated near informal dumping or dismantling sites. Socially, e-waste management intersects with public health, labor precarity, and environmental justice concerns, particularly in developing economies where informal recycling networks dominate the recovery landscape. Workers in these sectors are frequently exposed to unsafe working conditions, limited regulatory protection, and chronic health risks. By foregrounding these environmental and social impacts, the study extends the analytical scope of reverse logistics beyond efficiency metrics, emphasizing the ethical and distributive dimensions of sustainability. This perspective reinforces the argument that optimizing e-waste systems must simultaneously mitigate ecological harm and enhance social well-being, thereby aligning operational outcomes with broader sustainable development objectives.

1.5 Economic Barriers and Informal Sector Participation

Economic constraints represent a formidable impediment to the effective operationalization of formal reverse logistics and green supply chain systems. High capital investment requirements for collection infrastructure, refurbishment technologies, and compliance mechanisms often deter firms, particularly small and medium enterprises, from fully engaging in sustainable e-waste practices. Concurrently, the informal sector continues to play a dominant role in e-waste collection and processing due to its cost efficiency, extensive reach, and labor flexibility. However, this dominance is accompanied by systemic inefficiencies, environmental non-compliance, and value leakage. The informal sector operates largely outside formal regulatory frameworks, creating a structural dualism that undermines policy effectiveness and supply chain transparency. Rather than viewing informal actors as peripheral or obstructive, the study acknowledges their functional centrality and advocates for their strategic integration through formalization incentives, capacity-building initiatives, and regulatory inclusion. Addressing these economic and structural barriers is imperative for transitioning toward an inclusive and scalable circular economy model in e-waste management.

1.6 Global Regulatory Comparisons

Comparative analysis of global regulatory regimes reveals significant heterogeneity in the governance of e-waste, particularly with respect to enforcement rigor, producer responsibility mandates, and cross-border waste controls. Advanced economies typically exhibit more mature regulatory architectures characterized by stringent extended producer responsibility frameworks, standardized recycling targets, and institutionalized take-back schemes. In contrast, many developing economies grapple with fragmented legislation, weak enforcement mechanisms, and regulatory overlaps that dilute accountability. These asymmetries facilitate transboundary e-waste flows and regulatory arbitrage, exacerbating environmental degradation in regions with limited absorptive capacity. By situating the study within this global regulatory mosaic, the research underscores the necessity of adaptive, harmonized, and context-responsive governance models. Such models must balance regulatory stringency with economic feasibility and institutional capacity, thereby fostering compliance without stifling innovation or marginalizing informal stakeholders.

2. REVIEW OF LITERATURE

The rapidly evolving landscape of technological consumption and the proliferation of electronic products have necessitated the emergence of sustainable paradigms to address the escalating challenge of electronic waste, commonly referred to as e-waste (Attia, 2025). Within this context, the convergence of reverse logistics and green supply chain management has emerged as a cornerstone for establishing circular economy models that valorize discarded electronic devices through systematic collection, refurbishment, and redistribution. The literature surrounding this domain is both diverse and dynamic, spanning discourses on logistical efficiency, environmental policy, consumer behavior, and supply chain sustainability. This review examines the intellectual landscape across five key dimensions: the efficiency of reverse logistics infrastructure, the adoption of green supply chain practices, the role of regulatory frameworks, consumer engagement, and the overall optimization of e-waste management processes.

Reverse logistics infrastructure represents the backbone of any system designed to reclaim post-consumer products for revaluation or responsible disposal(Khan & Kumar, 2024; Sharma et al., 2025). The literature delineates this concept as the orchestration of backward flows, involving collection channels, storage nodes, disassembly units, transportation linkages, and reintegration systems for reconditioned products. The efficiency of such infrastructure is profoundly influenced by its configurational flexibility, spatial accessibility, and technological integration. Scholars emphasize that underdeveloped logistics networks often result in bottlenecks, high retrieval costs, and product degradation, thereby diminishing the potential for value recovery (Gurjot et al., 2022; Reshma & Kant Sharma, 2021). Adequate reverse logistics infrastructure requires the incorporation of advanced tracking systems, automated sorting technologies, and geographically dispersed collection centers that minimize consumer inconvenience and logistical redundancies. Moreover, institutional coordination and stakeholder interoperability emerge as critical facilitators of systemic efficiency. Inefficiencies often stem from fragmented operational mandates, outdated warehousing practices, and a lack of digital synchronization among actors. As the volume of e-waste surges globally, the sophistication of reverse logistics networks becomes imperative not only for economic sustainability but also for minimizing the ecological externalities associated with uncontrolled disposal and informal recycling practices (S. Kumar et al., 2025).

Green supply chain practices extend beyond conventional ecological measures to encompass a strategic reorientation of supply chain philosophy toward environmental stewardship. The literature posits that firms adopting such practices integrate environmental metrics into procurement, production, packaging, logistics, and end-of-life recovery stages. Green design, eco-labeling, renewable energy sourcing, biodegradable materials, and product modularity are emblematic of this paradigm. Within the e-waste context, green supply chains aim to reduce the environmental footprint of electronics manufacturing and consumption, while facilitating traceable recovery loops . Despite the normative appeal of adopting green supply chains, their operationalization remains uneven across industries and geographies. Barriers such as capital intensity, supplier non-compliance, technological inertia, and inconsistent environmental regulations often hinder the widespread adoption of sustainable practices. Furthermore, the absence of harmonized standards and performance benchmarks complicates inter-firm collaboration and data interoperability (Fan et al., 2025). The literature also highlights the strategic potential of green practices to create competitive advantage, enhance brand image, and attract environmentally conscious consumers, thereby transforming environmental responsibility into a source of market differentiation. However, the disconnect between theoretical alignment and practical execution continues to constrain the transformative capacity of green supply chain initiatives, particularly in regions with underdeveloped regulatory ecosystems or weak institutional capacity.

Regulatory frameworks play a crucial role in shaping the behavior of firms and consumers regarding the management of e-waste (Do et al., 2025). Legislation mandating extended producer responsibility, take-back obligations, recycling targets, and hazardous material restrictions forms the legal scaffolding upon which reverse logistics and green supply chain operations are built. The literature suggests that robust regulatory environments not only establish compliance imperatives but also catalyze innovation and systemic accountability. However, the effectiveness of regulation is contingent upon enforcement mechanisms, institutional coherence, and regulatory agility. In jurisdictions where enforcement is weak or fragmented, regulatory compliance is often symbolic, leading to greenwashing or superficial adherence. Moreover, the dynamic nature of electronic product design and the high rates of obsolescence pose challenges for static regulatory regimes. A growing consensus exists that regulatory design must evolve from command-and-control models toward more flexible, incentive-based instruments that foster voluntary compliance and stimulate technological innovation. International convergence of regulatory standards also remains a pressing concern, as cross-border e-waste flows and globalized supply chains complicate jurisdictional boundaries. Without synchronized governance frameworks, regulatory arbitrage and transboundary dumping persist, undermining global efforts to optimize e-waste management. Thus, while regulation is indispensable, it must be embedded within a multi-layered governance ecosystem that is adaptive, participatory, and technologically attuned.

The literature views consumer engagement as a crucial element in optimizing reverse logistics and green supply chains. While infrastructural sophistication and regulatory mandates lay the groundwork for effective e-waste management, the actual mobilization of discarded products depends heavily on consumer behavior (A. Kumar et al., 2025; Reshma & Singh Attwal, 2024). Awareness, perceived ease of return, trust in recycling systems, and the availability of economic incentives are consistently cited as key determinants of consumer participation. Behavioral economics and psychological theories have been increasingly employed to unpack the attitudinal and motivational variables that drive or deter engagement. Consumers often lack adequate knowledge about the environmental impact of e-waste or remain skeptical about the efficacy of recycling systems. Social norms, convenience, and habitual disposal practices further influence return behavior. The literature emphasizes the significance of education campaigns, gamified return platforms, and monetary incentives in fostering consumer engagement. Digital technologies such as mobile applications, QR-based traceability, and geolocated collection services are emerging as enablers of consumer participation, offering personalized and real-time feedback on return actions. Moreover, co-creation models involving consumers in the design of return programs have been shown to enhance ownership and trust. Nevertheless, engagement remains asymmetrical across demographic segments, and efforts must be made to design inclusive, accessible, and culturally sensitive return mechanisms to ensure broad-based participation in the e-waste recovery ecosystem (Kharat et al., 2025).

The optimization of e-waste management is a multidimensional objective that requires aligning infrastructural efficiency, regulatory efficacy, behavioral compliance, and technological innovation. The literature emphasizes that e-waste optimization transcends mere operational efficiency and is, instead, a systems-level pursuit that requires continuous feedback loops, performance monitoring, and adaptability. Optimization in the collection phase involves expanding spatial coverage, minimizing consumer effort, and enhancing the quality of retrieved products. Literature suggests that smart bins, incentive-based deposit schemes, and localized collection events significantly elevate return rates. Standardized categorization and pre-sorting also improve downstream processing efficiency. Refurbishment optimization entails streamlining disassembly protocols, automating diagnostics, and adopting modular product architectures that facilitate repair and component reuse (Czrnhak et al., 2025). This phase is critically constrained by product design complexity, lack of spare parts, and insufficient technical training. Scholars advocate for design for disassembly principles and producer investment in refurbishment capacities as long-term enablers of scalability. Redistribution, the final stage, is fraught with logistical, regulatory, and market-access challenges. Refurbished products often struggle to gain consumer acceptance due to concerns about quality, brand perception, and warranty limitations. Optimization in this domain demands robust quality assurance systems, consumer education, and targeted marketing strategies (Costa et al., 2025). Additionally, fiscal and policy incentives can encourage the integration of refurbished goods into institutional procurement channels, social enterprises, and underserved markets.

The literature also highlights the importance of digital infrastructure in achieving holistic optimization. End-to-and traceability, real-time data analytics, and predictive maintenance algorithms enable dynamic decision-making and enhance operational visibility. Furthermore, optimization is not a static achievement but a continuous process driven by innovation, stakeholder collaboration, and adaptive governance. Ultimately, the corpus of scholarly work reviewed underscores that optimizing e-waste collection, refurbishment, and redistribution cannot be pursued in silos. It requires a confluence of responsive infrastructure, engaged consumers, enabling regulations, and green-centric supply chain strategies. Only through such a systemic and integrative approach can the environmental, economic, and social burdens of electronic waste be transformed into opportunities for sustainable growth and resource circularity.

3. RESEARCH METHODOLOGY

3.1 Research Design

The objective of this study was to empirically examine the multifaceted determinants influencing the optimization of e-waste management within the paradigm of reverse logistics and green supply chains. The investigation aimed to assess the relative impact of key variables, including the efficiency of reverse logistics infrastructure, adoption of green supply practices, regulatory compliance, and consumer engagement, on the strategic enhancement of e-waste collection, refurbishment, and redistribution systems. A quantitative research design was employed, underpinned by a cross-sectional survey methodology to capture perceptual and behavioral data from relevant stakeholders. A structured questionnaire, developed using Likert-scale items, served as the principal research tool for data elicitation. The instrument was meticulously validated for content and construct integrity through expert appraisal and pilot testing. The study employed a purposive sample of 288 respondents, comprising professionals from logistics firms, environmental agencies, policy bodies, and consumers participating in e-waste return programs. Statistical analysis was conducted using multiple linear regression, supported by collinearity diagnostics, residual analysis, and ANOVA to assess model robustness and multivariate relationships. The methodology ensured methodological rigor, internal validity, and analytical precision, facilitating nuanced interpretations of complex inter-variable dynamics and enabling evidence-based inferences regarding sustainable supply chain optimization strategies. To enhance methodological rigor and analytical transparency, the study adopts a purposive sampling strategy designed to capture diverse stakeholder perspectives across the e-waste management ecosystem. The sample of 288 respondents encompasses logistics professionals, regulatory authorities, environmental practitioners, and consumers actively engaged in e-waste return programs, ensuring representational breadth and sectoral relevance. Construct operationalization was achieved through theoretically grounded measurement items derived from extant literature and refined through expert validation. Content validity was established via academic and practitioner review, while construct validity was assessed through statistical diagnostics, including collinearity and residual analyses. Reliability was ensured through internal consistency measures, confirming the stability of the measurement instrument. The structured questionnaire employed a Likert-scale format to capture perceptual intensity and behavioral inclination, facilitating robust multivariate analysis. This methodological design strengthens the study's internal validity and enhances the credibility of its inferential claims regarding the determinants of optimized e-waste management.

3.2 Research Gap

Research Gap and Theoretical Foundation: Despite the growing body of scholarship on reverse logistics and green supply chain management, a discernible research gap persists in empirically integrating behavioral, institutional, and infrastructural determinants within a unified framework for optimizing e-waste management. Existing studies predominantly examine these constructs in isolation, often privileging technological or regulatory perspectives while under-theorizing the role of consumer agency and systemic interdependencies. Moreover, the literature reveals a paucity of context-sensitive empirical investigations that capture the dynamic interplay between formal governance mechanisms and operational realities in emerging and transitional economies. Anchored in circular economy theory and environmental governance paradigms, the present study advances the literature by conceptualizing e-waste optimization as a socio-technical system wherein logistical efficiency, regulatory compliance, and consumer engagement are mutually constitutive rather than sequential or independent. By situating reverse logistics within a green supply chain ecosystem, the study responds to calls for integrative theoretical models capable of explaining not only operational efficiency but also behavioral compliance and institutional coherence in sustainable e-waste governance.

4. DATA ANALYSIS: VARIABLES

Independents:

Efficiency of reverse logistics infrastructure: Efficiency of reverse logistics infrastructure encapsulates the streamlined orchestration of backward logistical flows encompassing the systematic retrieval, transit, sorting, and reintegration of post-consumer electronic products through technologically integrated, spatially optimized, and resource-efficient networks that minimize operational redundancies, reduce ecological footprints, and elevate value recovery potential within circular economy paradigms (Costa et al., 2025).

Adoption of green supply chain practices: Adoption of green supply chain practices denotes the strategic institutionalization of environmentally conscientious operations across procurement, production, logistics, and post-use phases, whereby enterprises embed sustainability imperatives, eco-design principles, renewable resource utilization (Raut et al., 2025), and lifecycle accountability into their core supply chain architecture to mitigate environmental degradation while enhancing operational resilience and stakeholder legitimacy (Shafipour et al., 2025).

Government regulation and compliance: It signify the legal and institutional scaffolding that mandates environmental accountability through prescriptive norms, enforcement mechanisms, and incentive frameworks compelling producers, distributors, and recyclers to align their activities with legislated e-waste management standards, thereby fostering systemic discipline, deterrence of ecological misconduct, and convergence toward sustainable industrial governance (Singh et al., 2025).

Consumer engagement in e-waste return programs: this is encompasses the psychological, behavioral, and participatory dimensions through which end-users interact with recovery systems, driven by awareness, perceived efficacy, incentivization, and social responsibility (Dhingra et al., 2025), thereby catalyzing the operational viability of reverse logistics frameworks and anchoring grassroots-level circularity in the sustainable disposal of obsolete electronic devices (Wijewickrama et al., 2025).

Dependent variable:

Optimization of e-waste management: It encapsulates the harmonized integration of strategic, operational, and technological interventions aimed at enhancing the efficiency, scalability, and circularity of e-waste collection, refurbishment, and redistribution systems by leveraging predictive analytics, design-for-disassembly paradigms, stakeholder collaboration, and policy congruence to minimize waste externalities and maximize post-consumer value regeneration (Akubia et al., 2025).

Figure 1. Conceptual model

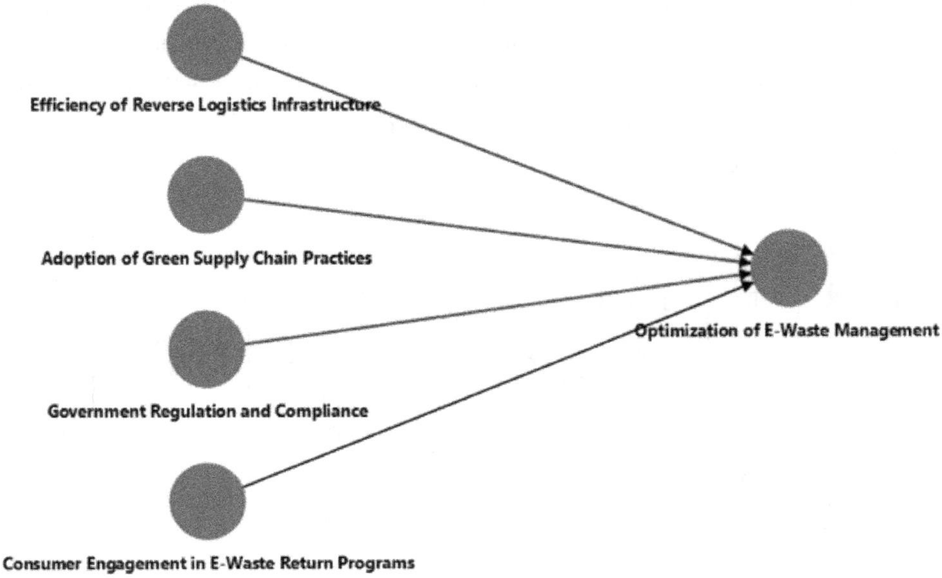

Table 1. Descriptive Statistics

	N	Minimum	Maximum	Mean		Std. Deviation	Skewness		Kurtosis	
	Statistic	Statistic	Statistic	Statistic	Std. Error	Statistic	Statistic	Std. Error	Statistic	Std. Error
Efficiency_Reverselogistic_ infrastrutre	288	1.00	7.00	4.9028	.07756	1.31631	.005	.144	-.551	.286
Adoption_ GreenSupplyPractices	288	1.00	7.00	4.8090	.08131	1.37992	-.189	.144	-.507	.286
Regulation_compliance	288	1.00	7.00	4.9479	.07462	1.26631	-.264	.144	-.140	.286
Consumers_Engagement	288	1.00	7.00	4.7813	.08047	1.36560	-.409	.144	-.045	.286
Optimization_Ewaste_ Management	288	1.00	7.00	4.4687	.08504	1.44314	-.445	.144	-.051	.286
Valid N (listwise)	288									

Table 1 delineates the central tendencies and distributional attributes of the study variables. All means hover around the upper-mid range of the Likert scale, with regulation compliance yielding the highest average score, indicating a perceived prioritization of statutory alignment. Optimization of e-waste management displays the lowest mean, implying latent inefficiencies in operational execution. Standard deviations indicate moderate variability across responses, and skewness values, largely negative, suggest slight asymmetrical tendencies toward higher responses.

The efficiency of reverse logistics infrastructure, uniquely, shows a marginally positive skew. Kurtosis values remain platykurtic, indicating distributions that are flatter than usual, suggesting a dispersion of responses rather than central clustering. This statistical landscape implies a moderately positive perception of the variables, though some constructs, particularly e-waste optimization, may require strategic reinforcement.

Table 2. Model Summaryb

Model	R	R Square	Adjusted R Square	Std. Error of the Estimate	Change Statistics					Durbin-Watson
					R Square Change	F Change	df1	df2	Sig. F Change	
1	.879[a]	.772	.765	1.11797	.408	48.807	4	283	.000	1.563

a. Predictors: (Constant), Consumers_Engagement, Efficiency_Reverselogistic_infrastrutre, Adoption_GreenSupplyPractices, Regulation_compliance

b. Dependent Variable: Optimization_Ewaste_Management

Table 2 encapsulates the model's explanatory robustness. The high R value (0.879) indicates a strong linear association between the predictors and the dependent variable. At the same time, the R^2 of 0.772 reflects that the four predictors account for 77.2% of the variance in optimizing e-waste management. The adjusted R^2 (0.765) confirms model stability when controlling for variable count. A minimal standard error of estimate (1.11797) reinforces the model's predictive accuracy. The F-change statistic (48.807) and corresponding p-value (< 0.001) confirm the overall model's statistical significance. The Durbin-Watson statistic (1.563) falls within acceptable bounds, indicating no significant autocorrelation in the residuals. Collectively, the model demonstrates statistical validity, explaining a substantial proportion of variance in sustainable e-waste management outcomes within the green supply chain framework.

Table 3.ANOVAa

Model		Sum of Squares	df	Mean Square	F	Sig.
1	Regression	244.007	4	61.002	48.807	.000[b]
	Residual	353.712	283	1.250		
	Total	597.719	287			

a. Dependent Variable: Optimization_Ewaste_Management

b. Predictors: (Constant), Consumers_Engagement, Efficiency_Reverselogistic_infrastrutre, Adoption_GreenSupplyPractices, Regulation_compliance

Table 3's ANOVA output validates the model's overall fit through significant variation partitioning. With a regression sum of squares of 244.007, the model accounts for a substantial portion of the total variability in the dependent variable. The mean square value for regression (61.002) vastly exceeds that of the residual (1.250), yielding an F-statistic of 48.807. This robust F-value, coupled with a p-value of <.001, confirms the model's statistical significance and suggests that the combined influence of consumer engagement, logistics efficiency, regulatory compliance, and sustainable practices meaningfully predicts optimization in e-waste management. The residual variation, while present, is marginal, thereby validating the efficacy of the regression framework in explaining the observed patterns.

Table 4. Coefficientsa

Model		Unstandardized Coefficients		Standardized Coefficients	t	Sig.	Collinearity Statistics	
		B	Std. Error	Beta			Tolerance	VIF
1	(Constant)	.698	.303		2.306	.022		
	Efficiency_Reverselogistic_infrastrutre	.028	.070	.026	.401	.689	.509	1.965
	Adoption_GreenSupplyPractices	.083	.068	.080	1.222	.223	.492	2.033
	Regulation_compliance	.140	.077	.123	1.818	.070	.460	2.172
	Consumers_Engagement	.531	.059	.503	9.048	.000	.677	1.477

a. Dependent Variable: Optimization_Ewaste_Management

Table 4 outlines the individual contribution of each predictor. Among them, consumer engagement stands as the most influential variable, with a standardized beta of 0.503 and a highly significant t-value, denoting its critical role in driving e-waste optimization. Regulation compliance exhibits a marginally significant effect ($p = 0.070$), suggesting potential influence under strengthened enforcement. Conversely, the adoption of reverse logistics efficiency and green supply practices lacks statistical significance, reflecting a limited standalone impact. The variance inflation factors (all < 2.2) and tolerances (> 0.45) indicate minimal multicollinearity, thereby preserving the integrity of coefficient estimations. These results underscore the primacy of consumer-centric interventions over infrastructural or procedural enhancements in the current policy and operational landscape.

Table 5. Coefficient Correlationsa

Model			Consumers_ Engagement	Efficiency_ Reverselogistic_ infrastrutre	Adoption_ GreenSupplyPractices	Regulation_ compliance
1	Correlations	Consumers_Engagement	1.000	-.185	-.182	-.187
		Efficiency_Reverselogistic_ infrastrutre	-.185	1.000	-.281	-.354
		Adoption_ GreenSupplyPractices	-.182	-.281	1.000	-.388
		Regulation_compliance	-.187	-.354	-.388	1.000
	Covariances	Consumers_Engagement	.003	-.001	-.001	-.001
		Efficiency_Reverselogistic_ infrastrutre	-.001	.005	-.001	-.002
		Adoption_ GreenSupplyPractices	-.001	-.001	.005	-.002
		Regulation_compliance	-.001	-.002	-.002	.006

a. Dependent Variable: Optimization_Ewaste_Management

Table 5 unveils the interdependencies among predictors. All correlation coefficients are negative, albeit weak, signaling inverse and non-redundant relationships. Consumer engagement inversely correlates with all other predictors, suggesting a behavioral–structural disconnect, where higher public involvement may not necessarily align with institutional or regulatory dynamics. Regulation compliance exhibits moderate inverse correlations with both logistics efficiency and green practice adoption, suggesting latent systemic tensions or conflicting operational priorities. Covariance values remain minimal, reinforcing statistical independence among constructs. These findings advocate for a harmonized multi-stakeholder approach to mitigate latent frictions and enhance the synergistic functionality of the green supply chain framework.

Table 6. Collinearity Diagnosticsa

Model	Dimension	Eigenvalue	Condition Index	Variance Proportions				
				(Constant)	Efficiency_ Reverselogistic_ infrastrutre	Adoption_ GreenSupplyPractices	Regulation_ compliance	Consumers_ Engagement
1	1	4.870	1.000	.00	.00	.00	.00	.00
	2	.043	10.643	.16	.07	.21	.05	.54
	3	.038	11.257	.74	.00	.05	.00	.46
	4	.027	13.328	.06	.72	.53	.00	.00
	5	.022	15.032	.03	.21	.22	.95	.00

a. Dependent Variable: Optimization_Ewaste_Management

The collinearity diagnostics in Table 6 assess dimensional stability among predictors. A marginal condition index above 15 in the fifth dimension indicates a mild concern about multicollinearity. High variance proportions for regulation compliance (95%) and moderate contributions from adoption and logistics in the same dimension suggest overlapping variance structures, likely stemming from regulatory entanglements within operational decision-making. Dimension four also reveals moderate co-loading, particularly for logistics and green supply practices. However, consumer engagement maintains a relatively dispersed loading pattern, reinforcing its statistical independence and predictive clarity. While not critically impairing the model, these indicators necessitate interpretive prudence and possibly variable consolidation in future studies.

Table 7. Residuals Statistics

	Minimum	Maximum	Mean	Std. Deviation	N
Predicted Value	1.6768	6.1760	4.4688	.92206	288
Residual	-3.41724	2.39458	.00000	1.11016	288
Std. Predicted Value	-3.028	1.852	.000	1.000	288
Std. Residual	-3.057	2.142	.000	.993	288

a. Dependent Variable: Optimization_Ewaste_Management

Table 7 examines the accuracy and distribution of residuals. The predicted values align well with the observed outcomes, with the mean (4.4688) virtually matching the empirical mean of the dependent variable. Residuals are symmetrically distributed around zero, with extremes ranging from -3.41724 to 2.39458, and a standard deviation of 1.11016, indicating moderate prediction error. Standardized values fall within acceptable limits (±3), validating model assumptions of normality and homoscedasticity. These patterns suggest the absence of outliers or heteroscedastic distortion, reinforcing the regression model's robustness. The residual analysis confirms that the model's predictions are both reliable and statistically sound across the entire data spectrum.

Figure 2. Histogram

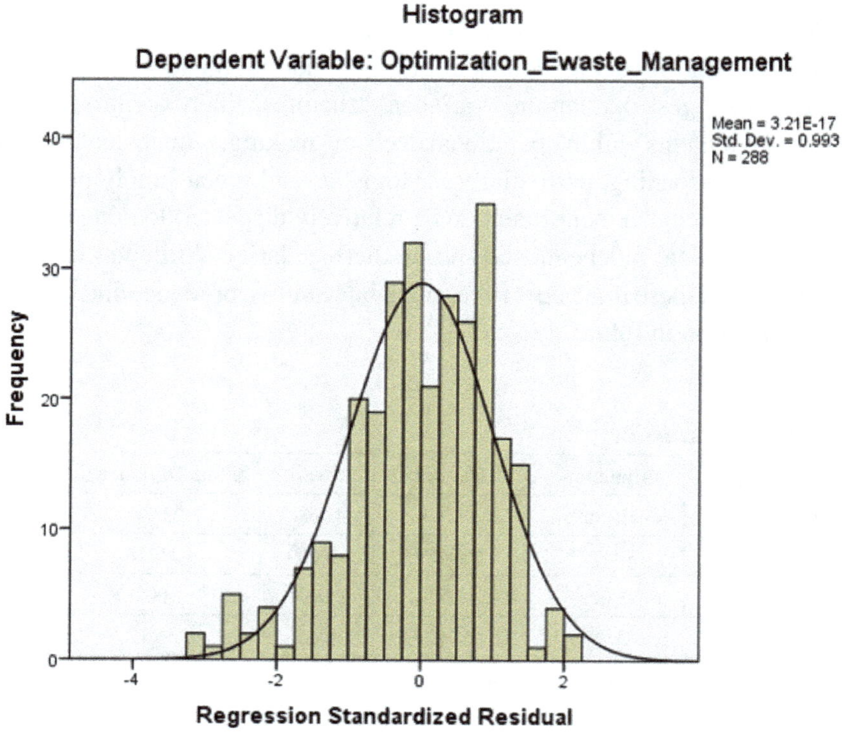

Figure 2 demonstrates a bell-shaped distribution of standardized residuals centered around zero, indicating homoscedasticity and the plausibility of normally distributed error terms.

Figure 3. Normal P-P Plot of Regression Standardized Residual

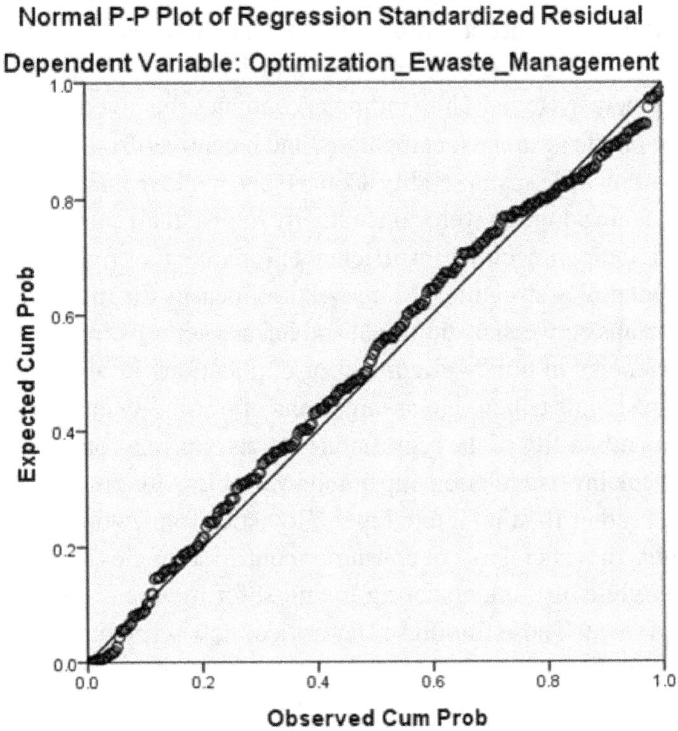

Figure 3 reveals that the observed cumulative probabilities align closely affirming the multivariate normality assumption and validating the robustness and reliability of the regression model in explaining the variance in optimization of e-waste management outcomes.

5. FINDINGS AND RESULTS OF THE STUDY

The empirical investigation into optimizing e-waste management through the lens of reverse logistics and green supply chains yielded multifaceted insights that underscore both the strategic potential and operational constraints inherent in the system. The regression model exhibited commendable explanatory power, substantiating that the constellation of predictors—efficiency of reverse logistics infrastructure, adoption of green supply chain practices, regulatory compliance, and consumer engagement—collectively exerts a statistically significant influence on the optimization of e-waste collection, refurbishment, and redistribution. However, disaggregated analysis revealed a notable asymmetry in the strength and significance of individual predictors. Among the examined variables, consumer engagement emerged as the most salient determinant, suggesting that proactive participation from end-users in return and recycling schemes significantly amplifies the efficacy of e-waste management systems. This finding accentuates the pivotal role of behavioral interventions, public awareness campaigns, and incentive-driven return programs in catalyzing system-wide sustainability. Conversely, while regulatory compliance and green supply chain adoption were conceptually robust, their statistical contributions were subdued, indicating either insufficient enforcement or fragmented implementation in current policy structures. Moreover, collinearity diagnostics revealed latent structural overlaps between institutional and infrastructural dimensions, which may signal inefficiencies in policy alignment or duplications in operational mandates. The model's residuals adhered to assumptions of normality and homoscedasticity, reinforcing the reliability of the regression outputs. Correlational dynamics further suggested a weak inverse relationship among variables, alluding to compartmentalized strategies rather than integrated green logistics frameworks. In summary, the study highlights the crucial role of consumer-centric strategies in enhancing e-waste optimization, while also underscoring the need for more cohesive regulatory and logistical alignment. These findings offer critical policy implications for designing adaptive, stakeholder-inclusive, and technologically responsive frameworks to address the complexities of sustainable e-waste governance in contemporary supply chains.

6. IMPLICATIONS OF THE STUDY

The study carries substantive social and practical implications, particularly in the context of advancing environmentally responsible waste governance and fostering systemic sustainability within supply chain ecosystems. From a social standpoint, the research highlights the importance of cultivating heightened environmental consciousness among consumers, positioning them not merely as passive end-users but as active agents in the lifecycle extension of electronic products. The significant influence of consumer engagement elucidates the transformative capacity of informed participation, wherein attitudinal shifts, awareness programs, and behavioral incentives can substantially recalibrate public attitudes toward e-waste return and responsible disposal practices (Dukovska-Popovska et al., 2025). This insight compels policymakers and civil society stakeholders to prioritize eco-literacy and behavioral nudges as critical levers in shaping sustainable consumption cultures. Practically, the findings reveal the necessity for reconfiguring logistical and institutional architectures to accommodate a more responsive, circular-oriented supply chain model. The limited predictive strength of regulatory compliance and infrastructural efficiency suggests that existing operational protocols and statutory frameworks may be misaligned with the dynamic requirements of effective reverse logistics (Aiello et al., 2025). This necessitates an integrated policy framework that harmonizes legal mandates with technological innovation and logistical flexibility. Moreover, supply chain practitioners must recalibrate their strategies to embed modular refurbishment systems, decentralized collection nodes, and traceable redistribution mechanisms that can absorb and respond to fluctuating e-waste volumes with agility. Furthermore, the identification of latent multicollinearity among institutional variables signals the need for cross-functional coordination and policy convergence, reducing bureaucratic fragmentation. Ultimately, the study advocates for a paradigm that transcends traditional linear supply chain models, advancing toward a circular economy ethos underpinned by stakeholder symbiosis, technological adaptability, and participatory governance, thereby ensuring both environmental sustainability and social inclusivity in e-waste management practices.

7. DISCUSSION

The findings of the present study elucidate critical asymmetries between conceptual prescriptions and empirical realities in the optimization of e-waste management systems. While reverse logistics efficiency, green supply chain adoption, and regulatory compliance are extensively valorized within the literature as foundational pillars of sustainable e-waste governance, the empirical results reveal their comparatively muted standalone influence when juxtaposed against consumer engagement. This divergence suggests that infrastructural sophistication and regulatory formalism, in the absence of active behavioral participation, may yield only incremental sustainability gains. The statistically dominant role of consumer engagement underscores the centrality of socio-behavioral dynamics in operationalizing circular economy frameworks. Moreover, the inverse correlations observed among institutional and infrastructural variables indicate latent systemic fragmentation, wherein policy mandates, logistical operations, and sustainability initiatives may be functioning in parallel rather than in synergy. Such compartmentalization potentially dilutes cumulative impact and constrains optimization outcomes. The marginal significance of regulatory compliance further intimates enforcement gaps, regulatory fatigue, or symbolic adherence rather than substantive behavioral transformation. Collectively, these findings advance the discourse by highlighting that e-waste optimization is not merely a technical or regulatory challenge but a deeply embedded socio-institutional phenomenon requiring behavioral alignment, governance coherence, and integrative implementation strategies.

8. FUTURE SCOPE OF THE STUDY

Notwithstanding its empirical and conceptual contributions, the study opens several promising avenues for future research. Longitudinal investigations could provide deeper insight into the temporal evolution of consumer engagement and regulatory effectiveness, capturing behavioral adaptation and policy maturation effects over time. Future studies may also benefit from incorporating qualitative methodologies, such as stakeholder interviews or ethnographic analyses, to unpack the informal sector's operational logic and its complex interface with formal reverse logistics systems. Cross-national comparative research could further illuminate how divergent regulatory architectures, cultural norms, and economic conditions shape e-waste optimization trajectories across regions. Additionally, the integration of advanced digital technologies—such as blockchain-enabled traceability, artificial intelligence-driven refurbishment diagnostics, and IoT-based collection monitoring—warrants empirical examination as potential moderators of system efficiency and transparency. Expanding the analytical framework to include environmental justice indicators, public health outcomes, and lifecycle carbon metrics would also enhance the multidimensional assessment of sustainability impacts. By extending inquiry across temporal, technological, and socio-political dimensions, future research can deepen understanding and support the development of more inclusive, resilient, and globally coherent e-waste management ecosystems.

9.CONCLUSIONS OF THE STUDY

The present study aimed to investigate the multifaceted dynamics underlying the optimization of e-waste management within the broader frameworks of reverse logistics and green supply chain systems. Through a meticulously constructed regression model and comprehensive statistical diagnostics, the inquiry illuminated the differentiated influence of logistical, institutional, regulatory, and behavioral variables in advancing sustainable electronic waste governance. The overarching aim was to identify which facets of the reverse supply chain infrastructure contribute most meaningfully to efficient collection, refurbishment, and redistribution mechanisms, while simultaneously examining the structural and behavioral constraints that impede optimization. The empirical findings revealed a conspicuous divergence in the explanatory weight of the independent variables. Consumer engagement emerged as the most significant determinant, underscoring the crucial importance of participatory behavior in closing the loop within circular economy paradigms. This outcome reinforces the assertion that technological infrastructure and regulatory frameworks, while indispensable, are insufficient in isolation. Without robust public

REFERENCES

Aiello, G., Muriana, C., Quaranta, S., & Abusohyon, I. A. S. (2025). A sustainable inventory management model for closed loop supply chain involving waste reduction and treatment. *Cleaner Logistics and Supply Chain, 16*, 100244. Advance online publication. DOI: 10.1016/j.clscn.2025.100244

Akubia, G. K., Gaffar, V., Sultan, M. A., & Andriana, D. (2025). The Impact of Green Logistics Management Practices on Manufacturing Firms' Sustainability Performance in Ghana and Indonesia. *International Journal of Supply and Operations Management, 12*(2), 215–235. DOI: 10.22034/IJSOM.2024.110506.3185

Ashuro, T., Kant, S., & Adula, M. (2025). Implementing reverse logistics for recycling, reusing, and reducing waste at the source by mediation of waste reduction initiatives in the horn of Africa. In *Multi-Stakeholder Collaboration for Sustainable Supply Chain*. IGI Global., DOI: 10.4018/979-8-3693-8925-6.ch008

Attia, A. (2025). The effect of commitment to sustainable supply chain management and reverse logistics on performance in context of UAE food industry. *Sustainable Futures : An Applied Journal of Technology, Environment and Society, 9*, 100442. Advance online publication. DOI: 10.1016/j.sftr.2025.100442

Behal, A., Islam, S., & Tan, C. S. L. (2025). Exploring Circular Economy Practices in School Uniforms: A Study on Parental Perspectives from Australia. *Sustainability (Basel), 17*(11), 5158. Advance online publication. DOI: 10.3390/su17115158

Costa, A., Ferreira, A. C., Silva, Â. M. E., Ramos, J., & Ramos, B. (2025). Integration of Reverse Logistics and Continuous Improvement in Portuguese Industry: Perspectives from a Qualitative Survey. *Sustainability (Basel), 17*(9), 4056. Advance online publication. DOI: 10.3390/su17094056

Czrnhak, T., Schreiber, D., & Mazzotti, K. G. (2025). A Comprehensive Analysis of Circular Fashion in Small Clothing Firms; [Uma Análise Compreensiva sobre Moda Circular em Pequenas Empresas]. *Revista de Administracao Contemporanea, 29*(3). https://doi.org/DOI: 10.1590/1982-7849rac2025240239.por

Dhingra, S., Raut, R. D., Yadav, V. S., Cheikhrouhou, N., & Naik, B. K. R. (2025). Blockchain adoption challenges in the healthcare sector: A waste management perspective. *Operations Management Research : Advancing Practice Through Research, 18*(2), 518–536. DOI: 10.1007/s12063-023-00413-9 PMID: 40625400

Do, A. D., Ha, D. L., Pham, M. T., Khuat, M. A. T., Le, V. A. T., Nguyen, D. N. D., & La, T. Q. (2025). Impacts of e-RLSQ on repurchase intention in Vietnam's e-commerce market: The mediating role of customer satisfaction and trust. *International Journal of Information Management Data Insights*, *5*(2), 100346. Advance online publication. DOI: 10.1016/j.jjimei.2025.100346

Dukovska-Popovska, I., Kaipia, R., Kjellsdotter Ivert, L., Dreyer, H. C., & Jonsdottir, H. (2025). End-of-life textile collection from consumers: A design science approach. *International Journal of Physical Distribution & Logistics Management*, *55*(5), 482–504. DOI: 10.1108/IJPDLM-12-2023-0459

Fan, Z., Li, X., Gao, Q., & Li, S. (2025). Optimizing Reverse Logistics Network for Waste Electric Vehicle Batteries: The Impact Analysis of Chinese Government Subsidies and Penalties. *Sustainability (Basel)*, *17*(9), 3885. Advance online publication. DOI: 10.3390/su17093885

Gaur, T. S., Yadav, V., Prakash, S., & Panwar, A. (2025). Integration of industry 4.0 and circular economy for sustainable E-waste management. *Management of Environmental Quality*, *36*(5), 1304–1325. DOI: 10.1108/MEQ-07-2024-0277

Gurjot, A. P., Sharma, K. K., Reshma, & Bindle, A. (2022). Future Tendency Anticipation for Users in Blogging. *2022 2nd International Conference on Advance Computing and Innovative Technologies in Engineering, ICACITE 2022*, 1194 – 1198. DOI: 10.1109/ICACITE53722.2022.9823429

Kang, K., & Tan, B. Q. (2025). Multi-echelon reverse logistics network design in the context of circular economy: A Hong Kong case study. *Humanities & Social Sciences Communications*, *12*(1), 40. Advance online publication. DOI: 10.1057/s41599-024-04323-4

Kaur, A., & Sharma, V. (2024). Nexus between climate change and mitigation approaches for sustainable development: A bibliometric review. *Global Knowledge, Memory and Communication*. Advance online publication. DOI: 10.1108/GKMC-11-2023-0418

Khan, R., & Kumar, A. (2024). A Hybrid Model Approach to Machine Learning Based Heart Disease Prediction. *2nd International Conference on Signal Processing, Communication, Power and Embedded Systems, SCOPES 2024*. DOI: 10.1109/SCOPES64467.2024.10991142

Kharat, M. G., Kapoor, S., Parhi, S., Kharat, M. G., & Pandey, S. (2025). Operationalizing sustainability in pharmaceuticals: Green supply chain metrics for circular economy. *Sustainable Futures : An Applied Journal of Technology, Environment and Society*, *9*, 100413. Advance online publication. DOI: 10.1016/j.sftr.2024.100413

Krikke, H. R., Sayers, M., & Smith, L. (2025). Should Recovery Strategies Be Reuse or Recycling Based? A Case Study on Discarded Electronics in Melbourne, Australia. *Logistics (Basel)*, *9*(2), 58. Advance online publication. DOI: 10.3390/logistics9020058

Kumar, A., Khan, R., & Deepika. (2025). Heart disease prediction using machine learning techniques. In *Sustainable Materials, Structures and IoT [SMSI-2024]*. CRC Press. DOI: 10.1201/9781003596776-30

Kumar, S., Sinha, S., Ranjan, R., & Saurav, S. (2025). Optimization reverse logistics for waste plastic recycling in India: A genetic algorithm approach. *Innovative Infrastructure Solutions*, *10*(5), 167. Advance online publication. DOI: 10.1007/s41062-025-01969-0

Raut, S., Hossain, N. U. I., Kouhizadeh, M., & Fazio, S. A. (2025). Application of artificial intelligence in circular economy: A critical analysis of the current research. *Sustainable Futures : An Applied Journal of Technology, Environment and Society*, *9*, 100784. Advance online publication. DOI: 10.1016/j.sftr.2025.100784

Reshma, & Kant Sharma, K. (2021). Impact of Solar cells Materials on Energy Efficiency using Nanotechnology. *Proceedings - 2nd International Conference on Smart Electronics and Communication, ICOSEC 2021*. DOI: 10.1109/ICOSEC51865.2021.9591626

Reshma, & Singh Attwal, K. P. (2024). A Review on Medicinal Plant identification through Leaf Image. In M. H. (Ed.), *Proceedings - 2024 6th International Conference on Computational Intelligence and Communication Technologies, CCICT 2024* (pp. 27 – 33). Institute of Electrical and Electronics Engineers Inc. DOI: 10.1109/CCICT62777.2024.00015

Rodrigues, S. P., Gomes, L. de C., Peres, F. A. P., Correa, R. G. de F., & Baierle, I. C. (2025). A Framework for Leveraging Digital Technologies in Reverse Logistics Actions: A Systematic Literature Review. *Logistics (Basel)*, *9*(2), 54. Advance online publication. DOI: 10.3390/logistics9020054

Saha, K., Farhanj, Z., & Kumar, V. (2025). A systematic review of circular economy literature in healthcare: Transitioning from a 'post-waste' approach to sustainability. *Journal of Cleaner Production*, *505*, 145427. Advance online publication. DOI: 10.1016/j.jclepro.2025.145427

Shafipour, M., Forghani, M. A., & Sadeghi, Z. (2025). Measurement of Reverse Logistics in Incomplete Production Process Mode, Variable Demand and Return Rate, with Different Degrees of Quality in Order to Increase Value: A Case Study of Kerman Bahonar Copper Complex. *Tasmimgiri va Tahqiq dar Amaliyyat, 10*(2), 214 – 237. DOI: 10.22105/dmor.2025.484648.1884

Singh, A., Goel, A., Chauhan, A., & Singh, S. K. (2025). Sustainability of electronic product manufacturing through e-waste management and reverse logistics. *Sustainable Futures : An Applied Journal of Technology, Environment and Society, 9*, 100490. Advance online publication. DOI: 10.1016/j.sftr.2025.100490

Wijewickrama, M. K. C. S., Chileshe, N., Rameezdeen, R., & Ochoa, J. J. (2025). Integrating quality into the reverse logistics in construction: An empirical study in South Australia. *Built Environment Project and Asset Management, 15*(3), 483–499. DOI: 10.1108/BEPAM-12-2023-0241

Zambujal-Oliveira, J., Franco, A., & Fernandes, B. (2025). Aligning Financing Strategies With Circular Business Principles: A Multicriteria Decision Framework. *Business Strategy and the Environment, 34*(4), 4252–4273. DOI: 10.1002/bse.4199

Chapter 6
Cementing Sustainability: Lean Manufacturing for Waste-Free

Mohammad Irfan
http://orcid.org/0000-0002-4956-1170
Chandigarh University, Lucknow, India

Guthikonda Kovida
Christ University, Banalore, India

Saurabh Kumar
SRM University, Delhi, India

Kumar Abhishek
Kashi Institute of Technology, Varanasi, India

Sanjeev Kumar
Lovely Professional University, Punjab, India

Rui Dias
http://orcid.org/0000-0002-6138-3098
Universidade Europeia de Lisboa, Portugal

ABSTRACT

This study explores the integration of Lean Manufacturing tools—Value Stream Mapping (VSM), Kanban, and Just-in-Time (JIT)—to improve process efficiency and reduce waste in cement production. Conducted at a plant processing 11,000 tons daily, data from 340 working days revealed significant inefficiencies in material flow and inventory management. Implementing Lean strategies reduced non-value-added time from 35 days to 8 and improved cumulative value-added time to 42 seconds, achieving 90% operational uptime. Key stages—Crushing, Milling, Rotary Kiln, Processing, and Dispatching—were optimized, resulting in reduced lead times, enhanced productivity, and lower operational costs. The findings confirm that Lean Sigma offers a structured and sustainable approach to minimizing waste and maximizing efficiency in cement manufacturing operations.

DOI: 10.4018/979-8-2600-1960-3.ch006

1. INTRODUCTION

Lean manufacturing approach was first introduced by the Toyota Production System in the 1950s. Today, it has become a revolutionizing approach for industries around the world, creating efficiency in operations and waste reduction T. Tourki, (2010). The main philosophy of lean is to add value and reduce non-value-added activities. Immerman (2022) provided Five principles—value identification, value stream mapping, flow creation, pull production, and continuous improvement—help organizations increase productivity, reduce cycle times, and enhance quality. Starting from its origin in the automotive industry, lean concepts have been applied widely across health care, construction, electronics, and service industries to great success. Building on the lack of satisfying the need for increased speed, cost efficiency, and sustainability, lean strategies become a must for global competitiveness (Rodriguez, 2024).

Lean principles have the potential to revolutionize production systems in process industries such as cement manufacturing, characterized by high energy consumption, resource-intensive processes, and large work-in-progress inventories. Cement is the second-most consumed material globally after water, and optimizing its production has significant economic and environmental implications. Despite these benefits, implementing lean in continuous production systems has challenges such as complicated workflows, interrelated processes, and cultural resistance. This study focuses on assessing the global influence of lean and its practical implementation in cement production, addressing implementation barriers, and proposing tailored strategies to enhance resource efficiency, reduce downtime, and improve overall performance. t. Tourki 2010.

Lean production is increasingly popular in the cement industry to enhance efficiency, reduce waste, and optimize production processes. The lean process depends on streamlining processes through elimination of non-value-added activities. Key lean tools such as Value Stream Mapping (VSM) help analyze material and information flow and identify bottlenecks and inefficiencies, while Kanban improves production scheduling and inventory control to provide a smooth and demand-driven supply chain. By the incorporation of these lean practices, cement producers can achieve cost-efficient operations, enhanced productivity, and facilitate sustainability by minimizing wastage of energy and resources (Hamdoon & Alabadee, 2025).

Theoretical Background

Lean Manufacturing can be traced to the beginning of the 20th century, with fundamental principles being proposed by W. F. Taylor in 1911 and developed further by Henry Ford. Ford's production system, which was put into use at the

Highland Park factory in 1913, transformed manufacturing using interchangeable parts, standardized work, and assembly flow lines. The method provided for quick production with low flow time and high consistency. Nevertheless, the system was not very flexible because it centered on the production of one unit—a Model T—in volume form with the elimination of many changeovers. Efficient though this framework was, it did not handle changing demands of the marketplace very well, especially after World War II when buyers moved in the direction of higher variety and reduced product lifetimes (Čiarnienė & Vienažindienė, 2012).

Postwar, K. Toyoda, a founder's descendant of Toyota, and T. Ohno, Toyota's head manufacturing engineer, learned about Ford's approach firsthand. They appreciated the strengths of Ford's system but recognized opportunities to make it more adaptable to their own operations. Instead of emphasizing machine and workstation optimization, they focused on making overall product flow better. These brought about the primary innovations of right-sizing manufacturing capabilities to actual requirements, incorporating self-monitoring processes for quality management, configuring process layouts for maximum workflow, and creating rapid setup methods for frequent changeovers. They also incorporated the kanban system to enhance production flow by coordinating workstations and suppliers and enabling Just-in-Time (JIT) inventory management. These innovations allowed Toyota to provide a broader product line while cutting lead times and inventory needs dramatically—an imperative benefit considering Japan's limited space for manufacturing and warehousing. These concepts eventually became the basis of the Toyota Production System (TPS), which became a worldwide model for efficiency and continuous improvement in production.

Value Stream Mapping

Saputra, D. I. (2019d). Value Stream Mapping (VSM) is a tool used to identify value-added and non-value-added activities within the manufacturing industry, helping to pinpoint the root causes of process issues. It provides a visual representation of the current state of the system, which serves as a foundation for defining optimal conditions.

Figure 1. Current Value Stream Mapping

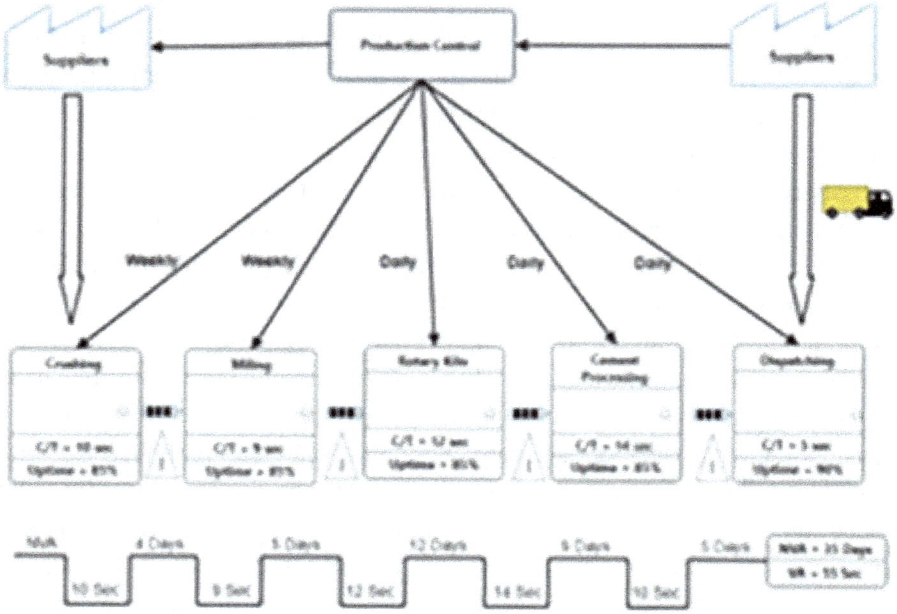

The study was undertaken at a cement factory based on a daily requirement of 11,000 tons, data was observed over 340 working days through surveys and site visits. The Value Stream Mapping (VSM) diagram represents the flow of materials and information in a cement production process, starting from suppliers and ending with product dispatching. It highlights key production stages, including crushing, milling, rotary kiln, cement processing, and dispatching, with corresponding cycle times and uptime percentages. Crushing and milling operations are supplied weekly, while other stages operate daily.

The diagram emphasizes the disparity between total non-value-added (NVA) time, which amounts to 35 days, and value-added (VA) time, which is only 55 seconds. This visualization identifies inefficiencies and delays in the process, offering a foundation for applying lean manufacturing techniques to reduce waste, improve process flow, and enhance overall production efficiency.

A push system is a production strategy where goods are produced based on forecasted demand, and then pushed through the supply chain to the customer. This approach relies heavily on demand predictions and often leads to excess inventory, as products are made ahead of actual customer orders.

Figure 2. Push Production System

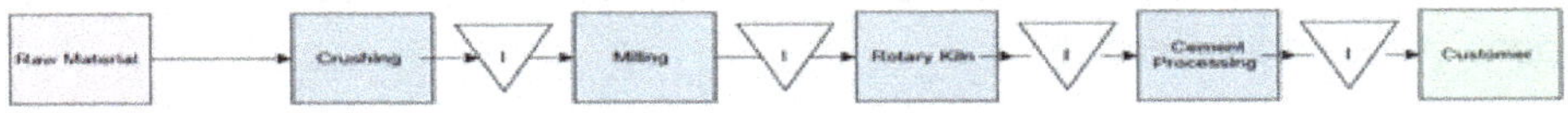

Kanban is an inventory control system used in just-in-time (JIT) manufacturing to track production and to order new shipments of parts and materials. Kanban uses visual cues to prompt the action needed to keep a process flowing (Rahman et al., 2013). In 1999, a group of ten major cement producers came together to form the Cement Sustainability Initiative (CSI) under the umbrella of the World Business Council for Sustainable Development. Over time, this platform expanded to include 24 companies operating in more than 100 countries, accounting for almost one-third of the world's cement output. The main purpose of CSI is to encourage sustainable practices and environmental responsibility across the cement industry. Since its establishment, the initiative has contributed significantly to shaping emission reduction strategies and improving environmental performance in cement manufacturing. One of the main goals of kanban is to limit the buildup of excess inventory at any point on the production line. Kanban also strives to limit bottlenecks by promoting communication and information sharing between individuals and departments. Successful implementation of kanban may lead to reduced expenses, greater customer satisfaction, more efficient processes, and minimized risk due to unforeseen problems.

Figure 3. Proposed Kanban

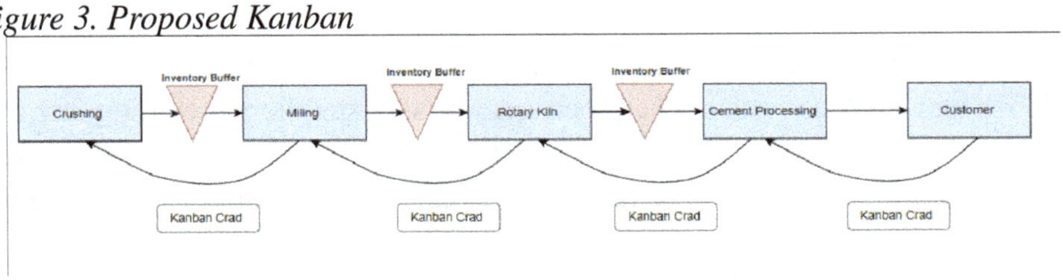

This demonstrates a Lean Manufacturing method in cement production with the aid of Kanban and Just-in-Time (JIT) strategies to limit waste and inventory. Inventory buffers between each step—Crushing, Milling, Rotary Kiln, and Cement Processing—are kept to a minimum, maintaining control over flow but to be eliminated for a smooth continuous process. The Kanban cards are signals for demand, pulling the materials only as and when required, avoiding overproduction and surplus inventory. By streamlining these buffers and putting flow efficiency into practice, the system is able to lower lead times, improve productivity, and facilitate Lean objectives such as reducing waste, process standardization, and greater responsiveness to demand.

Figure 4. Future Value Stream Mapping

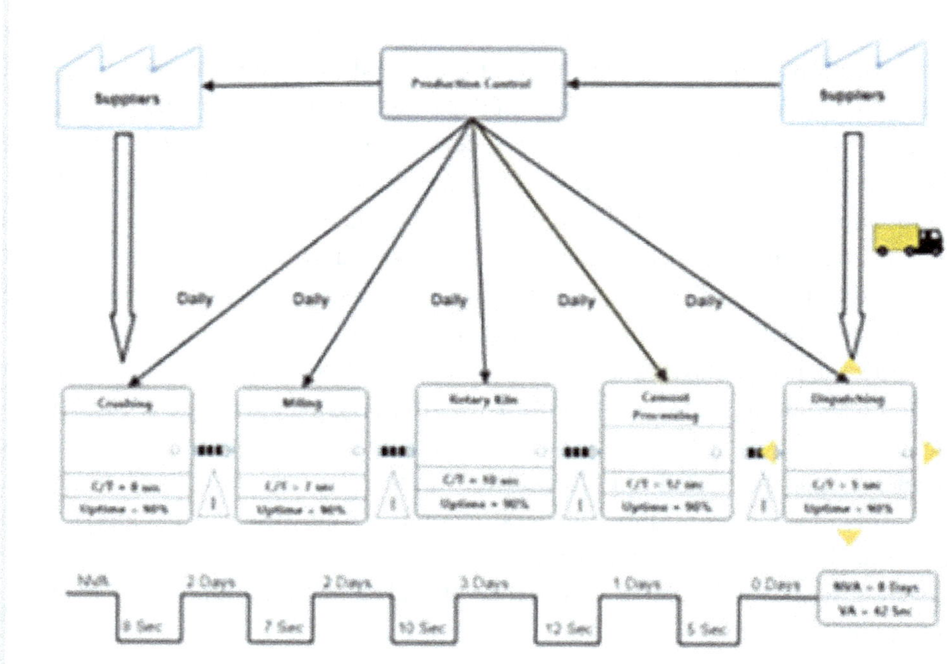

Figure 7 shows a future VSM process for cement production, which includes both VA and NVA activities. The process starts with suppliers providing raw materials to production control, which manages operations across five main stages: crushing, milling, rotary kiln, cement processing, and dispatching. Each stage has a specific cycle time (C/T), ranging from 5 to 12 seconds, with an uptime of 90% for each operation. Cumulative Value-Added time 42 sec while non-value added time extends 8 days showing great NVA times between the stages. Crush and mill take 2-day NVAs each rotary kiln has 3 days and cement processing another day, all that's mapping with some areas that are very clear on ways of NVA elimination in optimization of inter-stage workflow to ensure effective use of uptimes to complete a whole cycle in the process of production.

Objective:

This study aims to assess the global influence of Lean Manufacturing and its practical implementation in cement production. The CSI roadmap highlights several methods for lowering carbon emissions. A key approach is the gradual replacement of traditional fossil fuels with alternative energy sources such as biomass and different types of industrial waste. In addition, the use of alternative raw materials instead of conventional limestone and marl has been promoted. Further efforts focus on enhancing energy efficiency during production and advancing carbon capture and storage (CCS) technologies. It focuses on identifying implementation barriers, optimizing resource efficiency, minimizing downtime, and improving overall performance through lean strategies like Value Stream Mapping (VSM) and Kanban.

Conceptual Framework for the Impact of Waste Factors on Cement Production Performance

Figure 5. Sources: Author's Compiled

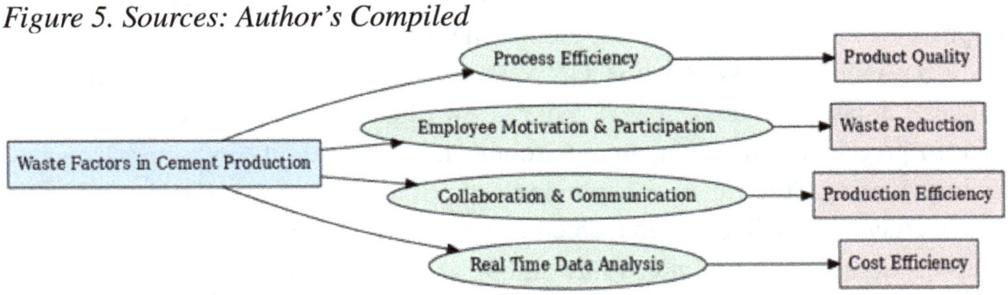

The conceptual framework is important key of waste factors in cement production, and their influence on performance outcomes. It identifies Waste Factors in Cement Production as the independent variable in this research, also affecting four mediating variables. Like, Process Efficiency, Employee Motivation & Participation, Collaboration & Communication, and Real-Time Data Analysis. These mediators, in turn, impact the dependent variables: Product Quality, Waste Reduction, Production Efficiency, and Cost Efficiency. This framework emphasizes the importance of optimizing internal processes and employee engagement, supported by real-time data. To enhance quality and efficiency while reducing waste and costs—offering a structured approach to lean implementation in the cement industry.

2. LITERATURE REVIEW

There were several studies published on the lean sigma strategy. This study focuses on minimizing waste in the cement production process using a lean manufacturing approach. It applies various tools such as the Waste Relationship Matrix (WRM), Waste Assessment Questionnaire (WAQ), and Value Stream Analysis Tools (VALSAT) to identify and reduce wastes. The findings highlight three primary wastes in cement production: defects, overproduction, and inventory. Process Activity Mapping (PAM) revealed that most activities in cement production are non-value-added but necessary. The study identified the raw mill process as having the highest failure mode, with low material quality posing a significant risk. Suggested improvements include reducing production cycle time and maintaining material quality to enhance production performance and competitiveness in the cement industry.

Masmali (2021) research applies lean manufacturing to cement production, using Value Stream Mapping (VSM) to identify and reduce waste. It found excessive inventory between workstations as a key issue. The study proposed two pull production strategies: Kanban to manage inventory and CONWIP to control work in process. The improvements reduced non-value-added time from 23 days to 4 days with Kanban and 2 days with CONWIP, with CONWIP being the most efficient. The study also suggests areas for future research. Wu & Low (2011) This study explores the role of lean production in enhancing environmental sustainability in precast-concrete factories, using CO2 emissions as an indicator. It focuses on two stages: site layout management and delivery management, analyzing 17 precasters in Singapore. The findings highlight several non-value-adding activities, such as large storage areas, excess supply base, lack of just-in-time (JIT) sourcing, and a lean workforce, that increase carbon emissions. The study offers practical guidance for identifying these activities to reduce emissions and supports regulatory agencies in recommending measures to achieve low-carbon production in the precast-concrete industry.

Deo, S. (2023-b) This project develops a methodology for applying lean manufacturing principles in the cement industry to reduce waste in processes shared with discrete manufacturing. Value Stream Mapping (VSM) is used to identify current waste, and a simulation model is created for both current and future states. Although these measures offer potential environmental benefits, they also involve technical and regulatory challenges. This report places particular emphasis on the use of alternative fuels, as this practice has raised concerns related to public health, environmental safety, and strong community responses. Lean principles are applied in the simulation, which is validated for accuracy. Data from the production department informs the model. The future state VSM shows improved throughput, reduced waiting time, inventory, and lead time. For systems with high volume, low product mix, and significant machine downtime, running low setups at the pacemaker process is recommended for profitability. De Souza, J. A. (2020b) This study applies Lean Manufacturing principles and quality tools to reduce packaging waste in an industrial unit in Sergipe. A case study approach led to a 12.1% reduction in annual packaging loss and improved communication between company sectors, along with better visibility of loss indicators. The methodology used is adaptable to other manufacturing contexts for waste reduction and process improvement.

Naeemah, A. J., & Wong, K. Y. (2022) Researcher investigates the positive effects of Lean Manufacturing (LM) tools on sustainability and addresses gaps in existing research. Using a systematic literature review (SLR) of 101 articles from 2010 to 2021, it identifies 36 LM tools impacting sustainability, with Six Sigma and Kaizen among the top 10. It also highlights 20 sustainability metrics, including eight wastes and air emissions, with the greatest impact observed in cost reduction under the economic aspect. The study emphasizes the need for further exploration of the negative effects of LM tools on sustainability and encourages future research to deepen understanding and application.

Muhammad. (2017) he applies lean production methods to reduce waste in an automotive company using Waste Relationship Matrix (WRM), Waste Assessment Questionnaire (WAQ), and Value Stream Mapping (VSM). The analysis identifies waiting, motion, and inventory as the major waste types in the assembly trimming process of the Daihatsu SIGRA 1.5L at PT. Astra Daihatsu Motor. Implementing WRM and VSM resulted in waste reductions: motion decreased from 17.65% to 15.75%, waiting from 15.88% to 13.12%, and lead time from 14.202 to 12.322 minutes. Suggested improvements include adopting a FIFO system, optimizing worker movements, and balancing the assembly line. Klabusayová, N. (2014) paper addresses improving production efficiency and labor productivity through lean manufacturing principles. It emphasizes the importance of systematic waste reduction and cost efficiency using appropriate methods. Value Stream Mapping (VSM) is highlighted for identifying process inefficiencies, while the Overall Equipment Efficiency (OEE) method evaluates equipment performance. Both methods are analyzed as key tools for enhancing manufacturing processes.

Vadodaria, D. (2014) discusses Lean Manufacturing, a production control method focused on eliminating waste and enhancing efficiency by optimizing flow and preserving value with minimal effort. Known as the Toyota Production System (TPS), it helps manufacturers improve productivity, quality, and cost-efficiency. Lean Manufacturing saves time and labor costs while reducing overall production time. The paper also highlights Overall Equipment Effectiveness (OEE) as a practical tool for monitoring and improving manufacturing efficiency by categorizing productivity losses into three key areas. The study concludes that applying Lean Manufacturing and OEE leads to maximum production with reduced time. Moreover, putting sustainability measures into practice requires strong policy backing, practical technological solutions, and reliable monitoring systems. Government agencies and environmental authorities play an important role by enforcing emission limits and promoting cleaner production through regulations and incentive-based programs. At the industry level, ongoing research and innovation are needed to improve the effectiveness of fuel substitution and to reduce possible environmental and operational risks. Clear reporting practices and active engagement with stakeholders also help build public confidence and ensure that sustainability targets reflect broader social and environmental concerns. When these efforts are combined, the cement sector can move toward a more sustainable growth path that balances industrial development with long-term environmental protection. (2009, November) discusses Bechtel Mining & Metals' (M&M) Six Sigma initiative to optimize water use in concrete operations at arid and semi-arid project sites in countries like Australia, Chile, Peru, and Argentina. While water consumption for concrete mixing is monitored, peripheral uses—such as aggregate washing, truck cleaning, dust suppression, and curing—lack control measures. The initiative aimed to reduce peripheral water usage by 30% (approximately 250 L per cubic meter of concrete), potentially conserving 4,000 million liters of water and saving about $8 million USD. This strategy supports sustainable resource management in large-scale construction projects.

Jakarta, Indonesia. (2018) paper highlights the importance of lean production in manufacturing to meet growing customer demand and remain competitive. Lean production eliminates unnecessary processes to enhance productivity, reduce production costs, increase output, and shorten lead times. Inefficient processes can cause delays and inventory buildup, making lean implementation essential for effective production management.

Research Gap

While Lean Manufacturing concepts have been widely implemented across various manufacturing industries to minimize waste and improve efficiency, their implementation specifically for the cement production process is still not well researched. Most research work focuses mainly on generic waste reduction methods without properly examining the interdependencies between various forms of waste or the overall effect on production efficiency and sustainability.

Although earlier works have aimed to research lean tools such as Waste Relationship Matrix (WRM), Waste Assessment Questionnaire (WAQ), Process Activity Mapping (PAM), and Overall Equipment Efficiency (OEE) for the purpose of detecting inefficiencies, not many works have taken a close look at the interaction of VSM and Kanban for cement manufacture. The majority of research has addressed sustainability practice or studied overall waste reduction techniques in general, yet few have looked at the specific means by which VSM and Kanban affect inventory control, process flow optimisation, and production efficiency within the cement sector. Apart from that, although CONWIP (Constant Work-in-Progress) and Kanban have been experimented upon in the context of lead time reduction, comparative case studies analyzing their long-term performance in cement plants are limited. Along with technological upgrades, building skills and training the workforce play a key role in adopting sustainable practices in the cement sector. Workers at different operational levels need proper training to operate modern equipment, handle alternative fuels safely, and follow environmental guidelines effectively. Awareness programs within organizations can further promote responsible behavior and strengthen commitment toward sustainability goals. When employees are actively involved in environmental initiatives, productivity tends to improve and the chances of regulatory violations are reduced. Therefore, investing in people and knowledge is just as critical as investing in advanced technologies for achieving long-term sustainability in the industry. Past research has also noted saving energy and water, but they do not elaborate much on the extent to which lean techniques may be combined with sustainable practices to maximize the use of resources. In addition, empirical research on actual implementation issues, flexibility of workers, and the financial effect of lean implementation in cement plants is still limited.

This study tries to fill such gaps by analyzing the implementation of VSM and Kanban in actual cement manufacturing and their value addition towards better process efficiency, elimination of inventory-related wastes, and smooth production flow. In the process, it adds to advancing the overall body of knowledge on lean approaches applied to cement manufacturing and their usability for application throughout the industry.

3. RESEARCH METHODOLOGY

The research is analytical and exploratory with emphasis on the application of Lean Manufacturing philosophy in the cement industry, with special interest in Value Stream Mapping (VSM) and Kanban. The research aims at establishing inefficiency, examining waste reduction measures, and ensuring production flow in an optimal manner in the production of cement. A quantitative and qualitative approach will be used to assess how VSM and Kanban can optimize inventory management, reduce lead times, and enhance overall process efficiency. The study includes data collection for cement manufacturing plants, process mapping, waste reduction analysis before and after the implementation of lean tools, and case studies and comparative analysis to ascertain the effectiveness of these lean practices.

By integrating both theoretical understanding and real-world solutions, this study will provide actionable solutions for cement manufacturers to raise productivity, lower waste, and adopt lean principles for cost-effective and sustainable operations.

Statement of Problem

The cement manufacturing sector also has non-efficient processes, excessive inventories, long lead times for manufacturing, and wasteful manufacturing. The traditional processes of manufacturing cause non-value-added activities that increase the cost, tardiness, and wastage of resources. While Lean Manufacturing has been effectively used in several sectors to get more efficient, it is rarely used and researched in the manufacturing of cement. Past research has confirmed waste in cement production, but few focused researches have been conducted on the use of Value Stream Mapping (VSM) and Kanban to enhance production flow and inventory management. The majority of past research also investigated lean tools separately without accounting for their synergistic impact on reducing lead time, material handling, and process efficiency improvement. This study aims to fill these gaps by the implementation of VSM and Kanban in cement manufacturing to identify their efficacy in reducing waste, simplifying workflow, and maximizing production efficiency.

Scope of the study

This research is keen on the use of Lean Manufacturing methods, namely Value Stream Mapping (VSM) and Kanban, in the cement sector. The research is keen to explore and improve production efficiency, inventory control, and waste minimization in cement production. Exploring various types of wastes (e.g., waiting time, defects, overproduction, and excess inventory) which are efficiency-affecting. Usage of Value Stream Mapping (VSM) Creation of the present process to reveal inefficiencies and a future-state map to create an enhanced flow. Kanban Implementation in Inventory Management Analyzing to what degree Kanban can eliminate waste inventory, optimize material flow, and avoid non-value-added operations. Comparative Analysis Assessing before and after impacts of lean implementation on production lead times, wastage savings, and operational efficiency. Data Collection and Case Study Approach Leverage actual industry data from cement plants to support findings and offer tangible recommendations.

Objectives of Study

To enhance productivity and reduce waste in the cement manufacturing process, this research considers the utilization of Lean production principles, Value Stream Mapping (VSM), and Kanban.

1. To improve productivity in cement manufacturing using Lean tools like VSM and Kanban.
2. To reduce waste and optimize inventory flow by evaluating pre- and post-Lean implementation.

The research offers actionable guidelines for increasing cost savings, sustainability, and efficiency in the entire process by examining production efficiency post-adoption against pre-adoption.

Hypotheses of the Study

1. Evaluate the different types of waste generated in the cement production process.
    Null Hypothesis (H0): Various types of waste (overproduction, defects, waiting) do not significantly contribute to inefficiencies in the cement production process.
    Alternative Hypothesis (H1): Various types of waste (overproduction, defects, waiting) significantly contribute to inefficiencies in the cement production process.
2. Categorize types of waste (overproduction, defects, waiting) into high, medium, and low priority to indicate which ones must be tackled first.
    Null Hypothesis (H0): There is no meaningful difference in priority between overproduction, defects, and waiting as far as their effect on production efficiency.
    Alternative Hypothesis (H1): Overproduction and defects are higher priority types of waste compared to waiting, and solving them first will lead to higher improvements in production efficiency.
3. Analyze the quantitative correlation between waste types and production performance measures like cost, productivity, and quality.
    Null Hypothesis (H0): There is no substantial correlation between waste types (overproduction, defects, waiting) and production performance measures (cost, productivity, quality).
    Alternative Hypothesis (H1): There is a substantial negative correlation between waste types (overproduction, defects, waiting) and production performance indicators (cost, productivity, quality)

4. Assess the impact of Value Stream Mapping (VSM) on identifying inefficiencies and improving production flow.

   Null Hypothesis (H_0): The application of Value Stream Mapping (VSM) does not significantly help in identifying inefficiencies or improving production flow in the cement industry.

   Alternative Hypothesis (H_1): The application of Value Stream Mapping (VSM) significantly helps in identifying inefficiencies and improving production flow in the cement industry.

5. Evaluate the effect of Value Stream Mapping (VSM) on reducing non-value-added activities in the cement production process.

   Null Hypothesis (H_0): The implementation of VSM does not significantly reduce non-value-added activities in cement production.

   Alternative Hypothesis (H_1): The implementation of VSM significantly reduces non-value-added activities in cement production, leading to higher efficiency and cost savings.

6. Measure the impact of Kanban on inventory management and work-in-process (WIP) reduction.

   Null Hypothesis (H_0): Implementing Kanban does not significantly improve inventory management or reduce work-in-process (WIP) in cement production.

   Alternative Hypothesis (H_1): Implementing Kanban significantly improves inventory management and reduces work-in-process (WIP) in cement production, enhancing material flow and reducing holding costs.

Data Types: This study explores how Value Stream Mapping (VSM) and Kanban Lean Manufacturing methods are applied in the cement sector.

Primary Data Collection:

Primary data were gathered by using questionnaires, site visits, and direct observations in a cement factory. Our data consist of both quantitative details and personal observations:

Quantitative Data: We collected data pertaining to lead times, stock levels, cycle times for production, and wastage percentages. This was conducted to compare against and after applying Lean techniques.

Qualitative Data: We heard from employees' suggestions, took experts' opinions, and interviewed individuals to learn about operational issues and assess the performance of adopting Lean practices.

Secondary Data Sources:

In addition, we collected secondary information from research studies, industry reports, academic journals, and case studies on implementing Lean Manufacturing in the production of cement. These were: Quantitative Data: We utilized industry standards, past production figures, and efficiency levels to lend perspective in terms of understanding improvements. Qualitative Data: We researched best practices, theoretical models, and typical challenges encountered when implementing Lean to inform our analysis and recommendations. This research integrates primary data from observations and site visits with secondary data from industry reports to assess Lean Manufacturing in cement manufacturing. By bringing together quantitative measures and qualitative findings, it offers a thorough analysis of VSM and Kanban in streamlining processes and minimizing waste.

Data Analysis Tool and Techniques

For analyzing the data gathered for this research, different statistical and analytical tools and methodologies were utilized. These tools and methodologies assist to detect inefficiencies, maximize production functions, and establish the effect of Lean Manufacturing values in the cement industry. The most important data analysis methodologies used for this research are discussed below: Descriptive Statistics and Inferential Statistics: Inferential statistics are applied while making inference from the population to samples. The following were used: Hypothesis Testing – Determination of significance of relationships among variables. t-tests – To determine comparison between two groups' means in order to know how one affects the other. ANOVA (Analysis of Variance) – Comparison of differences between more than two groups, e.g., efficiency of production prior and subsequent to Lean. Chi-square Tests – Measures how categorical variables are related to each other, e.g., production results and waste control actions.

1. Regression Analysis

Regression analysis was used to examine the relationship between waste reduction and production efficiency. The method is used to examine the extent to which independent variables (e.g., Lean actions) affect dependent variables (e.g., cost reductions and enhanced efficiency).

2. Principal Component Analysis - PCA (Factor Analysis)

Principal Component Analysis (PCA) was used as a factor analysis to retain significant variables affecting the production efficiency. PCA minimizes the data by bundling together variables that are closely associated with each other, and thus it is easier to recognize the most significant factors of change in operation.

3. Value Stream Mapping (VSM)

Value Stream Mapping (VSM) has been applied such that information flow and material flow within the period of production have been analyzed. Through the application of the future and current state production mapping, non-value-added activities and inefficiencies were mapped and minimized.

4. Kanban Analysis

The application of the kanban analysis technique assisted in the assessment of process improvement and the efficiency of inventory management. Material flow was made more efficient using a Kanban system with fewer unnecessary inventories aside from enhancing just-in-time (JIT) productivity. The incorporation of these statistical and analytical methods provided an accurate view of the impact of Lean Manufacturing on cement production efficiency. Using both qualitative and quantitative approaches, the research provides evidence-based results for effective resource utilization, waste minimization, and overall operating performance.

4. DATA ANALYSIS

Data analysis using for the collected information in the research.
Where ethical decision making will come in picture.

Table 1. Data analysis

Reliability Statistics	
Cronbach's Alpha	**0.94**
N of Items	**20**

Sources: Author's Complied

Cronbach's Alpha = 0.940: This suggests very good reliability. Greater than 0.9 → Very good reliability, N of Items = 20: The scale has 20 items, which means that the internal consistency is calculated on these 20 questions and 9 variables.

As Cronbach's Alpha is 0.940, it implies that the scale items are very consistent with one another and are measuring the same underlying construct very effectively. This high reliability implies that the questionnaire or measurement instrument is ready for research and decision-making purposes. Yet, a very high alpha (nearly 1) may imply redundancy of items, i.e., some questions may be too redundant.

Table 2. KMO and Bartlett's Test

KMO and Bartlett's Test	
Kaiser-Meyer-Olkin Measure of Sampling Adequacy	**0.95**
Bartlett's Test of Sphericity	
Approx. Chi-Square	**4603.924**
df	**190**
sig	**0.000**

Sources: Author's Complied

1. Kaiser-Meyer-Olkin (KMO) Measure of Sampling Adequacy:

KMO = 0.950 The KMO value lies between 0 and 1. A value of more than 0.9 is regarded as excellent, and it implies the dataset is greatly appropriate for factor analysis.

It means that variables have a great deal of correlation with each other, and therefore factor extraction would be significant.

2. Bartlett's Test of Sphericity:

Chi-Square = 4603.924, df = 190, Sig. = 0.000:

Bartlett's test verifies if the correlation matrix is an identity matrix (i.e., there are not significant relationships between variables).

A significant p-value (0.000) indicates that the correlation between variables is statistically and not random. The data is extremely good for factor analysis as both KMO value is superb (0.950) and Bartlett's test is significant ($p < 0.05$).

This indicates that the variables have high correlations, and factor analysis for dimension reduction is suitable.

Scree Plot

Figure 6. Sources: Author's Complied

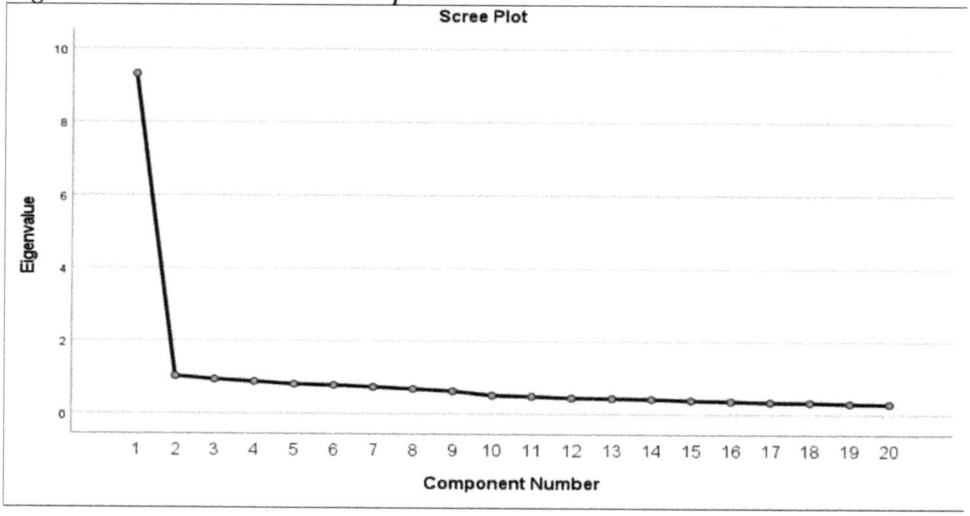

Table 3. Factor Analysis

Items	Factor 1	Factor 2	Factor 3	Factor 4	Factor 5	Factor 6	Factor 7
WF1	0.761						
WF2	0.742						
WF3	0.716						
PF1		0.749					
PF2		0.697					
PF3		0.787					
RT1			0.759				
RT2			0.777				
PQ1				0.786			
PQ2				0.741			
WR1					0.726		
WR2					0.766		
PP1						0.732	
PP2						0.761	

continued on following page

Table 3. Continued

Items	Factor 1	Factor 2	Factor 3	Factor 4	Factor 5	Factor 6	Factor 7
CE1							0.756
CE2							0.733

Sources: Author's Complied

Principal Component Analysis (PCA) or Factor Analysis employing Varimax rotation with Kaiser Normalisation yields this Rotated Component Matrix. It assists to figure out how various extracted components are loaded with the original variables (WF1, PF1, etc.). Greater Numbers = Stronger Connections, The strength of a variable's relationship to a factor (or component) is indicated by each number in the table. Stronger relationships are indicated by values greater than 0.6. Component 1 is loaded by PF1, PF2, and PF3 **Process Efficiency**, Component 2 is loaded by WF1, WF2, and WF3. **Waste Factors in Cement Production** Component 3 is loaded by RT1 and RT2, indicating that they might be a **Real-Time Data Analysis**. Component 4 is loaded by PP1 and PP2, which most likely measure a **Production Efficiency** Component 5 is loaded by PQ1 and PQ2 **Product Quality**. Component 6 is loaded by WR1 and WR2 **Waste Reduction**. Component 7 is loaded by CE1 and CE2 **Cost Efficiency**.

Descriptive Statistics

Table 4.

Variable	N	Min	Max	Mean	Stdv
WF1	458	1	5	3.32	1.22
waste_factors	458	1	5	3.32	1.03
process_efficiency	458	1	5	3.29	1.03
real_time	458	1	5	3.25	1.10
product_quality	458	1	5	3.31	1.06
waste_reduction	458	1	5	3.28	1.09
production_efficiency	458	1	5	3.30	1.07
cost_efficiency	458	1	5	3.29	1.02

Sources: Author's Complied

Descriptive statistics for some of the factors related to operations are presented in this table.

458 respondents scored each component on a scale of 1 to 5.

Mean Values:

- The majority of factors have a mean of approximately 3.2 to 3.35, reflecting a neutral to slightly positive attitude among the respondents.
- The highest average is for teamwork (3.35), which indicates employees consider teamwork most important.
- The lowest mean is for employee motivation (3.27), which suggests there is scope for improvement in motivating and engaging employees.

Standard Deviation (Spread of Answers):

- All the standard deviations range from 1.02 to 1.09, so responses are mildly spread out but not very heterogeneous.
- Real-time operations (1.09) and production efficiency (1.07) exhibit comparatively higher variability, indicating responses that varied across the respondents.

The statistics would indicate that staff in general look at operational variables with a negative to positive lean. Team Working is regarded as the strongest aspect, whereas the motivation of the employees can improve. The near uniform answers confirm that employees do share the same view, however, operations on the ground and efficiency of production are more diverse, reflecting variances of experience or difficulty experienced in these issues. In order to improve overall effectiveness, organizations need to concentrate on increasing employee motivation and resolving inconsistencies in real-time operations and production effectiveness.

Regression Analysis

1. R Value: Strength of Relationship The R value of 0.565 shows there is a moderate positive link between waste factors and the dependent variable, such as cost efficiency or performance. This means that waste factors do affect the dependent variable, but they are not the only things that matter. Other factors also have an impact on it.
2. R SquareVariance Explained An R Square value of 0.619 means that waste factors account for 61.9% of variation in the dependent variable. Waste factors have an impact, it indicates. Yet, it indicates

68.1% variation is due to other factors left out of this model.

3. Adjusted R Square

Adjusted R Square stands at 0.618, close to R Square of 0.619. The close indication demonstrates that waste factors are a consistent predictor and not an outcome of mere coincidence. It supports the argument that even in view of the predictors' numbers, the predictive power of the model is great.

4. Standard Error of Estimate

A standard error of 0.87678 indicates that the predictions made by the model do contain some degree of uncertainty, but the error is acceptable. A lower number would mean more precision, but as this error is less than 1, the model is still somewhat useful.

CONCLUSION, IMPLICATION AND LIMITATIONS

Conclusion

The study on Lean Manufacturing adoption in the cement sector validates the huge impact Lean

Sigma methods have on refining the production process. By utilizing Value Stream Mapping (VSM), Kanban, and Just-in-Time (JIT) methods, the study was able to identify areas of inefficiency and remove non-value-added activities. The results validated an astonishing reduction in lead time, inventory wastage, and production backlog, which consequently led to enhanced operating efficiency and sustainability.

By reducing excess buffer stocks and increasing flows through key stages of production—Crushing, Milling, Rotary Kiln, Cement Processing, and Dispatching—the research attained greater uptime, reduced wastage of resources, and greater cost-effectiveness. The results validate the fact that Lean Manufacturing offers a scientific and realistic approach for cement plants to become more productive and reduce their carbon footprint. Future studies need to consider long-term sustainability of Lean practice in cement production, and the use of digital technologies to monitor and automate processes in real-time. Generally, this research suggests the need for Lean implementation for efficiency, cost reduction, and sustainable production in the cement industry.

Recommendation

the recommendations suggested based on the findings of the study for the successful implementation of Lean Manufacturing in the cement industry: To enhance Lean Manufacturing in the cement industry, firms should adopt additional tools like Kaizen, JIT, and Six Sigma beyond VSM and Kanban to improve efficiency. Implementing a Kanban-based pull system helps minimize excess inventory and ensures smoother material flow. Regular employee training on Lean principles boosts engagement and eases implementation. Leveraging digital technologies such as automation, real-time monitoring, and analytics enhances decision-making and process control. Continuous improvement reviews should be conducted to monitor Lean effectiveness, identify gaps, and guide timely updates. Together, these steps create a more sustainable, efficient, and responsive production environment in the cement industry.

Social Implication

The research offers crucial information on the use of Lean Manufacturing methods, especially Value Stream Mapping (VSM) and Kanban, in the cement sector. From the research, one can see how lean practices improve production efficiency, minimize waste, and streamline inventory management, ultimately making costs more effective and sustainable. Through highlighting major inefficiencies and suggesting lean-based solutions, this study presents actionable advice to cement manufacturers in order to eliminate inefficiencies and implement a better-organized effort at reducing wastage. Second, the work adds to existing knowledge on Lean Manufacturing in process industries, enabling future research as well as its application in the industry.

Limitations

Even though it has made great contributions, there are some shortcomings in the research. First of all, this research relies upon data collected in a single cement plant, thereby restricting the extrapolation of its findings to production units of differing operational scenarios. Secondly, while the research highlights VSM and Kanban significantly, other lean tools like Kaizen and Six Sigma are kept out of sight, and considering them may enable further boosting in production efficiency. Third, Lean practices take long-term observation, but the current study only captures a short period, so it is challenging to evaluate long-term effects. Finally, employee resistance to change and adaptability were not extensively examined, which might be an important factor in the success of Lean implementation.

REFERENCES

Čiarnienė, R., & Vienažindienė, M. (2012). Lean manufacturing: Theory and practice. *Economics and Management, 17*(2), 359–366. DOI: 10.5755/j01.em.17.2.2205

Deo, S. (2023). Methodology to implement lean manufacturing in cement industry. *ProQuest.* https://www.proquest.com/openview/8df736e5ec74d93920526b5bbe38063e/?pq-origsite=gscholar&cbl=18750

Hamdoon, S., & Alabadee, S. (2025). The effect of strategic monitoring on lean manufacturing system: A case study of Northern Cement Company, Iraq. *International Journal of Social Science Management and Economics Research, 3*(1), 59–75. DOI: 10.61421/IJSSMER.2025.3106

Immerman, G. (2022, December 13). Lean manufacturing: The principles, wastes, benefits, and tools. *MachineMetrics.* https://www.machinemetrics.com/blog/lean-manufacturing

Klabusayová, N. (2014). The possibilities for increasing production processes efficiency by utilizing the lean manufacturing principles. *Applied Mechanics and Materials, 693*, 483–488. . DOI: 10.4028/www.scientific.net/AMM.693.483

Masmali, M. (2021). Implementation of lean manufacturing in a cement industry. *Engineering Technology & Applied Science Research, 11*(3), 7069–7074. DOI: 10.48084/etasr.4087

Muhammad. (2017). Implementation of lean manufacturing system to eliminate wastes on the production process of line assembling electronic car components with WRM and VSM method. *SciSpace.* https://typeset.io/papers/implementation-of-lean-manufacturing-system-to-eliminate-5023wabkhr

Rahman, N. A. A., Sharif, S. M., & Esa, M. M. (2013). Lean manufacturing case study with Kanban system implementation. *Procedia Economics and Finance, 7*, 174–180. DOI: 10.1016/S2212-5671(13)00232-3

Rodriguez, D. (2024, August 23). Lean tools, principles, and applications. *Invensis Learning Blog.* https://www.invensislearning.com/blog/lean-methodology-tools-principles-applications/

Tourki, T. (2010). *Implementation of lean within the cement industry* [Doctoral dissertation, De Montfort University]. https://hdl.handle.net/2086/4562

Vadodaria, D. (2014). Optimizing time with increasing productivity using lean manufacturing and overall equipment effectiveness. *SciSpace.* https://typeset.io/papers/optimizing-time-with-increasing-productivity-using-lean-1r3ccdxre1

Wu, P., & Low, S. P. (2011). Lean management and low carbon emissions in precast concrete factories in Singapore. *Journal of Architectural Engineering*, *18*(2), 176–186. DOI: 10.1061/(ASCE)AE.1943-5568.0000053

Chapter 7
Waste Reduction in Public Hospitals:
Integrating Digital Solutions and Sustainable Practices

Flaka Hoti
http://orcid.org/0000-0002-4780-5776
University of Gjakova "Fehmi Agani", Kosovo

Kamel Mouloudj
http://orcid.org/0000-0001-7617-8313
University of Medea, Algeria

Hasime Arif Bacaj Terziqi
http://orcid.org/0009-0001-3411-8272
University of Gjakova "Fehmi Agani", Kosovo

Ahmed Chemseddine Bouarar
http://orcid.org/0000-0001-8300-9833
University of Medea, Algeria

Basima Kzar Hasan
http://orcid.org/0000-0002-9298-0784
Center for Basra and Arab Gulf Studies, University of Basrah, Iraq

Antigona Ukëhaxhaj
http://orcid.org/0000-0003-4128-0769
University of Gjakova "Fehmi Agani", Kosovo & National Institute of Public Health, Kosovo

ABSTRACT

The growing complexity of healthcare waste management has prompted increased interest in leveraging digital technologies to enhance efficiency, safety, and sustainability. This study explores the integration of digital tools—such as blockchain and data analytics—into medical waste management systems, particularly in the context of emerging economies. The purpose is to examine how digital transformation can optimize waste handling processes, reduce environmental and health risks, and promote stakeholder engagement. Drawing on recent literature and case-based evidence, the study highlights how digital technologies support real-time tracking, improve route planning, enable effective segregation, and enhance traceability. Key

DOI: 10.4018/979-8-2600-1960-3.ch007

findings suggest that such technologies not only reduce operational costs and infection risks but also contribute to biofuel production and the reuse of e-waste materials. The paper concludes that cross-sector collaboration and increased environmental awareness are essential to fully realize the potential of digital transformation in medical waste management.

INTRODUCTION

Healthcare systems across the globe are confronting mounting pressures from the intensifying effects of climate change (Islam et al., 2025). These include not only the disruption of food systems and the proliferation of zoonotic, vector-, food-, and water-borne diseases but also the increasing frequency of extreme weather events such as heat waves, storms, and floods. Alongside these physical threats, climate change is also exacerbating mental health issues and undermining key social determinants of health, such as equity, employment, access to services, and social support networks. As these impacts extend into both the natural and socio-economic domains, the resilience and capacity of healthcare institutions are being fundamentally tested (Affordofe et al., 2025).

Among the growing environmental challenges facing healthcare systems, the issue of waste management stands out as both urgent and complex (Mouloudj et al., 2025). The unchecked generation of waste, driven by human activity and intensified by industrial, urban, and healthcare operations—poses a critical barrier to sustainable development (Al-Hassan et al., 2026a; Rittl et al., 2025). Healthcare facilities, in particular, are becoming significant contributors to environmental degradation through the continuous production of hazardous and non-hazardous waste (Kenny & Priyadarshini, 2021). This situation calls for innovative and sustainable strategies to mitigate waste and reduce the environmental footprint of the health sector (Manimaran et al., 2025).

Incineration remains one of the most commonly employed waste disposal methods, capable of rapidly reducing waste volume by up to 90% while generating heat that can be converted into energy. However, in the absence of modern control technologies, incineration carries substantial risks—most notably, the release of toxic emissions and the generation of hazardous by-products (Chisholm et al., 2021). These risks underscore the need for more environmentally responsible waste treatment approaches that align with the principles of the circular economy (Al-Hassan et al., 2026a; Shin et al., 2020).

Both high-income and low-income countries are grappling with rising levels of healthcare waste, and this trend is expected to intensify with population growth, medical innovations, and increased consumption of disposable medical products

(Sharma et al., 2025). As such, addressing healthcare waste is no longer a peripheral issue but a central pillar in the movement toward sustainable, climate-resilient healthcare systems (Kenny & Priyadarshini, 2021).

Encouragingly, advances in digital technology offer promising pathways for mitigating healthcare waste. Real-time data analytics, smart monitoring systems, and AI-enabled logistics are making waste collection more efficient and environmentally sound (Al-Hassan et al., 2026a). In some instances, these innovations have supported the development of alternative energy sources such as biofuels and solar batteries derived from waste materials (Farjana et al., 2023).

A successful transition toward sustainability depends largely on changing both individual and institutional behaviors to encourage more accountable waste management practices (Al-Hassan et al., 2026b). In healthcare settings, fostering environmental awareness, supporting sustainable routines, and prioritizing waste minimization are essential steps in advancing wider sustainability objectives (Almokdad & Mouloudj, 2026; Patwa et al., 2026). The healthcare workforce, regulatory frameworks, digital tools, and sustainable procurement practices all represent vital levers for transforming the health sector into a model of environmental responsibility (Mouloudj et al., 2023).

Given the pressing need to mitigate the environmental and public health consequences of healthcare-related waste, this chapter seeks to address several core questions: What are the main types and sources of waste generated in public hospitals, and what foundational strategies can help reduce them? How can digital transformation tools and data-driven solutions contribute to more efficient and sustainable waste management? What role do waste segregation, recycling, and disposal methods play in promoting sustainability within healthcare systems? And finally, how can staff training and awareness initiatives foster a culture of waste reduction and environmental responsibility in public hospitals?

Methodologically, this chapter is grounded in a conceptual and integrative review approach. It synthesizes findings from academic studies, industry reports, and policy documents to provide a comprehensive understanding of current challenges and innovative solutions related to healthcare waste management. This method allows for a multidimensional analysis of waste reduction strategies by incorporating perspectives from environmental sustainability, digital innovation, and organizational behavior, without relying on primary data collection. Through this lens, the chapter offers evidence-informed insights and practical guidance for healthcare practitioners, policymakers, and administrators.

The rest of this chapter is organized as follows. The first section, "*Reduce Waste in Public Hospitals: Concepts and Foundations*," outlines the various types and sources of waste, including clinical, pharmaceutical, food, and administrative waste, and explores core principles for effective waste minimization. The second section,

"*Digital Transformation for Waste Reduction*," examines how technologies such as IoT, AI, cloud computing, and blockchain can optimize waste tracking, management, and prevention. The third section, "*Waste Segregation, Recycling, and Disposal*," highlights best practices in handling different categories of medical waste and the importance of adopting environmentally sound disposal methods. The fourth section, "*Staff Training and Awareness Campaigns*," emphasizes the human element—promoting environmental literacy, staff engagement, and institutional culture shifts. The chapter concludes with a summary of key insights and recommendations for strengthening sustainable waste management practices in public healthcare systems.

REDUCE WASTE IN PUBLIC HOSPITALS: CONCEPTS AND FOUNDATIONS

Reducing waste effectively depends on the active participation of various stakeholders (Brotosusilo & Naldi, 2021), including pharmaceutical manufacturers, distributors, healthcare providers, and pharmacists, among others (Mouloudj et al., 2023). Hospitals are among the most resource-intensive public institutions (e.g., energy, food, and water), and their environmental footprint is both vast and multifaceted (Torrejón-Ramos et al., 2025). According to Health Care Without Harm (2025), hospitals generate over 42 pounds of waste per bed per day, contributing to more than five million tons of waste each year globally. This alarming figure underscores the urgency of implementing sustainable, systematic waste reduction strategies in public healthcare settings (Manimaran et al., 2025). Waste reduction is no longer an optional improvement; it is a critical component of hospital operations that intersects with health outcomes, environmental responsibility (Torrejón-Ramos et al., 2025), cost containment, and compliance with international safety standards (Kumar et al., 2016).

At the heart of any effective hospital waste strategy lies a clear understanding of what constitutes healthcare waste (HCW), how it is classified, and where it originates. Healthcare waste refers to all waste generated by healthcare facilities, including hospitals, clinics, laboratories, research centers, blood banks, and mortuaries. Although approximately 85% of HCW is considered non-hazardous, similar in nature to household waste, the remaining 15% is hazardous and requires specialized handling due to its potential to cause infections, injuries, environmental degradation, or long-term public health impacts (WHO, 2024a).

Types of Hazardous Healthcare Waste

The World Health Organization (2024a) identifies several categories of hazardous healthcare waste:

- Infectious Waste: Waste suspected to contain pathogens, such as blood-contaminated materials, bodily fluids, bandages, and surgical waste. This type of waste poses significant risks to patients, healthcare workers, and waste handlers if not properly treated (Alighardashi et al., 2025).
- Pathological Waste: Human or animal tissues, organs, and body parts, including fetuses and surgical specimens.
- Sharps Waste: Items capable of causing cuts or punctures, such as needles, scalpels, blades, and broken glass. This waste carries a high risk of injury and transmission of blood-borne infections.
- Pharmaceutical and Cytotoxic Waste: Expired, unused, or contaminated medications, and substances used in cancer treatment that are potentially mutagenic, teratogenic, or carcinogenic.
- Chemical Waste: Disinfectants, solvents, reagents, and other toxic chemicals, including heavy metals like mercury from thermometers or batteries.
- Radioactive Waste: Materials contaminated with radionuclides used in nuclear medicine or radiotherapy.

Each of these waste categories requires unique containment, treatment, and disposal approaches to mitigate risks to human health and environmental integrity. Improper disposal, such as open dumping, burning, or indiscriminate landfilling, can result in air and water pollution, injuries, radiation exposure, and the spread of resistant pathogens (Alighardashi et al., 2025).

Sources of Healthcare Waste in Public Hospitals

An effective waste management system involves proper segregation and disposal processes, along with adequate provisions for transportation, storage, and training for personnel handling the waste (Chisholm et al., 2021; Zulfa & Nugraheni, 2020). The production of healthcare waste is closely tied to the nature of hospital activities, which range from clinical care to administrative support. Key sources of waste include:

- Medical Waste: Arising from patient care, surgery, diagnostics, and laboratory activities (Saha et al., 2025). This includes soiled dressings, disposable gloves, tubing, IV sets, and biological specimens (Islam et al., 2025).

- Pharmaceutical Waste: Originating from expired drugs, improperly stored medications, and leftover doses (Limenh et al., 2025). Overprescription, poor inventory management, and fragmented supply chains often contribute to pharmaceutical waste.
- Administrative Waste: Paper files, packaging, and disposable office supplies. Although generally non-hazardous, this waste represents a significant volume and environmental burden.
- Food Waste: Generated by hospital kitchens and patient meal services, often including organic waste, plastic packaging, and unused food portions (Manimaran et al., 2025; Torrejón-Ramos et al., 2025).
- Energy Waste: Resulting from inefficient lighting, HVAC systems, or medical equipment left idle. While less tangible, energy waste contributes to the overall carbon footprint of the facility.

Notably, public hospitals in low- and middle-income countries face the additional challenge of poor segregation practices, leading to the mixing of hazardous and non-hazardous waste. As a result, the total volume of waste requiring expensive and specialized treatment is significantly increased (WHO, 2014; Taslimi et al., 2020). Furthermore, improper labeling and lack of infrastructure exacerbate the risks for hospital personnel, patients, and local communities.

Waste Management Hierarchy in Healthcare

Leadership bodies, including local authorities, generally hold responsibility for the supervision of waste management systems. However, their effectiveness is frequently limited by insufficient strategic planning, restricted financial and technical resources, and difficulties related to operational oversight and contract administration (Chisholm et al., 2021; Saha et al., 2025). Effective cold chain management is crucial for maintaining the safety and quality of perishable products from production to end use. Cold chains involve a network of coordinated processes and infrastructure that ensure strict temperature control, which is vital for preserving the integrity of temperature-sensitive items like pharmaceuticals (Fatorachian & Pawar, 2025; Limenh et al., 2025). A foundational framework for sustainable waste practices is the waste management hierarchy, a model adopted and promoted by the World Health Organization (WHO, 2024b). The hierarchy consists of five core principles ranked in order of environmental preference:

- Prevention: Avoiding the generation of waste at the source by changing processes, using less hazardous materials, and optimizing consumption.

- Reduction: Minimizing the volume or toxicity of waste through efficient use of materials and better procurement practices.
- Reuse: Employing durable, sterilizable, and reusable tools, such as surgical instruments and textiles.
- Recycling: Processing materials such as plastics, glass, or paper into new products, when contamination risk is absent.
- Disposal: Safe, compliant final disposal methods—used only when the other options are not feasible.

This hierarchy shifts the focus from waste "treatment" to waste "avoidance," encouraging healthcare systems to integrate sustainability into planning, procurement, and daily operations. The aim is not only to reduce waste outputs but to do so in a way that preserves patient safety, reduces operational costs, and safeguards environmental quality.

The global disparity in waste management capacity is evident when comparing waste generation and treatment between high- and low-income countries. In high-income countries, hospitals produce more waste per bed but typically have infrastructure for waste segregation, collection, and environmentally safe disposal. In contrast, in low-income regions, while overall waste production may be lower, poor classification and mixing of waste types result in more waste being treated as hazardous, raising disposal costs and environmental risks unnecessarily (WHO, 2014).

In some facilities, infectious waste is incinerated in outdated, inefficient incinerators, releasing toxic emissions such as dioxins and furans into the atmosphere. Others rely on unregulated open burning or dumping, which can contaminate local water supplies or soil. Thus, building foundational waste literacy and improving classification systems are crucial first steps in strengthening overall waste management capabilities.

Beyond environmental concerns, waste reduction in hospitals also touches on ethical and legal obligations. Hospitals have a duty to prevent harm, not only to patients but also to workers, communities, and future generations (Limenh et al., 2025). Exposure to improperly handled waste may result in occupational injuries, long-term illnesses, or outbreaks of disease.

Legally, various international and national regulations govern healthcare waste. In the United States, for instance, the "Resource Conservation and Recovery Act" (RCRA) impose cradle-to-grave accountability for hazardous waste management (EPA, 2021). The European Union implements the Waste Framework Directive (2008/98/EC) and Landfill Directive (1999/31/EC), mandating the reduction of hazardous waste disposal and prioritizing recycling and reuse (European Commission, 2008; 2020). These frameworks define not just environmental expectations but also institutional responsibilities, reporting obligations, and performance targets.

Establishing a culture of waste accountability begins with these foundations, an understanding of what waste exists, where it originates, what risks it presents, and how it can be minimized systematically through structured frameworks and informed decision-making.

DIGITAL TRANSFORMATION FOR WASTE REDUCTION

AI-driven technologies are increasingly reshaping organizational management and workplace dynamics by enhancing decision-making processes, improving operational performance, and transforming human resource practices (Ali & Mahmood, 2025). The integration of digital technologies into healthcare waste management is playing a transformative role in modern public hospitals, offering innovative tools to streamline processes, improve decision-making, and enhance environmental sustainability (Leow et al., 2026). The need for digital solutions that connect physical and digital infrastructure is steadily increasing, with a focus on building more effective systems for managing waste across its lifecycle (Rittl et al., 2025).

Digital transformation in this context refers to the use of technologies such as the Internet of Things (IoT), artificial intelligence (AI), blockchain, cloud computing, sensors, and radio frequency identification (RFID) to enhance the efficiency, accuracy, and transparency of waste management operations (Sharma et al., 2025). These tools facilitate the monitoring and analysis of waste generation, segregation, transport, and disposal in real time, enabling data-driven decision-making and proactive responses.

For instance, IoT sensors have been deployed to detect inefficiencies within temperature-sensitive supply chains, such as vaccine storage, allowing hospitals to intervene before spoilage occurs and thus reduce pharmaceutical waste (Fatorachian & Pawar, 2025; Limenh et al., 2025). Similarly, data analytics powered by AI can help identify patterns in waste production, optimize disposal schedules, and minimize resource consumption (Uysal et al., 2026).

Blockchain technology plays a vital role in ensuring traceability and accountability in waste disposal, particularly for hazardous medical waste. By providing immutable records of waste transactions and transport routes, blockchain builds trust among stakeholders and reinforces compliance with safety standards (Fatorachian & Pawar, 2025; Sharma et al., 2025).

Furthermore, rather than relying solely on rigid hierarchical structures, organizations are gradually adopting management approaches supported by intelligent systems that enable more flexible, data-informed processes (Ali & Mahmood, 2025). In parallel, advanced tools such as machine learning and explainable AI are being applied to address the rising challenge of electronic waste in healthcare settings

(Uysal et al., 2026). These technologies support the automated classification of discarded electronic materials, allowing valuable elements like metals and plastics to be efficiently recovered and reused or transformed into energy through methods such as pyrolysis, thereby supporting circular economy objectives (Farjana et al., 2023).

Infectious medical waste management has also benefited significantly from data-driven digital systems. Real-time data collection and analysis support quicker and more informed responses to potential biohazards, enhancing overall safety within hospital environments (Yaspal et al., 2023). Web-based geographic information systems (WebGIS), remote sensing, and geoprocessing tools are further contributing to these efforts by improving collection route planning, reducing landfill dependency, and encouraging public participation in waste governance (Rittl et al., 2025).

Despite these advancements, challenges remain, particularly in emerging economies where the adoption of digital technologies is uneven. Limited infrastructure, lack of digital literacy, and insufficient investment create significant barriers to the safe and effective management of hazardous healthcare waste (Islam et al., 2025; Sharma et al., 2025). Addressing these challenges is critical to achieving broader goals of environmental protection, operational efficiency, and public health.

Overall, digital transformation represents a powerful lever for waste reduction in public hospitals (Uysal et al., 2026). By leveraging a combination of real-time monitoring, automation, secure data sharing, and predictive analytics, healthcare systems can transition toward more sustainable, transparent, and efficient waste management models.

WASTE SEGREGATION, RECYCLING, AND DISPOSAL

Healthcare facilities, though essential for protecting human health, paradoxically contribute to environmental degradation through the generation of substantial quantities of medical waste (Kenny & Priyadarshini, 2021). Hospitals are directly implicated in climate-related harms, including greenhouse gas emissions, hazardous discharges, and energy-intensive operations (WHO, 2024). As climate change exacerbates public health risks—rising temperatures, pollution, and extreme weather events (Shin et al., 2020)—the urgency to mitigate the environmental footprint of healthcare institutions becomes more apparent (Mouloudj et al., 2025). Central to this effort is effective healthcare waste management, which includes three primary pillars: segregation, recycling, and safe disposal.

Waste Segregation

Segregation is the foundational step in healthcare waste management. It involves the immediate and systematic separation of different waste types at the point of generation, typically using color-coded containers. This process ensures that infectious or hazardous waste is not mixed with general or recyclable waste, thereby minimizing both health risks and disposal costs (Longe & Williams, 2006; Aseweh Abor, 2013).

The use of a standardized color-coding system is essential. For instance, black containers are used for non-hazardous general waste, yellow for infectious waste (including sharps), red for waste containing heavy metals or effluents, and white for recyclable medical glass such as ampoules or vials (Ramteke & Sahu, 2019). In low-resource facilities, red containers may be omitted, with heavy metals handled via yellow-coded systems.

Proper segregation is not only an operational necessity but also a regulatory and ethical obligation. Health workers are responsible for ensuring that waste is categorized and disposed of correctly. However, multiple studies have identified low compliance with segregation protocols. Ibrahim et al. (2023) found that segregation practices correlated with factors such as hospital ownership, training, knowledge, and the presence of clear signage and infrastructure. Inadequate segregation complicates waste classification and increases health and environmental risks (Mouloudj et al., 2023).

Recycling

Recycling in healthcare settings involves converting waste materials into reusable resources. This is especially crucial for plastic-based items, which make up a large portion of healthcare waste, including "personal protective equipment" (PPE), IV tubing, and syringes (Kenny & Priyadarshini, 2021). Recent studies underscore that while healthcare systems still rely heavily on single-use materials, innovation in recycling can significantly reduce the environmental burden (Ebrahimzadeh-sarvestani et al., 2025).

Different recycling methods are applicable depending on the material type. Mechanical recycling, which involves melting and remolding plastics, is a feasible short-term solution when waste is properly segregated and decontaminated. Long-term solutions such as chemical recycling, where polymers are depolymerized and repolymerized, retain material value and align with circular economy principles (Ahamad et al., 2025).

Beyond plastics, a variety of other materials in the healthcare sector can be recycled: aluminum from blister packs, stainless steel from surgical tools, mercury from dental amalgams, and prosthetics from cremation processes (Shukla et al.,

2022; Lidgren et al., 2020; Van Straten et al., 2021). Despite its benefits, recycling often requires substantial energy input and centralized infrastructure. Therefore, AI-powered initiatives implemented on-site or near-site, particularly for non-hazardous waste like cardboard, tubing, and packaging, can reduce costs and environmental impact (Uysal et al., 2026; WHO, 2014).

Public engagement is also a key enabler of successful recycling programs. Ionescu and Cazan (2024) report strong community support for pharmaceutical take-back schemes and stress the importance of public education and institutional accountability in boosting recycling rates. Additionally, digital innovations such as AI-supported decision models can help hospitals select optimal recycling strategies based on cost, safety, and environmental impact (Li & Qiao, 2025).

Disposal: Effective Strategies

While segregation and recycling reduce waste volume, disposal remains necessary for certain hazardous materials. Effective disposal strategies are vital for safeguarding environmental and human health, especially when dealing with toxic, infectious, or chemical waste.

One of the most promising innovations in this domain is plasma pyrolysis; a high-temperature thermal process that converts hazardous waste into syngas, bio-oil, and inert slag. This method drastically reduces waste volume and toxicity, and when combined with IoT systems, it allows real-time monitoring for optimal performance. The resulting slag can even be repurposed in anaerobic digestion systems to support biogas production, contributing to circular economy goals (Li & Qiao, 2025).

In decentralized or residential healthcare settings, community participation plays a critical role. It can be said that community-based education and infrastructure investments can led to improvement in hazardous waste identification and a marked reduction in unsafe disposal behaviors (Brotosusilo & Naldi, 2021; Zulfa & Nugraheni, 2020). These outcomes highlight the importance of behavior change communication, intersectoral collaboration, and stakeholder empowerment (Almokdad & Mouloudj, 2026).

Surveillance systems and safety protocols are also essential in managing chemical hazards. Wijnands et al. (2025), in their review of HAZMAT incidents in the Netherlands, emphasized the prevalence of chemical exposures and the need for targeted training and preparedness, especially in fixed healthcare facilities where such risks are most concentrated.

Finally, institutional capacity building is key to successful waste disposal. Training programs focused on occupational safety and waste handling can dramatically improve compliance and outcomes. Ebrahimzadehsarvestani et al. (2025) and Kumar et al.

(2016) both highlight the need for systematic training, especially in low-resource settings where incineration remains the dominant, yet problematic, disposal method.

By integrating segregation, recycling, and advanced disposal technologies, healthcare systems can significantly reduce their ecological footprint. These strategies not only protect public health but also align healthcare institutions with global sustainability goals, transforming them from contributors to climate change into leaders of environmental stewardship.

STAFF TRAINING AND AWARENESS CAMPAIGNS

In addition to addressing economic and environmental concerns, the operational workforce plays a pivotal role in protecting public health and implementing effective waste management strategies (Patwa et al., 2026). As frontline actors in healthcare settings, staff members are not only responsible for delivering medical services but also for ensuring that waste generated during these processes is handled safely and sustainably. Their active participation is essential for translating institutional sustainability goals into day-to-day practices (Leow et al., 2026). Without their involvement in waste segregation, safe disposal, and adherence to environmental protocols, broader waste reduction efforts may falter. Therefore, equipping the workforce with appropriate training and empowering them to understand the environmental impact of their roles is crucial for the success of any sustainable healthcare initiative (Sharma et al., 2025).

In this context, promoting sustainable practices within healthcare institutions is no longer optional; it is essential. Unsustainable environmental practices in healthcare, whether in critical care, primary care, dental settings, or social and veterinary services, are increasingly recognized as contributing to environmental degradation (Huffling & Schenk, 2014; Nicolet et al., 2022). As a result, training and awareness campaigns have emerged as crucial strategies for embedding sustainability into everyday healthcare practices.

Environmental education and training programs are key tools for cultivating a culture of sustainability within healthcare organizations. These programs equip staff with the knowledge and skills needed to align operational activities with broader environmental goals (Affordofe et al., 2025; Jasim et al., 2026). As noted in recent research, the commitment of staff to the organization's sustainability vision, combined with targeted training, can form the backbone of a greener healthcare infrastructure, without compromising the quality of care (Al-Hassan et al., 2026; Arous et al., 2025; Sepetis et al., 2024). Education not only fosters informed decision-making but also encourages behavioral change among different segments of society (employees, customers, patients, etc.), leading to more sustainable practices in different settings (Almokdad & Mouloudj, 2026; Alyousuf et al., 2026).

The importance of training is underscored by several empirical studies. For instance, Altassan (2024) found that training and development programs significantly enhanced employees' environmental commitment in the manufacturing sector, particularly when aligned with green human resource management (HRM) practices. Similarly, Lee and Lee (2022) emphasized that leadership and organizational support are vital for fostering environmental engagement among staff, highlighting the importance of structured education on sustainability policies and procedures.

Within the healthcare sector, however, gaps remain. A survey of 379 healthcare professionals across Greece revealed that environmental sustainability often remains a peripheral concern for administrators. The study found low levels of staff participation in sustainability initiatives, with key barriers including poor institutional communication, limited awareness, and insufficient education (Sepetis et al., 2024). This disconnects between administrative priorities and on-the-ground practices poses a significant hurdle to environmental reform in healthcare settings.

However, targeted and periodic training programs have shown measurable success. A study by Bannour et al. (2024), conducted over a two-year period, demonstrated that consistent training significantly improved healthcare waste management practices among medical staff. Despite progress, the study also uncovered persistent issues such as improper labeling of containers and waste bags, underscoring the need for ongoing capacity building. Similarly, Guihenneuc et al. (2024) found that healthcare professionals in France cited inadequate methodological support, lack of leadership endorsement, and limited access to training as barriers to implementing sustainability projects. Interestingly, the presence of a long-standing sustainability committee was positively correlated with staff knowledge, perceptions of hospital transformation efforts, and proactive engagement in eco-friendly initiatives.

Sustainability in healthcare requires not only technical solutions but also behavioral shifts (Leow et al., 2026). Medical professionals and support staff must be empowered to integrate sustainability into their daily routines, from energy and resource use to waste sorting and disposal. Raising awareness of the environmental impacts of clinical and non-clinical activities is critical for promoting a culture of responsibility and innovation. Staff and waste handlers who are well-informed can help reduce the institution's carbon footprint by making environmentally conscious decisions in purchasing, procedures, and patient care (Affordofe et al., 2025).

Moreover, healthcare workers are central to implementing waste reduction policies. Their understanding of relevant guidelines, combined with their active participation, ensures the safe and efficient handling of medical waste (Mouloudj et al., 2025). This is particularly crucial in under-resourced healthcare systems, such as in many African nations, where inadequate training and low awareness contribute to widespread mishandling of hazardous waste (Chisholm et al., 2021). Equipping frontline staff with the necessary knowledge and resources is a fundamental step toward safeguarding both environmental and public health.

Finally, awareness campaigns that highlight the ethical implications of waste and promote environmental responsibility can shape positive attitudes and behavioral intentions (Alyousuf et al., 2026; Jasim et al., 2026). As Mouloudj et al. (2023) emphasized, fostering awareness around the risks of improper medication disposal, and encouraging environmentally responsible behavior, plays a key role in reducing healthcare-related waste.

In sum, comprehensive staff training and awareness-building is indispensable for embedding sustainability into healthcare institutions. These efforts help bridge the gap between policy and practice, foster environmental stewardship, and ensure that sustainability becomes a shared responsibility across all levels of the healthcare workforce.

FUTURE RESEARCH DIRECTIONS

As the integration of digital solutions into public healthcare waste management continues to evolve, future research can play a pivotal role in evaluating, refining, and expanding these approaches. One promising avenue involves assessing the effectiveness of AI and machine learning technologies in minimizing waste across various operational dimensions of healthcare institutions. For instance, empirical studies could explore how predictive analytics and AI-powered inventory management systems help reduce expired pharmaceuticals, optimize supply chains, or forecast waste generation patterns. Additionally, there is a need to investigate the synergistic impact of combining digital transformation strategies with sustainability-oriented practices, such as green procurement policies or circular economy models in healthcare settings. Future studies could explore how integrating digital tools with institutional policies improves outcomes not only in waste reduction but also in energy efficiency, patient safety, and cost savings.

Another important research direction involves conducting field investigations and case studies to identify the practical barriers and facilitators associated with implementing digital waste management solutions in public hospitals. These could include infrastructural limitations, staff resistance, financial constraints, or gaps in digital literacy. Such context-specific research would be valuable in tailoring interventions to different healthcare environments, especially in low- and middle-income countries. Further studies could also explore the role of interdisciplinary collaboration, between healthcare professionals, environmental scientists, and technology developers, in designing and implementing waste-reduction strategies. In addition, research into behavioral change and training programs using gamification, e-learning, or mobile applications could yield insights into the most effective ways to promote sustainable practices among hospital staff.

Finally, future research could contribute by developing standardized frameworks and performance indicators for measuring the success of digital waste reduction initiatives in the healthcare sector. This would help in benchmarking progress across institutions and guiding evidence-based policymaking at local, national, and international levels.

MANAGERIAL IMPLICATIONS

From a managerial standpoint, the implications are wide-ranging. Healthcare administrators are encouraged to prioritize investments in digital infrastructure and ensure the interoperability of waste tracking and clinical management systems. Policy-makers and regulatory bodies should support the development of national standards and incentives for sustainable hospital operations, including funding schemes for digital waste management solutions. Environmental and public health officials must collaborate with hospital managers to monitor compliance and promote circular economy principles across healthcare supply chains. Equally important, hospital executives and department leaders must invest in continuous staff training and awareness campaigns, ensuring that sustainability becomes embedded in the organizational culture. Technology providers and innovators, meanwhile, are called upon to design scalable, user-friendly, and context-sensitive solutions that meet the evolving needs of healthcare institutions, particularly in under-resourced settings.

CONCLUSION

Waste reduction in public hospitals is no longer a peripheral concern but a central pillar of sustainable healthcare management. As healthcare systems worldwide face increasing pressure from climate change, resource constraints, and rising medical waste, integrating digital technologies with sustainable practices emerges as a vital response. This chapter has outlined the core concepts and sources of healthcare waste, highlighted the transformative potential of digital tools such as AI and EHR systems, and emphasized the significance of staff engagement and effective waste segregation practices. Together, these dimensions form a comprehensive framework for rethinking waste management in healthcare institutions. Digital transformation, when aligned with environmental responsibility and supported by a trained workforce, can lead to more efficient operations, cost savings, and improved health outcomes.

In sum, the integration of digital solutions with sustainable waste management in public hospitals requires coordinated, multi-stakeholder action. With the right mix of leadership, innovation, and policy support, healthcare systems can significantly reduce their environmental footprint while safeguarding both public health and institutional resilience.

ACKNOWLEDGMENT

This research received no specific grant from any funding agency in the public, commercial, or not-for-profit sectors

REFERENCES

Affordofe, M., Udofia, E. A., Akyeampong, E., Koranteng, F. O., Tettey, P., Botwe, P. K., Andoh, T., Peprah, E. K., Attah, H. Y., Tabong, P., Boateng, G., & Quansah, R. (2025). Knowledge and practice of solid healthcare waste management among waste handlers in hospitals in Southern Ghana: A qualitative study. *BMC Public Health*, *25*(1), 702. DOI: 10.1186/s12889-025-21874-6 PMID: 39979880

Ahamad, A., Kumar, P., & Kumar, B. (2025). Treatment and Recycling of Plastic Waste: A Dire Need for a Sustainable Future. In Kumar, P., & Kumar, B. (Eds.), *Global Waste Management* (pp. 359–381). John Wiley & Sons., DOI: 10.1002/9781394318414.ch14

Al-Hassan, R. A., Alyousuf, A. N., Jasim, Z. A., Kamil, K., & Yaseen, A. (2026a). Fashion Waste Management in the Age of Artificial Intelligence: Strategies and Challenges. In Mouloudj, K., & Arantes, L. (Eds.), *Ethical Consumption and Responsible Innovation in Fashion through AI*. IGI Global., DOI: 10.4018/979-8-3373-7200-6.ch009

Al-Hassan, R. A., Dirawi, W. M., Almosafri, W. H., Alabduladheem, M. F., Jahloul, I. H., & Alyousuf, A. N. (2026b). Green HRM in Hospitals as a Strategic Approach to Promoting Employees' Pro-Environmental Behavior and Achieving Organizational Sustainability. In Rodrigues, R., Mohanty, V., & Figueiredo, P. (Eds.), *Organizational Culture and Employee Experience in the AI-Driven Organization* (pp. 269–308). IGI Global., DOI: 10.4018/979-8-2600-0610-8.ch010

Ali, Q., & Mahmood, M. A. (2025). Transforming Leadership: The Role of Artificial Intelligence in Modern Organizational Management. In Baig, S., Mohsin, M., Lodhi, R., & Batool, G. (Eds.), *Reshaping Organizational Management and Workplace Culture with AI* (pp. 223–262). IGI Global., DOI: 10.4018/979-8-3693-8960-7.ch009

Alighardashi, M., Mousavi, S. A., Almasi, A., & Mohammadi, P. (2025). A study of indoor air pollution caused by disinfection equipment as a consequence of infectious waste management. *International Journal of Environmental Health Research*, *35*(3), 634–644. DOI: 10.1080/09603123.2024.2368123 PMID: 38899861

Almokdad, E., & Mouloudj, K. (2026). Exploring Actual Sustainable Behavior of Tourists: Qualitative Analysis. In Mouloudj, K., & Bouarar, A. (Eds.), *Insights on Consumer Psychology in the Digital Landscape* (pp. 185–210). IGI Global., DOI: 10.4018/979-8-3373-2424-1.ch007

Altassan, M. A. (2024). The nexus between green HR practices and firm sustainable performance in Saudi Arabia manufacturing industry: The role of green innovation and green transformation leadership. *Uncertain Supply Chain Management, 12*(4), 2207–2220. DOI: 10.5267/j.uscm.2024.6.010

Alyousuf, A. N., Sabeeh, A. O., Alnaemat, N. D., Mouloudj, K., & Hussein, S. A. (2026). Understanding Gen Z's Adoption of AI Technologies in Green Product Purchases: An Extension of the Technology Acceptance Model. In Mouloudj, K., & Bouarar, A. (Eds.), *Insights on Consumer Psychology in the Digital Landscape* (pp. 369–392). IGI Global., DOI: 10.4018/979-8-3373-2424-1.ch013

Arous, J. A., ElRabbat, S. A., ElAwad, I. H., Hima, N. A., & El Hamdi, R. (2025). Assessing the impact of environmental training programs on awareness and job performance in institutions in the Kingdom of Saudi Arabia. *International Area Studies Review, 28*(1), 1–23. DOI: 10.69473/iasr.2025.28.1.1

Aseweh Abor, P. (2013). Managing healthcare waste in Ghana: A comparative study of public and private hospitals. *International Journal of Health Care Quality Assurance, 26*(4), 375–386. DOI: 10.1108/09526861311319591 PMID: 23795428

Brotosusilo, A., & Naldi, A. (2021, March). Policy on optimization of household waste and hazardous waste management based on community empowerment at the local level. In *IOP Conference Series: Earth and Environmental Science* (Vol. 716, No. 1, p. 012085). IOP Publishing. DOI: 10.1088/1755-1315/716/1/012085

Chisholm, J. M., Zamani, R., Negm, A. M., Said, N., Abdel daiem, M. M., Dibaj, M., & Akrami, M. (2021). Sustainable waste management of medical waste in African developing countries: A narrative review. *Waste Management & Research, 39*(9), 1149–1163. DOI: 10.1177/0734242X211029175 PMID: 34218734

Ebrahimzadehsarvestani, E., Sarvestani, E. M., Safarzadeh, H., Enayati, M., & Di Maria, F. (2025). Implementing circular economy in healthcare waste: Technological innovations and process management. *Journal of Solid Waste Technology Management, 51*(1), 125–134. DOI: 10.5276/jswtm/iswmaw/51SI1/2025.014

EPA. (2021). *Resource Conservation and Recovery Act (RCRA) overview*. U.S. Environmental Protection Agency. https://www.epa.gov/rcra

European Commission. (2008). Directive 2008/98/EC on waste and repealing certain directives. https://eur-lex.europa.eu/legal-content/EN/TXT/?uri=celex%3A32008L0098

European Commission. (2020). A new circular economy action plan: For a cleaner and more competitive Europe. https://environment.ec.europa.eu/strategy/circular-economy-action-plan_en

Farjana, M., Fahad, A. B., Alam, S. E., & Islam, M. M. (2023). An IoT- and cloud-based e-waste management system for resource reclamation with a data-driven decision-making process. *IoT*, *4*(3), 202–220. DOI: 10.3390/iot4030011

Fatorachian, H., & Pawar, K. (2025). Enhancing waste management in cold supply chains via digital transformation. *Procedia Computer Science*, *253*, 167–176. DOI: 10.1016/j.procs.2025.01.080

Health Care Without Harm. (2025). *Waste management*. https://us.noharm.org/waste-management

Huffling, K., & Schenk, E. (2014). Environmental sustainability in the intensive care unit: Challenges and solutions. *Critical Care Nursing Quarterly*, *37*(3), 235–250. DOI: 10.1097/CNQ.0000000000000028 PMID: 24896556

Ibrahim, M., Kebede, M., & Mengiste, B. (2023). Healthcare waste segregation practice and associated factors among healthcare professionals working in public and private hospitals, Dire Dawa, Eastern Ethiopia. *Journal of Environmental and Public Health*, *2023*(3), 1–7. DOI: 10.1155/2023/8015856 PMID: 36747502

Ionescu, A.-M., & Cazan, C. (2024). Pharmaceutical Waste Management: A Comprehensive Analysis of Romanian Practices and Perspectives. *Sustainability (Basel)*, *16*(15), 6571. DOI: 10.3390/su16156571

Islam, M. M., Zaman, S., Ferdous, J., & Sarker, M. H. (2025). Policy gaps, actor dynamics, and implementation challenges in sustainable medical waste management in Bangladesh. *Journal of Environmental Management*, *389*, 126203. DOI: 10.1016/j.jenvman.2025.126203 PMID: 40517630

Jasim, Z. A., Alameri, S. A., Almado, A. A., Abdulrahman, F. Y., & Alyousuf, A. N. (2026). Investigating the Impact of Ethical Organizational Culture on Employee Retention. In Rodrigues, R., Mohanty, V., & Figueiredo, P. (Eds.), *Organizational Culture and Employee Experience in the AI-Driven Organization* (pp. 85–110). IGI Global., DOI: 10.4018/979-8-2600-0610-8.ch004

Kenny, C., & Priyadarshini, A. (2021). Review of current healthcare waste management methods and their effect on global health. *Health Care*, *9*(3), 284. DOI: 10.3390/healthcare9030284 PMID: 33807606

Kumar, R., Somrongthong, R., & Ahmed, J. (2016). Impact of waste management training intervention on knowledge, attitude and practices of teaching hospital workers in Pakistan. *Pakistan Journal of Medical Sciences*, *32*(3), 705–710. DOI: 10.12669/pjms.323.9903 PMID: 27375718

Lee, S. M., & Lee, D. (2022). Effective medical waste management for sustainable green healthcare. *International Journal of Environmental Research and Public Health*, *19*(22), 14820. DOI: 10.3390/ijerph192214820 PMID: 36429539

Leow, N. X., Lin, W. L., Yong, J. Y., Chong, S. C., Pek, C. K., & Feranita, F. (2026). Optimizing Remote Work for a Sustainable Workplace: Strategies to Promote Employee Green Behavior. In Yusliza, M. Y., Renwick, D. W. S., Yong, J. Y., & Kamil, N. L. M. (Eds.), *Employee Green Behavior* (pp. 97–117). Springer., DOI: 10.1007/978-981-95-1422-9_6

Li, T., & Qiao, J. (2025). A novel q-rung orthopair fuzzy MAGDM method for healthcare waste treatment based on three-way decisions. *Pattern Recognition*, *157*, 110867. DOI: 10.1016/j.patcog.2024.110867

Lidgren, L., Raina, D. B., Tägil, M., & Tanner, K. E. (2020). Recycling implants: A sustainable solution for musculoskeletal research. *Acta Orthopaedica*, *91*(2), 125. DOI: 10.1080/17453674.2019.1706301 PMID: 31902268

Limenh, L. W., Simegn, W., Bizuneh, G. K., Chanie, G. S., Seid, A. M., Mengesha, A. K., Beyna, A. T., Melese, M., Gela, Y. Y., Esubalew, D., Mitku, M. L., & Ayenew, W. (2025). Management of pharmaceutical waste from hospitals and community pharmacies in Bahir Dar and Gondar cities, Ethiopia. *BMC Health Services Research*, *25*(1), 379. DOI: 10.1186/s12913-025-12555-6 PMID: 40087596

Longe, E. O., & Williams, A. (2006). A preliminary study of medical waste management in Lagos metropolis, Nigeria. *Journal of Environmental Health Science & Engineering*, *3*(2), 133–139. https://ijehse.tums.ac.ir/index.php/jehse/article/view/82

Manimaran, S., Razalli, N. H., Manaf, Z. A., & Shahar, S. (2025). Challenges and strategies to reduce food waste in malaysian hospitals from the perspective of multidisciplinary professionals: A qualitative descriptive study. *BMC Health Services Research*, *25*(1), 224. DOI: 10.1186/s12913-025-12365-w PMID: 39930430

Mouloudj, K., Bouarar, A. C., Hoti, F., Asanza, D. M., & Kabashi, D. (2025). Nurses' Intentions to Implement Sustainable Practices in Public Hospitals: An Empirical Investigation. In Carrasco-Garrido, C., Martínez Raya, A., & De-Pablos-Heredero, C. (Eds.), *Global Economic Challenges and Opportunities for the Workplace* (pp. 1–24). IGI Global., DOI: 10.4018/979-8-3693-6289-1.ch001

Mouloudj, K., Njoku, A., Asanza, D. M., Bouarar, A. C., Evans, M. A., Mouloudj, S., & Bouarar, A. (2023). Modeling predictors of medication waste reduction intention in Algeria: Extending the theory of planned behavior. *International Journal of Environmental Research and Public Health*, *20*(16), 6584. DOI: 10.3390/ijerph20166584 PMID: 37623170

Nicolet, J., Mueller, Y., Paruta, P., Boucher, J., & Senn, N. (2022). What is the carbon footprint of primary care practices? A retrospective life-cycle analysis in Switzerland. *Environmental Health : A Global Access Science Source*, *21*(1), 3. DOI: 10.1186/s12940-021-00814-y PMID: 34980135

Patwa, N., Paul, R., Vittal, S., Jadhav, V., & Ranjan, B. (2026). Harnessing Employee Engagement for Environmental Sustainability and CSR. In Yusliza, M. Y., Renwick, D. W. S., Yong, J. Y., & Kamil, N. L. M. (Eds.), *Employee Green Behavior* (pp. 259–276). Springer., DOI: 10.1007/978-981-95-1422-9_14

Ramteke, S., & Sahu, B. L. (2019). Health care waste management: A review. *International Journal of Trend in Scientific Research and Development*, *3*(5), 1235–1240. DOI: 10.31142/ijtsrd23109

Rittl, L. G. F., Zaman, A., & de Oliveira, F. H. (2025). Digital transformation in waste management: Disruptive innovation and digital governance for zero-waste cities in the global south as keys to future sustainable development. *Sustainability (Basel)*, *17*(4), 1608. DOI: 10.3390/su17041608

Saha, K., Patel, B., & Paladini, S. (2025). The role of leadership and cultural barriers in the adoption of lean six sigma in clinical pharmacy practice and medicine waste reduction. The case of NHS-UK. *International Journal of Quality & Reliability Management*, *42*(3), 809–836. DOI: 10.1108/IJQRM-02-2024-0069

Sepetis, A., Parlavatzas, I., Zaza, P. N., Platis, C., Fotios, R., & Nikolaou, I. E. (2024). The Role of Organizational Behavior to Sustainable Health Care: The Case of Greece. *Environmental Health Insights*, *18*, 1–15. DOI: 10.1177/11786302241298788 PMID: 39641108

Sharma, V., Jamwal, A., Agrawal, R., & Pratap, S. (2025). A review on digital transformation in healthcare waste management: Applications, research trends and implications. *Waste Management & Research*, *43*(6), 828–849. DOI: 10.1177/0734242X241285420 PMID: 39352741

Shin, S.-K., Um, N., Kim, Y.-J., Cho, N.-H., & Jeon, T.-W. (2020). New policy framework with plastic waste control plan for effective plastic waste management. *Sustainability (Basel)*, *12*(15), 6049. DOI: 10.3390/su12156049

Shukla, S., Halli, P., Khalid, M. K., & Lundström, M. (2022). Waste pharmaceutical blister packages as a source of secondary aluminum. *Journal of the Minerals Metals & Materials Society*, *74*(2), 612–621. DOI: 10.1007/s11837-021-05038-6

Taslimi, M., Batta, R., & Kwon, C. (2020). Medical waste collection considering transportation and storage risk. *Computers & Operations Research*, *120*, 104966. DOI: 10.1016/j.cor.2020.104966

Torrejón-Ramos, M., Medina-Salgado, M. S., & Ortiz-de-Urbina-Criado, M. (2025). A scoping review on reducing food waste and loss in hospitals. *Gaceta Sanitaria*, *39*, 102462. DOI: 10.1016/j.gaceta.2025.102462 PMID: 39985827

Uysal, İ., Altın, F. G., & Özcan, F. (2026). Predicting medical waste amounts and costs using machine learning and XAI for sustainable waste management in healthcare. *International Journal of Environmental Science and Technology*, *23*(1), 21. DOI: 10.1007/s13762-025-06895-3

Van Straten, B., Dankelman, J., Van Der Eijk, A., & Horeman, T. (2021). A circular healthcare economy; a feasibility study to reduce surgical stainless steel waste. *Sustainable Production and Consumption*, *27*, 169–175. DOI: 10.1016/j.spc.2020.10.030

WHO. (2014). Safe management of wastes from health-care activities. https://iris.who.int/bitstream/handle/10665/85349/9789241548564_eng.pdf

WHO. (2024b). Safe management of wastes from health-care activities. https://iris.who.int/bitstream/handle/10665/85349/9789241548564_eng.pdf

Wijnands, A. P. G., Koppen, A., de Vries, I., de Lange, D. W., & Rietjens, S. J. (2025). Poison center surveillance of occupational incidents with hazardous materials (2016–2023): Insights for risk mitigation and incident preparedness. *International Journal of Environmental Research and Public Health*, *22*(2), 158. DOI: 10.3390/ijerph22020158 PMID: 40003384

World Health Organization. (2024a). Health-care waste. https://www.who.int/news-room/fact-sheets/detail/health-care-waste

Yaspal, B., Jauhar, S. K., Kamble, S., Belhadi, A., & Tiwari, S. (2023). A data-driven digital transformation approach for reverse logistics optimization in a medical waste management system. *Journal of Cleaner Production*, *430*, 139703. DOI: 10.1016/j.jclepro.2023.139703

Zulfa, V., & Nugraheni, P. L. (2020, May). Effectiveness of community empowerment in waste management program to create sustainable tourism in Karawang, West Java. In *IOP Conference Series: Earth and Environmental Science* (Vol. 485, No. 1, p. 012087). IOP Publishing. DOI: 10.1088/1755-1315/485/1/012087

KEY TERMS AND DEFINITIONS

Disposal: The final and safe elimination of waste materials through approved methods, such as incineration or landfilling, in order to protect public health and the environment.

Healthcare Sustainability: The ability of healthcare systems to deliver quality care while minimizing environmental impact, ensuring economic viability, and supporting social well-being over the long term.

Healthcare Waste Management: The systematic process of collecting, handling, treating, and disposing of waste generated by healthcare activities in a manner that reduces health risks and environmental harm.

Recycling: The process of collecting and converting waste materials into new products to reduce resource consumption, environmental pollution, and the volume of waste requiring disposal.

Staff Awareness: The level of knowledge, understanding, and responsibility among healthcare workers regarding proper waste handling, environmental protection, and sustainable practices.

Waste Segregation: The practice of separating different types of waste at the point of generation based on their characteristics and risk levels to enable safe handling, recycling, or disposal.

Chapter 8

Household Waste and Everyday Practices:

Understanding Rural-Urban Differences in Kitchen Waste Disposal in Kerala

V. K. Aswathy
http://orcid.org/0000-0002-8895-8373
Lovely Professional University, India

Sanjeev Kumar
http://orcid.org/0000-0002-7375-7341
Lovely Professional University, India

ABSTRACT

Kerala's waste management practices,research shedding light on significant rural-urban disparities and emphasizing the importance of targeted interventions in rural regions, particularly addressing kitchen waste disposal. Drawing from a robust dataset of 46,334 respondents in Kerala, sourced from NFHS 5, the study uncovers a compelling commitment to responsible waste management. Both rural and urban areas exhibit a shared dedication to sustainability, with a consistent preference for closed drain disposal and reusing kitchen waste for gardening or farming. Result: Encouragingly, urban regions show higher levels of awareness and engagement in sustainable waste management, reflected in the prevalence of manual collection. However, nuanced variations in less eco-friendly methods require vigilant attention. This research underscores Kerala's resolute commitment to responsible waste disposal, buoyed by the success of eco-conscious initiatives, and highlights the necessity of tailoring strategies to address local variations for a more sustainable future for all residents.

DOI: 10.4018/979-8-2600-1960-3.ch008

INTRODUCTION

Effective waste management is a global imperative, and it holds particular significance in regions like Kerala, India, known for their environmental consciousness and unique waste disposal practices. Kitchen waste, comprising organic materials like food scraps and peels, represents a substantial portion of household waste (Nguyen, T. T. T., et al. 2022). Sustainable management of kitchen waste is crucial not only for environmental conservation but also for public health and resource optimization. Kerala's waste management practices have garnered attention for their eco-friendly and community-driven nature. However, understanding the variance in these practices between rural and urban regions is vital for tailoring effective waste management policies and interventions.

The state of Kerala, nestled in the southwestern part of India, is celebrated for its lush green landscapes, rich biodiversity, and a culture deeply rooted in sustainable living. As Kerala continues to experience urbanization and population growth, the dynamics of waste generation and disposal are evolving. Urban areas are expanding, and lifestyles are changing, influencing how residents dispose of their kitchen waste. In this context, it becomes imperative to assess the variations in approaches to kitchen waste disposal between rural and urban regions within Kerala.

This research embarks on a comprehensive exploration of the waste management practices in Kerala, with a specific focus on kitchen waste disposal. It is grounded in the recognition that waste management is not a one-size-fits-all solution but a complex interplay of geographical, cultural, and infrastructural factors. As Kerala continues to evolve and adapt to changing demographics and urbanization, this study strives to provide valuable insights that can inform targeted waste management policies and practices. It also serves as a testament to Kerala's commitment to responsible waste disposal and sustainability, highlighting the success of eco-conscious initiatives in the region. Ultimately, the research underscores the need to tailor strategies to address local variations and promote a more sustainable future for all residents of Kerala.

Need and significance of the study

This research is of paramount importance due to the pressing need for sustainable waste management practices in Kerala, a state renowned for its environmental consciousness. Understanding the variance in kitchen waste disposal methods between rural and urban regions is essential to address significant disparities and promote responsible waste management practices. The need for this study arises from the observed differences in waste disposal practices, which have implications for public health, sanitation, and the environment. Bridging the rural-urban gap in waste management is crucial to ensure equitable access to clean and safe living conditions for all residents. Furthermore, this research holds significance for policymakers, environmentalists, and community leaders seeking to design targeted interventions, awareness campaigns, and infrastructure improvements that can foster eco-friendly waste management practices and enhance Kerala's commitment to sustainability. By shedding light on these disparities and their underlying causes, this research contributes to the collective effort to create a more environmentally conscious and equitable future for Kerala's residents.

Kerala, known for its natural beauty and commitment to sustainability, faces distinct challenges in managing kitchen waste disposal practices, with significant rural-urban disparities. While Kerala's waste management practices have garnered praise for eco-consciousness, there remains a need to address these disparities in kitchen waste disposal methods. The problem at hand is the notable divergence in approaches to kitchen waste disposal between rural and urban regions in Kerala, which warrants in-depth investigation. Hence the study aims to identify potential opportunities to bridge the gap between rural and urban kitchen waste disposal practices in Kerala. By understanding and addressing these variations, Kerala can progress toward a more equitable and sustainable waste management landscape.

Data and Methods

This study relies on secondary data sourced from the National Family Health Survey 5 (NFHS 5), a comprehensive and nationally representative survey conducted in India. The National Family Health Survey 5 (NFHS 5) is a vital and extensive demographic and health survey program in India. It serves as an invaluable source of information for researchers, policymakers, and healthcare professionals. NFHS 5 is conducted under the aegis of the Ministry of Health and Family Welfare, Government of India, and is implemented by the International Institute for Population Sciences (IIPS) with support from various organizations. The survey is characterized by its comprehensive and nationally representative nature, making it a powerful tool for studying various aspects of public health, demography, and social indicators across India. The dataset comprises responses from 46,334 (41.6% from urban and 58.4% from rural areas) respondents residing in various districts of Kerala, representing a diverse cross-section of the population. The wide scope of the NFHS 5 survey and its commitment to providing high-quality, nationally representative data make it an ideal resource for studies like the one assessing kitchen waste disposal practices in rural and urban regions of Kerala.

Table 1. District-wise Sample Distribution for the Study

District	Number	Percent
Kasaragod	3743	8.1
Kannur	3815	8.2
Wayanad	3342	7.2
Kozhikode	3427	7.4
Malappuram	3793	8.2
Palakkad	3483	7.5
Thrissur	3021	6.5
Ernakulam	3221	7.0
Idukki	3106	6.7
Kottayam	3198	6.9
Alappuzha	3156	6.8
Pathanamthitta	2791	6.0
Kollam	3157	6.8
Thiruvananthapuram	3081	6.6
Total	46334	100.0

Analytical approach

The analytical approach in this study is structured to comprehensively investigate the differences in kitchen waste disposal methods between rural and urban settings using SPSS v26.0. The key variables of interest are the various kitchen waste disposal methods, categorized into options such as open drain disposal, closed drain disposal, reuse for gardening or farming, and manual collection. The study first employs a descriptive analysis to provide an initial overview of how these disposal methods are distributed within the population. Subsequently, the analytical focus shifts to bivariate analysis. Within this phase, cross-tabulations are used to compare the prevalence of different waste disposal methods separately for rural and urban regions, offering insight into the relative frequency of each method in both settings. Chi-square tests are then applied to assess the statistical significance of the observed variations, determining whether the differences between rural and urban regions are meaningful. This analytical approach allows the study to systematically explore, characterize, and confirm the disparities in kitchen waste disposal practices between rural and urban areas in Kerala, providing valuable insights into waste management strategies in these distinct contexts.

Result

Variance in Approaches to Kitchen Waste Disposal

In the realm of waste management practices in Kerala, a state known for its commitment to environmental sustainability, this study presents a comprehensive exploration of diverse approaches to kitchen waste disposal. The examined methods include 'Let out into drain/sewer,' 'Open drain,' 'Closed drain,' 'Reuse for garden or farming,' 'Domestic purposes,' and 'Manual collection.' Each approach carries its own implications in terms of environmental impact, sanitation, and resource conservation.

Figure 1.Distribution of kitchen waste disposal methods in Kerala

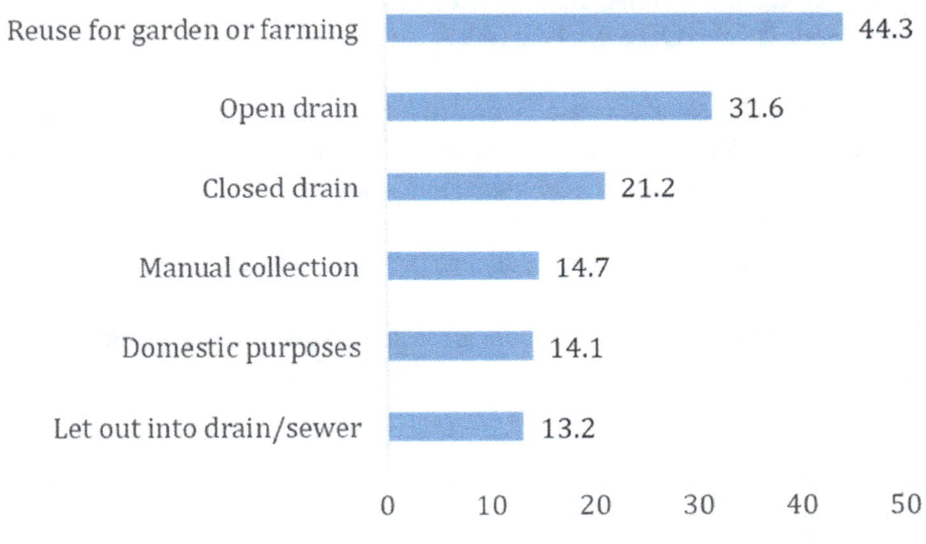

A significant portion of respondents (44.3%) demonstrate a strong commitment to sustainable waste management practices by choosing to reuse their kitchen waste for gardening or farming. This suggests a substantial level of awareness and successful promotion of composting and organic recycling methods in Kerala. While there is an overall positive trend toward sustainable practices, a substantial portion of respondents (31.6%) still opt for open drain disposal, which is less eco-friendly and can lead to sanitation and drainage issues. Addressing open drain disposal remains a challenge, requiring increased awareness efforts to promote responsible waste disposal. Approximately 21.2% of respondents use closed drains for kitchen waste disposal, which is an improvement over open drain disposal but still offers room for enhancement. Continued efforts are needed to encourage more environmentally friendly disposal methods like composting. A smaller percentage (13.2%) of respondents release their kitchen waste into drains or sewers, posing the risk of sewer blockages and environmental pollution. To combat this, educational and awareness campaigns should emphasize the adverse consequences of this disposal method and promote better alternatives. Notably, 14.7% of respondents engage in manual collection of kitchen waste, indicating an active interest in waste management. This reflects a willingness to participate in waste management efforts, and community-based initiatives can further support and enhance these practices. Around 14.1% of respondents repurpose kitchen waste for domestic purposes, which may involve creative and diverse reuse or recycling methods.

RURAL URBAN VARIATION IN KITCHEN WASTE DISPOSAL METHODS IN KERALA

In the context of Kerala, a state renowned for its commitment to environmental sustainability, understanding the divergence in kitchen waste disposal practices between rural and urban areas is of paramount importance. This study delves into the nuanced differences in waste management methods, exploring how residents in these distinct settings navigate the challenge of sustainable kitchen waste disposal. By examining the rural-urban variations, we aim to gain insights into the factors influencing these choices and contribute to the development of targeted waste management strategies tailored to the unique needs of each locality.

Table 2. Variance in Approaches to Kitchen Waste Disposal between Rural and Urban Regions

Kitchen waste disposal Method	Rural	Urban	Significance
Let out into drain/sewer	12.4	14.3	0.00
Open drain	36.5	24.7	0.00
Closed drain	20.4	22.3	0.00
Reuse for garden or farming	43.2	45.9	0.00
Reuse for Domestic purposes	13.9	14.3	0.12
Manual collection	11.7	18.9	0.00

In urban regions, 14.3% of the survey respondents indicated that they dispose of their kitchen waste into drains or sewers, while in rural areas, this figure is slightly lower at 12.4%. This data highlights a noticeable disparity in the adoption of this less environmentally friendly disposal method between rural and urban settings, with urban areas having a somewhat higher prevalence. The significant rural-urban difference in the practice of letting kitchen waste into drains or sewers suggests that various factors may be contributing to this variation. One potential factor could be the greater availability of sewer systems in urban areas, where some residents may not fully grasp the environmental consequences associated with this disposal method. This discrepancy underscores the need for targeted educational campaigns and awareness initiatives to promote responsible waste disposal practices, particularly in urban settings where sewer access might be more prevalent but where awareness of the environmental impact needs improvement. Open drain disposal is noticeably more widespread in rural areas, with 36.5% of respondents employing this method, whereas in urban areas, the practice is less common at 24.7%. This substantial difference between rural and urban areas underscores a significant rural-urban divide in the preference for open drain disposal. Rural communities, more than urban ones, tend to rely on this method, which is known for its potential environmental harm and sanitation challenges. Given the higher prevalence of open drain disposal in rural areas, there is a pressing need to prioritize addressing this issue. This situation highlights the importance of targeted interventions and improvements in infrastructure within rural communities to encourage more sustainable approaches to waste management. About 22.3% of urban respondents and 20.4% of rural respondents use closed drains for kitchen waste disposal. There is a slight difference in the usage of closed drains between rural and urban areas, although the Chi-Square test indicates that this difference is statistically significant but relatively small in magnitude. While there is a statistically significant difference between rural and urban areas in using closed drains for kitchen waste disposal, the magnitude of this difference is relatively small. Both rural and urban areas have a similar level of adoption of this method. This suggests that, overall, residents in Kerala tend to use closed drains for kitchen waste disposal at a consistent rate, regardless of their location. The practice of reusing kitchen waste for gardening or farming demonstrates remarkable consistency, with 45.9% of respondents in urban areas and 43.2% in rural areas adopting this eco-friendly approach. This data underscores a shared dedication among both rural and urban participants to sustainable waste management practices. The widespread adoption of this method in both settings not only reflects a shared commitment to sustainability across Kerala but also signifies a high level of awareness and the effective promotion of composting and organic recycling techniques in both rural and urban communities. A proportion of 14.3% of respondents in urban areas and 13.9% in rural areas employ kitchen waste for domestic purposes. Importantly, statistical analysis reveals no significant difference between rural and urban regions in the utilization of kitchen waste for domestic needs. This data indicates a consistent practice across both settings, where repurposing kitchen waste for various household requirements is a relatively uniform and stable practice observed throughout Kerala. Manual collection of kitchen waste is notably more widespread in urban areas, where it accounts for 18.9% of respondents, in contrast to rural areas where it is lower at 11.7%. This data reveals a clear and statistically significant rural-urban distinction in the preference for manual collection, with urban regions displaying a greater inclination to actively manage kitchen waste. The higher prevalence of manual collection in urban areas suggests a heightened awareness of sustainable waste management practices among urban residents, potentially underpinned by better access to resources for such practices.

District wise Rural-Urban variation in waste disposal method: Let out into drain/sewer

This study delves into the district-wise variation in kitchen waste disposal methods, focusing on the practice of letting kitchen waste out into drain/sewer in various districts of Kerala. Understanding how this practice varies across different districts is essential for tailoring effective waste management policies and interventions. The data will shed light on the prevalence of this disposal method, pinpointing districts where it is more common, and potentially revealing factors such as infrastructure, awareness, and local dynamics that influence these variations. This insight can guide targeted efforts to promote responsible and eco-friendly waste disposal practices in specific regions, contributing to overall sustainable waste management in Kerala.

Table 3. District-wise Rural-Urban Variation in Kitchen Waste Disposal Method: Let out into Drain/Sewer

District	Rural	Urban	Sig
Kasaragod	12.9	23.9	0.000
Kannur	4.4	16.5	0.000
Wayanad	18.5	35.8	0.000
Kozhikode	6.6	18.1	0.000
Malappuram	27.7	9.3	0.000
Palakkad	15.3	17.3	0.092
Thrissur	3.9	12.6	0.000
Ernakulam	4.6	8.6	0.000
Idukki	3.3	1.2	0.790
Kottayam	6.3	7.4	0.156

continued on following page

Table 3. Continued

District	Rural	Urban	Sig
Alappuzha	23.0	17.7	0.000
Pathanamthitta	9.8	4.1	0.000
Kollam	11.9	12.9	0.220
Thiruvananthapuram	18.1	13.3	0.000

The data from various districts in Kerala reveals a consistent urban-rural disparity in the practice of letting kitchen waste into drain/sewer. In Kasaragod and Kannur, significant differences indicate that urban regions have a higher prevalence of this method, potentially linked to better urban access to sewer systems. Wayanad displays a substantial urban-rural divide, underlining the influence of urbanization on disposal choices. Kozhikode aligns with the urban trend seen in other districts, with urban areas practicing this method more. Notably, Malappuram presents a unique pattern with rural areas showing a higher prevalence, possibly due to local factors. While Palakkad exhibits a difference, it's not statistically significant, suggesting similar prevalence in both rural and urban areas. Thrissur and Ernakulam show significant urban dominance in this disposal method. Idukki and Kottayam reveal relatively consistent practices in both settings. Alappuzha deviates, with rural areas practicing more. Pathanamthitta mirrors the common urban-rural divide. Kollam shows a difference, but not statistically significant, indicating a balanced prevalence. Thiruvananthapuram exhibits the expected urban-rural disparity. Overall, these findings underscore the need for tailored waste management strategies and awareness campaigns, particularly in urban areas, to promote responsible and eco-friendly waste disposal practices. The significant differences observed in Kasaragod and Kannur, with urban areas having a higher prevalence of drain/sewer disposal, suggest that urban regions often benefit from better access to sewer systems. This indicates that infrastructure plays a crucial role in shaping disposal practices. Efforts to improve waste management infrastructure in rural areas may be necessary to level the playing field. The substantial urban-rural divide seen in Wayanad highlights the significant influence of urbanization on disposal choices. As urban areas expand and develop, residents are more likely to adopt modern waste disposal methods, such as drain/sewer disposal. This emphasizes the need for waste management policies that adapt to changing urban landscapes. The urban dominance observed in Kozhikode, Thrissur, and Ernakulam suggests that urban residents may have higher awareness levels regarding the environmental consequences of their waste disposal practices. This underscores the importance of targeted awareness campaigns and educational initiatives, particularly in urban areas, to promote responsible waste disposal. Malappuram's unique pattern, where rural areas have a higher prevalence, indicates that local factors or specific infrastructure differences may influence disposal choices. Understanding these local dynamics is crucial for tailoring waste management strategies effectively. In districts like Idukki and Kottayam, where there is relatively consistent practice in both rural and urban areas, it may be due to uniform waste management practices and infrastructure across settings. These districts can serve as examples of successful waste management strategies that can be applied elsewhere. Alappuzha's deviation, with rural areas practicing this method more, suggests that unique factors specific to this district play a role in shaping waste disposal practices. Investigating these factors can provide insights for targeted interventions.

District wise Rural-Urban variation in waste disposal method: Open drain

The examination of rural-urban disparities in waste disposal practices, specifically focusing on open drain usage, provides valuable insights into the waste management landscape in various districts of Kerala. Open drain disposal is a critical aspect of environmental sanitation, and understanding the variations between rural and urban areas is essential for the development of targeted waste management strategies.

Table 4. District-wise Rural-Urban Variation in Kitchen Waste Disposal Method: Open drain

District	Rural	Urban	Sig
Kasaragod	38.0	34.9	0.027
Kannur	54.0	16.4	0.000
Wayanad	29.3	25.0	0.148
Kozhikode	32.4	31.4	0.284
Malappuram	33.4	36.3	0.035
Palakkad	45.0	38.6	0.001
Thrissur	29.7	22.8	0.000
Ernakulam	34.5	22.5	0.000
Idukki	41.9	14.6	0.000
Kottayam	31.5	19.5	0.000
Alappuzha	25.3	21.2	0.004
Pathanamthitta	44.7	28.7	0.000
Kollam	31.7	20.9	0.000
Thiruvananthapuram	32.2	15.2	0.000

The analysis of open drain disposal practices across various districts in Kerala reveals distinct rural-urban disparities. In Kasaragod, open drain disposal is slightly more prevalent in rural areas than in urban areas, albeit with a significant difference. In contrast, Kannur exhibits a striking divergence, with rural regions predominantly relying on open drain disposal, emphasizing the urgency of promoting sustainable waste management in rural Kannur. Wayanad and Kozhikode show no significant differences between rural and urban areas, signifying a relatively consistent practice of open drain disposal in both settings. Malappuram presents a unique scenario where urban areas display a slightly higher prevalence, indicating a somewhat urban inclination toward this method. Palakkad underscores the need for targeted interventions in rural areas to encourage eco-friendly waste disposal, while Thrissur and Ernakulam demonstrate more eco-conscious trends in urban regions. Idukki and Kottayam highlight substantial rural-urban divides, necessitating comprehensive waste management strategies for rural communities. Alappuzha and Pathanamthitta call for focused efforts in rural regions to improve waste disposal practices, and Kollam and Thiruvananthapuram underline the significance of awareness campaigns and infrastructure improvements in rural areas. These findings emphasize the need for tailored approaches to waste management and awareness initiatives, particularly in rural districts, to mitigate the prevalence of open drain disposal and promote responsible waste disposal practices more effectively.

District wise Rural-Urban variation in waste disposal method: Closed drain

The district-wise rural-urban variation in waste disposal methods, particularly the use of closed drains, provides critical insights into how different regions of Kerala manage their kitchen waste. Understanding these variations is essential for tailoring effective waste management strategies that align with the unique needs and preferences of each locale. Closed drain disposal is a prevalent method for handling kitchen waste, and analyzing its distribution across rural and urban areas sheds light on the influence of factors such as infrastructure, awareness, and cultural practices in shaping waste disposal habits.

Table 5. District-wise Rural-Urban Variation in Kitchen Waste Disposal Method: Closed drain

District	Rural	Urban	Sig
Kasaragod	15.3	13.9	0.129
Kannur	10.6	25.5	0.000
Wayanad	26.5	7.4	0.000
Kozhikode	33.4	20.3	0.000
Malappuram	27.2	35.3	0.000
Palakkad	12.8	16.0	0.011
Thrissur	21.9	17.7	0.003
Ernakulam	16.8	17.5	0.326
Idukki	14.0	14.0	0.359
Kottayam	27.9	35.4	0.000
Alappuzha	17.2	19.7	0.036
Pathanamthitta	24.2	35.5	0.000
Kollam	21.7	30.0	0.000
Thiruvananthapuram	16.2	17.0	0.292

The analysis of closed drain disposal practices across Kerala's districts reveals significant rural-urban disparities and offers valuable insights into waste management trends. In districts like Kannur, Malappuram, and Kottayam, urban areas exhibit a significantly higher prevalence of closed drain disposal compared to rural areas. This urban dominance suggests that urban residents are more inclined toward this waste disposal method, possibly due to better waste management infrastructure and awareness. Conversely, in districts such as Wayanad, Kozhikode, and Thrissur, rural areas show a notably higher prevalence of closed drain disposal. This indicates a rural preference for this method, possibly influenced by local factors, traditional practices, or limited access to alternative waste disposal options. Palakkad, Alappuzha, and Idukki exhibit rural-urban differences in closed drain disposal, albeit not as pronounced. These localized variations emphasize the importance of considering district-specific factors when designing waste management strategies. Few districts like Kasaragod and Ernakulam, there are no significant rural-urban disparities in closed drain disposal, suggesting consistent practices across both settings. These findings underscore the need for tailored waste management approaches that consider the unique waste disposal practices and preferences within each district. District-specific interventions, infrastructure improvements, and awareness campaigns should be implemented to promote responsible waste disposal effectively. The prevalence of closed drain disposal in urban areas across multiple districts indicates that urban residents may benefit from better waste management infrastructure and higher awareness levels regarding responsible waste disposal. These factors contribute to the observed urban dominance in certain districts. In districts where rural areas favor closed drain disposal, it is essential to address the underlying challenges, which may include limited access to modern waste management facilities and a need for awareness campaigns to promote eco-friendly practices.

District wise Rural-Urban variation in waste disposal method: Reuse for garden or farming

The exploration of waste disposal practices across Kerala's districts, with a specific focus on the reusing of kitchen waste for gardening or farming, unveils important facets of the state's waste management landscape. Kerala, celebrated for its unique blend of tradition and modernity, grapples with the challenge of sustaining environmentally responsible waste disposal practices while adapting to urbanization and changing lifestyles. Understanding how rural and urban areas differ in their commitment to repurposing kitchen waste for agricultural purposes is critical for tailoring effective waste management strategies.

Table 6. District-wise Rural-Urban Variation in Kitchen Waste Disposal Method: Reuse for garden or farming

District	Rural	Urban	Sig
Kasaragod	59.8	53.9	0.000
Kannur	36.7	46.8	0.000
Wayanad	40.5	37.8	0.291
Kozhikode	44.5	61.3	0.000
Malappuram	48.5	53.8	0.001
Palakkad	44.1	37.5	0.000
Thrissur	48.5	49.6	0.298
Ernakulam	47.3	33.1	0.000
Idukki	36.6	45.6	0.012
Kottayam	31.3	30.3	0.310
Alappuzha	38.7	41.3	0.075
Pathanamthitta	41.0	22.0	0.000
Kollam	53.6	46.7	0.000
Thiruvananthapuram	39.8	42.8	0.049

The analysis of rural-urban variation in the waste disposal method of reusing kitchen waste for gardening or farming in Kerala's districts reveals significant differences and provides valuable insights into the state's waste management practices. In districts like Kasaragod, Kannur, Kozhikode, Palakkad, and Pathanamthitta, urban areas exhibit a significantly higher prevalence of reusing kitchen waste for gardening or farming compared to their rural counterparts. This suggests that urban residents in these districts are more actively engaged in sustainable waste management practices, particularly by repurposing kitchen waste for agricultural purposes. The higher prevalence in urban areas could be attributed to better awareness, access to resources, and potentially more space for gardening or farming in urban settings. Conversely, in districts like Idukki, Kottayam, and Thiruvananthapuram, there are no significant rural-urban disparities in this disposal method, indicating relatively consistent practices across both rural and urban areas. This suggests that in these districts, the commitment to reusing kitchen waste for gardening or farming is consistent irrespective of the geographical location. Malappuram shows a unique pattern where rural areas have a slightly higher prevalence of this disposal method compared to urban areas, albeit with a significant difference. This indicates that in Malappuram's rural communities, there is a stronger inclination toward using kitchen waste for agricultural purposes, highlighting a potential area for promoting sustainable waste management practices in urban Malappuram. In districts like Wayanad, Thrissur, Alappuzha, and Kollam, there are disparities in prevalence between rural and urban areas, but the differences are not highly significant. This suggests that while there may be some variation, the commitment to reusing kitchen waste for gardening or farming is relatively balanced in both settings.

District wise Rural-Urban variation in waste disposal method: Reuse for Domestic purposes

The exploration of district-wise rural-urban variation in waste disposal methods, focusing on the reuse of kitchen waste for domestic purposes, unveils crucial insights into the waste management practices prevalent across different regions of Kerala. This analysis delves into the diverse approaches employed by rural and urban communities in repurposing kitchen waste within their households. Understanding these variations is pivotal in crafting targeted waste management strategies that cater to the distinct preferences, resources, and cultural dynamics that shape waste disposal practices. As kitchen waste reuse for domestic purposes represents a sustainable and eco-friendly approach, examining its prevalence and discrepancies across districts provides a comprehensive view of the state's commitment to responsible waste management.

Table 7. District-wise Rural-Urban Variation in Kitchen Waste Disposal Method: Reuse for Domestic purposes

District	Rural	Urban	Sig
Kasaragod	24.4	21.1	0.012
Kannur	19.0	16.6	0.033
Wayanad	11.6	8.1	0.114
Kozhikode	13.5	32.2	0.000
Malappuram	24.0	6.1	0.000
Palakkad	6.1	6.0	0.481
Thrissur	12.4	16.2	0.003
Ernakulam	14.8	11.1	0.002
Idukki	14.7	31.0	0.000
Kottayam	11.0	8.0	0.007
Alappuzha	17.8	10.4	0.000
Pathanamthitta	8.2	4.4	0.011
Kollam	11.5	10.7	0.275
Thiruvananthapuram	10.2	7.1	0.002

The analysis of rural-urban variation in the waste disposal method of reusing kitchen waste for domestic purposes in Kerala's districts reveals significant differences and provides valuable insights into the state's waste management practices. In several districts, urban areas exhibit a significantly lower prevalence of reusing kitchen waste for domestic purposes compared to their rural counterparts. These districts include Kozhikode, Malappuram, Idukki, and Alappuzha, where urban areas lag significantly behind rural areas in adopting this sustainable waste management practice. This may be attributed to factors such as limited space for household gardening, lesser awareness, or higher reliance on disposal methods in urban settings. In contrast, districts like Kasaragod, Kannur, Thrissur, Ernakulam, Kottayam, Pathanamthitta, and Thiruvananthapuram show urban areas with a slightly lower prevalence, but the differences are not highly significant. This suggests that in these districts, both rural and urban residents have relatively similar rates of reusing kitchen waste for domestic purposes. Wayanad exhibits no significant rural-urban difference, indicating a relatively consistent practice of kitchen waste reuse across the district. These findings emphasize the need for targeted awareness campaigns and interventions in urban areas of districts like Kozhikode, Malappuram, Idukki, and Alappuzha to promote responsible waste disposal practices. For districts with more balanced rural-urban prevalence, fostering a culture of kitchen waste reuse among urban populations is still essential to enhance overall sustainability in waste management.

District wise Rural-Urban variation in waste disposal method: Manual collection:

Understanding how waste disposal methods vary between rural and urban areas within different districts is crucial for effective waste management strategies. The choice of waste disposal methods can significantly impact environmental sustainability and public health. In this context, examining the rural-urban variation in the waste disposal method of manual collection across various districts in a region can provide valuable insights. This analysis can shed light on the prevalence of manual collection in different settings, helping policymakers, environmentalists, and local authorities tailor their efforts to promote responsible waste management practices. A comprehensive understanding of these variations can inform targeted interventions to enhance waste disposal practices, optimize resource allocation, and contribute to a cleaner and healthier environment.

Table 8. District-wise Rural-Urban Variation in Kitchen Waste Disposal Method: Manual collection

District	Rural	Urban	Sig
Kasaragod	15.4	30.5	0.000
Kannur	17.8	12.6	0.000
Wayanad	20.7	88.5	0.000
Kozhikode	14.3	26.8	0.000
Malappuram	14.1	12.5	0.078
Palakkad	10.2	9.1	0.196
Thrissur	9.0	16.8	0.000
Ernakulam	5.0	27.3	0.000
Idukki	7.2	21.1	0.000
Kottayam	10.2	12.1	0.071
Alappuzha	4.4	7.8	0.000
Pathanamthitta	10.5	16.6	0.002
Kollam	9.4	11.8	0.000
Thiruvananthapuram	6.8	23.9	0.000

The analysis of rural-urban variation in the waste disposal method of manual collection across different districts in Kerala highlights significant differences and offers valuable insights into waste management practices. In several districts, urban areas exhibit a significantly higher prevalence of manual collection compared to their rural counterparts. These districts include Kasaragod, Kannur, Wayanad, Kozhikode,

Thrissur, Ernakulam, Idukki, Alappuzha, Pathanamthitta, and Thiruvananthapuram. The substantial urban dominance in manual collection suggests that urban residents in these districts are more engaged in this method, possibly due to better waste management infrastructure and services. Conversely, Malappuram, Palakkad, Kottayam, and Kollam show no significant rural-urban disparities, indicating relatively balanced practices across both rural and urban areas. These findings emphasize the need for targeted interventions to promote responsible waste management practices in urban areas, where manual collection is more prevalent, and to maintain the existing balanced practices in districts where rural-urban differences are minimal.

DISCUSSION ON FINDINGS

A significant portion of respondents (44.3%) choose to reuse their kitchen waste for gardening or farming, showcasing a commendable dedication to sustainable waste management (Kuligowski et al. 2023). This high percentage suggests a considerable level of awareness and the successful promotion of composting and organic recycling methods in Kerala. Despite the overall positive trend toward sustainable practices, a substantial portion of respondents (31.6%) still opt for open drain disposal. This method is less environmentally friendly (Blom 2015) and can lead to sanitation and drainage issues (Jemberie et al., 2023). Addressing the prevalence of open drain disposal remains a challenge, necessitating increased awareness efforts to promote responsible waste disposal alternatives. Approximately 21.2% of respondents use closed drains for kitchen waste disposal, which is an improvement over open drain disposal (Department of Drinking Water and Sanitation, Ministry of Jal Shakti, Government of India, 2021). However, there is still room for enhancement in terms of adopting more environmentally friendly disposal methods like composting (Taiwo, 2011). Continued efforts are needed to encourage such practices and further reduce the reliance on less sustainable methods. A smaller percentage (13.2%) of respondents release their kitchen waste into drains or sewers. This poses the risk of sewer blockages and environmental pollution (Abubakar, 2017). To mitigate this issue, it is imperative to conduct educational and awareness campaigns that emphasize the adverse consequences of this disposal method and promote better, eco-friendly alternatives. Notably, 14.7% of respondents engage in manual collection of kitchen waste, indicating an active interest in waste management. This reflects a willingness to participate in waste management efforts, demonstrating the potential for community-based initiatives to support and enhance such practices. Around 14.1% of respondents repurpose kitchen waste for domestic purposes, which may involve creative and diverse reuse or recycling methods. This finding showcases a level of ingenuity in waste reduction and resource optimization within households.

The data indicates that urban areas in Kerala exhibit a higher prevalence of eco-friendly kitchen waste disposal methods, including manual collection (Sankar Cheela et al., 2021) and reuse for gardening or farming. This suggests a heightened awareness of sustainable waste management practices in urban settings. One prominent finding is the significantly higher prevalence of open drain disposal in rural areas, compared to urban areas (Berendes et al. 2020). Open drain disposal poses sanitation and environmental challenges, emphasizing the need for targeted interventions in rural regions. Initiatives to improve infrastructure and raise awareness about responsible waste disposal are essential in addressing this issue. Interestingly, both rural and urban areas show a similar level of adoption of the closed drain disposal method. This indicates a consistent practice of using closed drains for kitchen waste disposal, regardless of the geographical location. A positive finding is the shared commitment to reusing kitchen waste for gardening or farming across both rural and urban areas (Mohan et al. 2020). This practice reflects a strong inclination toward sustainability and signifies the successful promotion of composting and organic recycling methods in both settings (Mamun et al. 2023). The data reveals no significant rural-urban differences in using kitchen waste for domestic purposes. This suggests that repurposing kitchen waste for various household needs is a relatively consistent practice across Kerala (Kaithery, N. & Karunakaran, U. 2019). Urban areas stand out with a higher prevalence of manual collection of kitchen waste. This signifies a greater willingness among urban residents to actively manage their kitchen waste through manual collection (Schanes et al. 2018). It also suggests a higher awareness of sustainable waste management practices and potentially better access to resources for such practices in urban settings. The study highlights the necessity of recognizing and addressing rural-urban variations in kitchen waste management. It underscores the importance of tailoring waste management policies, infrastructure improvements, and educational campaigns to suit the unique needs and challenges of each locale. Despite the variations observed, the data reaffirms the shared commitment of both rural and urban populations in Kerala to sustainable waste management. This collective aspiration for eco-friendly practices serves as evidence of the success of eco-conscious initiatives in the state.

The consistent urban-rural disparity in the choice of letting kitchen waste into drain/sewer is evident across most districts. Urban areas tend to have a higher prevalence of this disposal method. This urban dominance can be attributed to factors such as better access to sewer systems, infrastructure development, and potentially higher awareness levels among urban residents regarding responsible waste management (Schellenberg et al. 2020). The significant differences observed in districts like Kasaragod and Kannur highlight the importance of infrastructure in shaping waste disposal practices. The better availability of sewer systems in urban regions suggests that infrastructure plays a crucial role in determining disposal methods (Kumar and Joseph 2012). This raises the need for infrastructure improvements in rural areas to create a level playing field. The wide urban-rural gap seen in Wayanad underscores the impact of urbanization on waste disposal choices. As urban areas expand and evolve, residents are more likely to adopt modern disposal methods, including drain/sewer disposal. This emphasizes the need for waste management policies that adapt to changing urban landscapes and accommodate the shifting preferences of urban populations. Districts like Kozhikode, Thrissur, and Ernakulam exhibit urban dominance in drain/sewer disposal. This suggests that urban residents may possess higher awareness levels regarding the environmental consequences of their waste disposal practices. To address this urban-rural disparity, targeted awareness campaigns and educational initiatives, particularly in urban areas, are essential to promote responsible waste disposal and sustainability. Malappuram's unique pattern, where rural areas show a higher prevalence of drain/sewer disposal, indicates that local factors or specific infrastructure differences may play a significant role in shaping disposal choices. Understanding these localized dynamics is crucial for tailoring waste management strategies effectively and accounting for regional variations. Districts like Idukki and Kottayam display relatively consistent disposal practices in both rural and urban areas, possibly due to uniform waste management practices and infrastructure. These districts can serve as models for successful waste management strategies that can be implemented in other regions. On the other hand, Alappuzha's deviation suggests that unique factors specific to this district influence waste disposal practices, highlighting the importance of investigating and addressing local variations.

The analysis of open drain disposal practices in various Kerala districts highlights significant discrepancies between rural and urban regions. While some districts exhibit a relatively consistent distribution of open drain disposal in both settings, others demonstrate significant discrepancies. The striking contrast observed in Kannur, with rural areas predominantly relying on open drains, underscores the urgency of implementing sustainable waste management practices in rural parts of this district (Centre for Science and Environment (CSE), 2017). In contrast, Palakkad, Thrissur, and Ernakulam display more eco-conscious trends in urban regions, indicating that urban residents may have a higher awareness of responsible waste disposal methods. However, the unique scenario in Malappuram, where urban areas exhibit a slightly higher prevalence of open drain disposal, highlights the need for localized interventions (Buttenheim 2008). Districts like Idukki and Kottayam reveal substantial rural-urban divides, calling for comprehensive waste management strategies tailored to rural communities. Alappuzha and Pathanamthitta emphasize the importance of focused efforts in rural regions to improve waste disposal practices, while Kollam and Thiruvananthapuram underscore the significance of awareness campaigns and infrastructure improvements in rural areas. Overall, these findings emphasize the necessity of adopting tailored approaches to waste management and awareness initiatives, especially in rural districts, to mitigate the prevalence of open drain disposal and promote responsible waste disposal practices effectively. Such efforts will be pivotal in advancing sustainable waste management across Kerala.

The findings from the analysis of closed drain disposal practices in Kerala's districts highlight the complex and diverse nature of waste management trends across the state. The significant rural-urban disparities observed underscore the need for nuanced approaches (Salvia et al. 2021) to address these variations effectively. In districts where urban areas dominate closed drain disposal, such as Kannur, Malappuram, and Kottayam, it is evident that urban residents benefit from better infrastructure and awareness, leading to their preference for this method. On the other hand, rural areas in districts like Wayanad, Kozhikode, and Thrissur show a distinct preference for closed drain disposal, possibly due to local factors or limited access to alternative waste disposal options (Kalonde et al. 2023). The localized variations in districts like Palakkad, Alappuzha, and Idukki highlight the importance of tailoring waste management strategies based on district-specific factors. Overall, these findings emphasize the necessity of customized interventions, infrastructure enhancements, and awareness campaigns to promote responsible waste disposal practices effectively. Additionally, addressing the challenges faced by rural areas in adopting eco-friendly disposal methods will be crucial in achieving more sustainable waste management practices across Kerala.

The findings from the analysis of rural-urban variation in the waste disposal method of reusing kitchen waste for gardening or farming in Kerala's districts offer valuable insights into the state's waste management practices and underscore the importance of tailored approaches to promote sustainable waste disposal. One prominent trend is the higher prevalence of this eco-friendly disposal method in urban areas of districts like Kasaragod, Kannur, Kozhikode, Palakkad, and Pathanamthitta. This reflects a positive aspect of urban areas in these districts being more actively engaged in environmentally conscious practices. This trend highlights the potential for urban areas to serve as models for sustainable waste management practices. Conversely, the unique pattern observed in Malappuram, where rural areas have a slightly higher prevalence of this disposal method, indicates a strong inclination toward agricultural reuse in rural communities (Wu et al. 2022). This finding presents an opportunity for targeted interventions and awareness campaigns in urban Malappuram to encourage similar eco-friendly practices. In districts like Idukki, Kottayam, and Thiruvananthapuram, where there are no significant rural-urban disparities in this disposal method, the data suggests a consistent commitment to reusing kitchen waste for gardening or farming across both rural and urban areas (Liu and

Nguyen 2020). This indicates a strong cultural or regional affinity for sustainable waste management practices, regardless of the geographical location. Districts like Wayanad, Thrissur, Alappuzha, and Kollam, which exhibit disparities in prevalence but not highly significant differences, demonstrate a relatively balanced adoption of this disposal method in both rural and urban settings. While there may be some variation, the overall commitment to eco-friendly waste disposal remains consistent.

The findings regarding the rural-urban variation in the waste disposal method of reusing kitchen waste for domestic purposes shed light on the diverse waste management landscape in Kerala. One of the striking observations is the significant urban-rural divide in several districts, with urban areas notably behind in adopting the sustainable practice of kitchen waste reuse (Schanes et al., 2018). This urban lag may result from challenges related to space constraints for household gardening (Pilone, di Santo, and Sisto 2023), lower awareness levels (Roy, Mohanty, Dick, and Misra 2023), or a preference for more convenient disposal methods in urban settings. On the other hand, districts like Kasaragod, Kannur, Thrissur, Ernakulam, Kottayam, Pathanamthitta, and Thiruvananthapuram demonstrate a more balanced prevalence of kitchen waste reuse between rural and urban areas. While this indicates a relatively positive scenario where both segments of the population are engaged in responsible waste management practices, there is still room for improvement in fostering a culture of kitchen waste reuse among urban residents. Wayanad stands out as a district where there is no significant rural-urban difference in kitchen waste reuse, reflecting a consistent practice across the entire region. This suggests a potential model for other districts to encourage this sustainable waste management method regardless of urbanization levels.

The findings regarding rural-urban variation in the waste disposal method of manual collection reveal intriguing disparities in waste management practices across Kerala's districts. The significant prevalence of manual collection in urban areas, particularly in districts like Kasaragod, Kannur, Wayanad, and Kozhikode, underscores the influence of urban infrastructure and services on residents' waste disposal habits. It suggests that urban areas benefit from more accessible waste management systems, leading to higher participation in manual collection (Cheela, Ranjan, Goel, John, and Dubey 2021). In contrast, districts such as Malappuram, Palakkad, Kottayam, and Kollam demonstrate a more uniform prevalence of manual collection across rural and urban areas, indicating consistent waste management practices. These findings highlight the importance of tailoring waste management strategies based on local infrastructure and service availability. In districts where urban dominance in manual collection is pronounced, there is an opportunity to further promote responsible waste management practices in rural areas. Conversely, in districts with balanced practices, efforts should focus on maintaining the existing level of engagement in manual collection and enhancing waste management infrastructure where needed.

IMPLICATIONS, LIMITATIONS, AND FUTURE DIRECTION

The results emphasize the significance of continuous awareness and educational initiatives, with a specific focus on rural areas. These efforts should aim to raise awareness about the environmental and health risks associated with open drain disposal while highlighting the advantages of closed drain disposal and the reuse of kitchen waste. The data indicates a willingness among respondents in both urban and rural settings to actively engage in waste management endeavors. This presents an opportunity for community-driven projects that can significantly contribute to enhancing responsible waste disposal practices and fostering sustainability. In the regions with a high prevalence of open drain disposal, especially in rural areas, there is an urgent need for infrastructural enhancements. This involves establishing proper waste disposal systems to ensure sanitation and prevent environmental contamination. Prioritizing research and initiatives aimed at improving waste management infrastructure, particularly in rural regions, is crucial. Such efforts are essential for ensuring the long-term sustainability of waste disposal practices and reaping the associated health and environmental benefits for the population.

CONCLUSION

The resounding commitment of a significant majority of respondents to reuse kitchen waste for gardening or farming reflects a commendable dedication to sustainable and eco-friendly waste management. While this high percentage underscores a substantial level of awareness and successful promotion of composting and organic recycling methods, the persistence of open drain disposal among a notable portion of respondents remains a concern. It underscores the need for continued efforts to raise awareness and promote responsible waste disposal methods. The adoption of closed drain disposal, although an improvement over open drains, presents an opportunity for further enhancement, emphasizing the ongoing importance of encouraging eco-friendly practices. Likewise, discouraging the release of kitchen waste into drains or sewers necessitates persistent education and awareness campaigns. The active involvement of respondents in manual waste collection and the creative repurposing of kitchen waste for domestic purposes demonstrate a willingness to participate in waste management efforts. These practices underscore the potential of community-based initiatives to further support and enhance responsible waste disposal practices. Notably, urban areas exhibit a higher prevalence of manual collection and, to a lesser extent, letting kitchen waste out into drains or sewers. These practices suggest a heightened awareness of sustainable waste management among urban residents. However, it is crucial to address the slight uptick in less eco-friendly disposal methods in urban settings. Conversely, rural areas grapple with a more significant reliance on open drain disposal, presenting sanitation and environmental challenges that demand immediate attention. Targeted interventions, including infrastructure improvements and awareness campaigns, are imperative to promote responsible waste disposal in rural regions. Despite these variations, the data reveals a heartening consistency in reusing kitchen waste for gardening or farming across both rural and urban areas. This shared commitment to sustainable practices serves as a testament to the success of eco-conscious initiatives in Kerala.

The analysis of waste disposal methods in Kerala reveals a clear urban-rural disparity, underlining the need for interventions in rural areas, particularly regarding open drain disposal. Encouragingly, both rural and urban areas consistently favor closed drain disposal and reusing kitchen waste for gardening or farming, reflecting a shared dedication to sustainability. Moreover, higher prevalence of manual collection in urban areas signals greater awareness and engagement in sustainable waste management. However, some areas exhibit a slight increase in less eco-friendly methods, warranting attention. Overall, the data reaffirms Kerala's commitment to responsible waste disposal and the success of eco-conscious initiatives. Tailoring strategies to address local variations is essential for fostering a more sustainable future for all residents.

REFERENCES

Abubakar, I. R. (2017). Household Response to Inadequate Sewerage and Garbage Collection Services in Abuja, Nigeria. *Journal of Environmental and Public Health, 5314840*. Advance online publication. DOI: 10.1155/2017/5314840 PMID: 28634496

Berendes, D. M., de Mondesert, L., Kirby, A. E., Yakubu, H., Adomako, L., Michiel, J., Raj, S., Robb, K., Wang, Y., Doe, B., Ampofo, J., & Moe, C. L. (2020). Variation in E. coli concentrations in open drains across neighborhoods in Accra, Ghana: The influence of onsite sanitation coverage and interconnectedness of urban environments. *International Journal of Hygiene and Environmental Health, 224*, 113433. DOI: 10.1016/j.ijheh.2019.113433 PMID: 31978730

Blom, K. (2015). Drainage systems, an occluded source of sanitation related outbreaks. *Archives of Public Health = Archives Belges de Santé Publique, 73*(8), 8. Advance online publication. DOI: 10.1186/s13690-014-0056-6 PMID: 25722855

Buttenheim, A. M. (2008). The sanitation environment in urban slums: Implications for child health. *Population and Environment, 30*(1-2), 26–47. DOI: 10.1007/s11111-008-0074-9 PMID: 25825551

Centre for Science and Environment (CSE). (2017). *SFD Report Kannur*. Retrieved from SFD Promotion Initiative.

Cheela, V. R., Ranjan, V. P., Goel, S., John, M., & Dubey, B. (2021). Pathways to sustainable waste management in Indian Smart Cities. *Journal of Urban Management, 10*(4), 419–429. DOI: 10.1016/j.jum.2021.05.002

. Department of Drinking Water and Sanitation, Ministry of Jal Shakti, Government of India. (2021). *Manual Greywater Management.*

Jemberie, M. A., Melesse, A. M., & Abate, B. (2023). Urban Drainage: The Challenges and Failure Assessment Using AHP, Addis Ababa, Ethiopia. *Water (Basel), 15*(5), 957. DOI: 10.3390/w15050957

Kaithery, N., & Karunakaran, U. (2019). Study on attitude of household waste management in a rural area of Northern Kerala. *International Journal of Community Medicine and Public Health, 6*(5), 2095. DOI: 10.18203/2394-6040.ijcmph20191826

Kalonde, P. K., Austin, A. C., Mandevu, T., Banda, P. J., Banda, A., Stanton, M. C., & Zhou, M. (2023). Determinants of household waste disposal practices and implications for practical community interventions: Lessons from Lilongwe. *Environmental Research: Infrastructure and Sustainability, 3*(1). Advance online publication. DOI: 10.1088/ersi/3/1/015001

Kuligowski, K., Konkol, I., Świerczek, L., Chojnacka, K., Cenian, A., & Szufa, S. (2023). Evaluation of Kitchen Waste Recycling as Organic N-Fertiliser for Sustainable Agriculture under Cool and Warm Seasons. *Sustainability (Basel)*, *15*(10), 7997. DOI: 10.3390/su15107997

Kumar, S. G., & Joseph, N. (2012). Drainage and sewerage system in urban India: Need for action. *Indian Journal of Occupational and Environmental Medicine*, *16*(3), 150–151. DOI: 10.4103/0019-5278.111767 PMID: 23776328

Mamun, A., Yang, Q., Naznen, F., Aziz, N. A., & Masud, M. M. (2023). Modelling the mass adoption potential of food waste composting among rural Chinese farmers. *Heliyon*, *9*(8), e18998. DOI: 10.1016/j.heliyon.2023.e18998 PMID: 37609413

Mohan, S. V., Hemalatha, M., Amulya, K., Velvizhi, G., Chiranjeevi, P., Sarkar, O., Kumar, A. N., Krishna, K. V., Modestra, J. A., Dahiya, S., Yeruva, D. K., Butti, S. K., Sravan, J. S., Chatterjee, S., & Kona, R. (2020). Decentralized Urban Farming Through Keyhole Garden: A Case Study with Circular Economy and Regenerative Perspective. *Materials Circular Economy*, *2*(12), 12. Advance online publication. DOI: 10.1007/s42824-020-00011-1

Nguyen, T. T. T., Malek, L., Umberger, W. J., & O'Connor, P. J. (2022). Household food waste disposal behavior is driven by perceived personal benefits, recycling habits, and ability to compost. *Journal of Cleaner Production*, *379*(1), 134636. DOI: 10.1016/j.jclepro.2022.134636

Pilone, V., di Santo, N., & Sisto, R. (2023). Factors affecting food waste: A bibliometric review on the household behaviors. *PLoS One*, *18*(7), e0289323. DOI: 10.1371/journal.pone.0289323 PMID: 37506105

Roy, P., Mohanty, A. K., Dick, P., & Misra, M. (2023). A Review on the Challenges and Choices for Food Waste Valorization: Environmental and Economic Impacts. *ACS Environmental Au*, *3*(2), 58–75. DOI: 10.1021/acsenvironau.2c00050 PMID: 36941850

Salvia, G., Zimmermann, N., Willan, C., Hale, J., Gitau, H., Muindi, K., Gichana, E., & Davies, M. (2021). The wicked problem of waste management: An attention-based analysis of stakeholder behaviors. *Journal of Cleaner Production*, *326*, 129200. DOI: 10.1016/j.jclepro.2021.129200 PMID: 34866810

Sankar Cheela, V. R., Ranjan, V. P., Goel, S., John, M., & Dubey, B. (2021). Pathways to sustainable waste management in Indian Smart Cities. *Journal of Urban Management*, *10*(4), 419–429. DOI: 10.1016/j.jum.2021.05.002

Schanes, K., Dobernig, K., & Gözet, B. (2018). Food waste matters - A systematic review of household food waste practices and their policy implications. *Journal of Cleaner Production*, *182*, 978–991. DOI: 10.1016/j.jclepro.2018.02.030

Schellenberg, T., Subramanian, V., Ganeshan, G., Tompkins, D., & Pradeep, R. (2020). Wastewater Discharge Standards in the Evolving Context of Urban Sustainability–The Case of India. *Frontiers in Environmental Science*, *8*, 30. Advance online publication. DOI: 10.3389/fenvs.2020.00030

Taiwo, A. (2011). Composting as a Sustainable Waste Management Technique in Developing Countries. *Journal of Environmental Science and Technology*, *4*(2), 93–102. Advance online publication. DOI: 10.3923/jest.2011.93.102

Wu, W., Li, L., Chen, H., Xu, M., & Yuan, Y. (2022). Farmers' Preference for Participating in Rural Solid Waste Management: A Case Study from Shaanxi Province, China. *International Journal of Environmental Research and Public Health*, *19*(21), 14440. DOI: 10.3390/ijerph192114440 PMID: 36361319

Chapter 9
Sustainable E-Waste Disposal Methods:
The Case of Discarded Mobile Phones in Zimbabwe

Paul Mukucha
http://orcid.org/0000-0002-6264-6574
Bindura University of Science Education, Zimbabwe

ABSTRACT

There is an exponential increase in e-waste material at most dump sites in Zimbabwe. The absence of proper guidelines for dealing with domestic e-waste material perhaps might account for its rampant presence at most dumpsites in Zimbabwean urban areas. This led to the need to empirically determine how individuals dispose of obsolete electronic gadgets in general and, more specifically, mobile phones. A phenomenological approach of qualitative research design was adopted, and 30 participants were recruited for in-depth interviews. Data was analysed thematically. The results indicated that most of the participants dispose of their obsolete mobile phones through toyification, incineration, landfilling, and using sanitary bins. The study therefore recommended the relevant statutory bodies tasked with environmental management provide comprehensive guidelines on how to domestically dispose of e-waste material.

1. INTRODUCTION

There has been a high penetration of mobile phones among the Zimbabweans that is well above 104% (Zimba et al., 2024). This high penetration has been determined by the massive reduction in production costs that is driven by mass

DOI: 10.4018/979-8-2600-1960-3.ch009

production emphasised by the mobile phone producers and suppliers (Tang et al., 2016). Mass production of mobile phones led to the reduction in prices that made mobile phones affordable to everyone and sundry. The existence of economy brands, which match the performance of premium brands, serves as evidence of this. The other factor that led to the rise in demand for mobile phones was the development of smartphones that enabled connectivity through social media platforms such as WhatsApp, Facebook, and YouTube (Ahad & Busch, 2024) and had some novel features like cameras (Blahnik & Schindelbeck, 2021).

The connectivity that mobile phone users attained through the use of smartphones enabled them to perform and access several remote services. These services include, but are not limited to, e-library, e-learning (Chaudhari & Shirude, 2025), mobile banking, and cyber security (Limna et al., 2023). The need to access several e-services through mobile phones led to their higher frequency of usage that triggers obsolescence. It has been suggested that, on average, a mobile phone lasts for only 2 years before it becomes obsolete. Considering that almost everyone in Zimbabwe, a nation of 16 million people, including babies, possesses a mobile phone, it means that every 2 years the nation produces about 16 million obsolete mobile phones. This translates to an average of 1600 tonnes of e-waste material from mobile phones. Obsolescence in general, and in the case of mobile phones, elicits the need for replacement (Oraee et al., 2024; Zhilyaev et al., 2021).

There are many types of obsolescence in the extant literature, but the most prominent is planned obsolescence. Planned obsolescence is driven by the marketers' need to induce frequent product replacement as a way of increasing their sales revenue (Malinauskaite & Erdem, 2021). This is particularly common with mobile phones that are made incompatible with some of the functions as a way of eliciting replacements (Bilici & Özdemir, 2024). For instance, recently, all smartphones that leverage Android version 23 were made incompatible with the WhatsApp application. This drove the consumers to buy new versions of smartphones as replacements for their current models. The replaced obsolete mobile phones needed to be disposed of responsibly in order to avoid disasters associated with e-waste material (Laeequddin et al., 2022).

The production of products is supposed to be accompanied by a plan for their end of life (Mukucha et al., 2025). However, most of the mobile phone brands in the Zimbabwean market do not provide the relevant information related to how they can be safely disposed of upon becoming obsolete. Without the prescribed guidelines of proper product disposal, the consumers are left with indiscriminate ways of disposing of their obsolete mobile phones, resulting in the creation of e-waste material. This raises the research question specified as follows:How do mobile phone users dispose of their obsolete mobile phones? With this research question in mind, this study seeks to determine how mobile phone users dispose of their obsolete phones.

This assessment will assist environmental advocates to come up with appropriate policies to regulate the disposal of e-waste material. Henceforth, the rest of the study is structured as follows: a literature review using the social cognitive theory (SCT) as the theoretical framework, methodology, results presentation and discussion, as well as policy recommendations.

2. LITERATURE REVIEW

2.1 Theoretical framework

The disposal of e-waste material, particularly from obsolete mobile phones, has no explicit guidelines from most of the manufacturers. This oversight from both the manufacturers and policymakers has left mobile phone users to indiscriminately dispose of their mobile phones at the end of their life cycle. The indiscriminate disposal of mobile phones at the end of their life cycle can be understood through the lens of the social cognitive theory (SCT) conceptualised by Bandura (1986). The SCT suggests that individuals learn through observing the appropriate or generally practised behaviour from others during social interactions and personal experiences (Bandura, 2014). Moreover, individuals also learn the popularised ideas in outside media messages or posts.

The application of the SCT highlights the interconnection between personal factors, environmental circumstances, and consumer behaviour (Husin et al., 2024). More specifically, this interconnectedness dictates how individuals deal with e-waste material in the absence of formal guidelines. The essential elements from the SCT that can be applied to dealing with sustainable e-waste material disposal relate to observational learning, environmental advocacy, self-efficacy, and reinforcement.

Observational learning

Observational learning refers to knowledge acquisition through observing the actions of others (Danwitz & von Helversen, 2025; Putra, 2024). This suggests that learning is not limited to direct experiences only. For observational learning to be effective, it should involve paying a lot of attention, putting more effort towards retention, attempting to do reproduction, and eliciting higher levels of motivation (Munguiko et al., 2025; Martín, 2024).

Observational learning in the context of e-waste material relates to how individuals adopt certain disposal practices through observing the actions of their peers and community members. Such imitations generally lack the understanding of the motivations behind the adoption of the chosen practices. This may lead to the adoption of irresponsible disposal of e-waste material practices that may pose some health hazards to a community. Health hazards resulting from indiscriminate disposal of e-waste material are well documented in the extant literature. For instance, in a recent study by Mukucha & Jaravaza (2025), it was found that informal waste pickers frequently contract diseases through exposures to e-waste material.

Environmental advocacy

Advocacy is another aspect of SCT that individuals can learn from. Advocacy relates to actions meant to support a cause. It involves various activities meant to impart persuasive knowledge on a particular cause. Several activities can be used in advocacy, such as lobbying, media campaigns, research publications and public speaking. Advocacy applies to various trending issues such as child labour (van Baar et al., 2024), ecofeminism (Özden, 2024; Aga et al., 2024), and environmental preservation issues (Mukucha et al., 2025).

Environmental advocacy is another key factor in the disposal of e-waste material. The availability of public information regarding how e-waste material should be handled has a greater influence on how individuals deal with their disused electronic gadgets (Laeequddin et al., 2022). Environmental advocacy has been found to be effective when individuals are nudged to adopt an acceptable practice. For instance, in a recent study by Mukucha et al. (2025a), nudging baby diapering mothers to adopt waste material segregation at source practices when handling soiled nappies was found to be effective in reducing indiscriminate disposals. Another study by Mukucha et al. (2025b) found various nudging strategies to be effective in reducing plate waste in the foodservice establishments.

Self-efficacy

Self-efficacy in general refers to beliefs that individuals hold with regards to their ability to perform certain tasks (Power et al., 2025). Self-efficacy emanates from four primary sources, which are mastery experiences, vicarious experiences, verbal persuasion, and physiological state. Mastery experiences relate to an individual's belief in her abilities. Vicarious experiences pertain to the boosting of one's beliefs that she can succeed as a result of seeing the successes of others. Verbal persuasion involves engaging others to boost one's confidence. Lastly, physiological states suggest that an emotional state can influence one's perception of capability to take certain action (Bandura, 1982). Such beliefs apply to general tasks in life (Stephen, 2024).

In the case of e-waste material, individuals can hold a belief that they can easily dismantle disused electronic devices and safely dispose of them. Their intentions usually lead to the actual behaviour. Thus, higher levels of self-efficacy may lead individuals to follow the prescribed e-waste material disposal methods. However, low levels of self-efficacy may lead individuals into indiscriminate disposal, particularly for complex products such as e-waste material. There is therefore a need for easy guidelines that can be followed by individuals with low self-esteem in order to avoid indiscriminate disposals.

Reinforcement

Individuals can also have their environmental behaviours shaped from learning through observing how others' actions are reinforced. Reinforcement may take two forms: positive and negative (Baron & Galizio, 2005). Positive reinforcement relates to the encouragement of behaviours that are deemed to be appropriate through the offering of incentives (Sari & Rahmani, 2024). In the case of waste material disposals, the relevant authorities may offer rebates for residents whose environment is always clean. Negative reinforcement suggests the imposition of penalties to discourage certain behaviours that are deemed to be retrogressive (Martínez-Torres et al., 2024). In the case of waste disposal, there are always fines imposed for behaviours such as public littering (Rauf et al., 2024; Lev et al., 2023).

2.2 Conceptual framework and empirical evidence

Product design

Under the marketing concept, new products find their way to the market as a result of responding to consumer needs and wants (Zhang et al., 2024; Bekele et al., 2024). This contrasts with the production concept, where a product is designed from a manufacturer's ingenuity and then forced onto the market (Ateeq et al., 2025; Poetz & Schreier, 2012). Such an approach results in the availability of cutting-edge innovations that surprisingly may not find takers in the market or have a short life cycle (Tao & Ucbasaran, 2025; Shah, 2024; Freisinger & McCarthy, 2024). This has resulted in the marketing literature being replete with epithets of failed products (Benson, 2024; Cefis et al., 2024).

The design and production of products should have a bird's-eye view of the entire product life cycle (Fazlollahtabar, 2024; Gulivindala et al., 2023). This suggests that producers must also pay some attention to the products' end-of-life. This is particularly important for electronic products, considering that they have relatively short life cycles (Xu & Yan, 2024). Product end-of-life planning should include taking into consideration how the product will be disposed of. In line with the circular economy approach, most products that have reached their end of life, in what is technically referred to as obsolescence (Alzaydi, 2024), should either be recycled or reused (Kastanaki & Giannis, 2022). However, such guidance seems to be unavailable for most products that are found in developing countries' markets.

Extant literature has shown various sources of e-waste, such as consumer electronics and industrial gadgets (Bhattacharjee et al., 2025; Seif et al., 2023). While the disposal of industrial electronic gadgets is regulated and enforced by legislation, the one generated by consumer electronics is largely chaotic. Consumer electronics include laptops, tablets, televisions, radios, and mobile phones (Bhattacharjee et al., 2025; Vishwakarma et al., 2023). A prominent research stream regarding e-waste suggests that indiscriminate disposal of such waste leads to several externalities of varying degrees of frequency and severity. E-waste is a public health concern that affects human beings and animals through dermal absorption, particularly for waste material pickers (Mukucha & Jaravaza, 2025); ingestion for scavengers; and inhalation for residents in the vicinity of dumpsites (Shrivastava et al., 2025; Ly et al., 2025).

Obsolescence and electronic waste material

Obsolescence refers to a condition where a product becomes irrelevant for its purpose (Lechelt et al., 2024; Kane et al., 2018). Obsolete products are characterised by being no longer in use, mostly as a result of being outdated (Guo & Yan, 2017). There are several types of obsolescence that affect both the manufacturers and consumers, but in this study we will focus on those related to the consumers as a result of manufacturers' premeditated actions and market dynamics. The most common types of obsolescence that affect the electronic gadgets market are technical, functional and planned obsolescence.

Technical obsolescence occurs when a new, improved product is introduced to the market (Reyes-Santias et al., 2024). This type of obsolescence has been seen to drive some of the obsolescence in the mobile phone industry, as new products are always introduced with improved features compared to the previous models (Oraee et al., 2024). Functional obsolescence relates to reduced usefulness of a product as a result of outdated designs and features (Costa et al., 2024). This type of obsolescence has been common since smartphones were introduced to cater to internet connectivity, something which the previous models of smartphones could not do. In a study by Proske & Jaeger-Erben (2019), functionally obsolete smartphones were found to have some ecological footprints.

Planned obsolescence occurs when manufacturers deliberately design product features that have a limited lifespan (Hunka et al., 2021). Limited life spans, which are common in electronic appliances (Hudomiet & Willis, 2022), are factored into a product's features so as to encourage frequent purchase replacements (Vail, 2022). A study by Corrocher & Paganuzzi (2025) revealed that frequent replacements are a result of fast-changing software updates that are incompatible with existing smartphone hardware. In a study by Barros and Dimla (2021), frequent replacements as a result of planned obsolescence in mobile phones is motivated by the need to stimulate sales, particularly in saturated markets. Irreparability of some smartphones was also identified in a study by Makov & Fitzpatrick (2021) to be the driver of e-waste under planned obsolescence conditions.

When products become obsolete, they turn into waste. There are various types of waste depending on the type of material used to make a product. For instance, in the textile industry they generate textile waste (Biyada & Urbonavičius, 2025), pharmaceutical waste in the pharmaceutical industry (Ajekiigbe et al., 2025), packaging waste in the packaging industry (Lorang et al., 2022), construction waste in the construction industry (Bonifazi et al., 2025), and in the electronic industry e-waste is generated (Bagwan, 2024).

E-waste is made up of discarded electric and electronic devices (Serpe et al., 2025; Mairizal et al., 2021). This type of waste usually contains toxic materials (Murthy & Ramakrishna, 2022), such as leaked acid from batteries, and heavy metals such as mercury, beryllium, and lead (Prabhu & Majhi, 2023). E-waste also contains dangerous chemical compounds and flame retardants.

E-waste disposal methods

There is a growing global consumption of electronic gadgets. This consumption has resulted in a parallel increase in e-waste that should be disposed of responsibly (Vishwakarma et al., 2023). However, there is a significant disparity between the global south and global north in terms of how e-waste is handled. There are several international best practices for disposing of e-waste in the extant literature originating from the global north. The extant literature covers the entire e-waste supply chain network spanning from collection centres, logistics infrastructure, and dismantling technologies (Noliya et al., 2025).

There are various types of e-waste: industrial and domestic. Industrial e-waste is generated through economic activities, while consumer e-waste is generated domestically. Generally, e-waste generated from smartphones can be regarded as domestic e-waste since most smartphones are for individual uses. Extant literature has documented several methods of how consumers handle e-waste from smartphones, such as donations, storing at home and recycling.

Storage at home has been found to be the most common method of disposing of obsolete mobile phones. The storage of mobile phones is necessitated by the nature of mobile phones, which is a small size that does not occupy much storage space within the domestic facilities (Prabhu & Majhi, 2023). Ideally, obsolete mobile phones should be discarded, but the reasons why consumers may decide to store them may be due to limited knowledge of how to get rid of the unwanted phones

(Yin et al., 2014) or the need to have them as a backup in the event that their new phone gets dysfunctional (Nowakowski, 2019).

The other methods of disposing obsolete mobile phones revealed in the extant literature include donating to less fortunate friends and relatives (Kwatra et al., 2014). The donation of mobile phones to those in need delays the movement of the e-gadgets to the dumpsites.

Finally, the international best practices for dealing with e-waste relate to the practice of circular economy practices.

Circular economy practices

The traditional handling of products follows a linear fashion where products are produced, used, and disposed of at the end of their useful life (Abdirahman et al., 2025). This type of production and consumption model thrived in an economy where the resources are in abundance (Sakthivelmurugan et al., 2022; Rashid & Malik, 2023). However, as the availability of virgin material for producing new products and externalities associated with disposing of obsolete products grew, manufacturers progressively leaned towards the circular economy (Abdirahman et al., 2025). A circular economy is a production and consumption model that aims at minimising waste (Agrawal et al., 2025; Chioatto & Sospiro, 2023). Minimisation of waste can be achieved through minimising the use of input resources and maximising the outputs (Anbarasu et al., 2025; Onyeaka et al., 2023). The circular economy model is characterised by the adoption of one or more of the following practices: reduce, reuse, or recycle (Bressanelli et al., 2022; Ghosh et al., 2021).

The circular economy model involves reducing the use of materials through efficient use of resources to achieve maximum output. (Kastanaki, 2025). This involves investments in technologies that reduce the consumption of input resources (Butturi et al., 2025). It also reduces the use of resources through innovative means of coming up with alternative resources that can be used to achieve the same or a higher output (Ghosh et al., 2021). For instance, in the spirit of reducing the extraction and use of resources, most businesses are increasingly leaning towards the use of renewable energy (Marciniuk-Kluska & Kluska, 2025). For instance, in a study by Marciniuk-Kluska & Kluska (2025), it was determined that countries such as Poland can drastically reduce the use of energy through turning to renewable energy from municipal waste.

Where reduction of the use of materials is not feasible, the circular economy promotes the reuse or refurbishment of materials in order to extend their shelf life. Extending the useful life of products reduces the production of new products, which in turn reduces the need for extracting virgin materials (Sonkusare & Shastri, 2025). Extending the useful life of products also reduces the volumes of obsolete products destined for the landfills (Mukucha et al., 2023a). Reusing also involves repairing the existing products in order to avoid the acquisition of new productive products (Oyenuga et al., 2025; Christensen et al., 2021).

The circular economy also promotes the recycling of the products that can no longer be reused (Khajuria et al., 2022). Recycling involves breaking down obsolete products to extract materials that can be used as raw material for the production of the same original product or new, different products (Pulido-Ocegueda et al., 2025). This practice has the same effect as the other elements of the circular economy of reducing the extraction of virgin materials and handling the disposal of obsolete products. In the extant literature recycling emerges as one of the sustainable ways of handling e-waste (Serpe et al., 2025). Recycling of e-waste material leads to the recovery of important components such as metals, rare-earth minerals, and semiconductor chips (Bhattacharjee et al., 2025).

2.3 Conclusion

The conceptual framework has revealed that e-waste can emanate from several sources of obsolescence, such as technical, functional, and planned obsolescence. Theoretically, the social cognitive theory, in summary, provides an all-encompassing framework for understanding how individuals deal with the disposal of a product that has reached its end of life in the absence of an official disposal guideline. Finally, empirical evidence suggested that there are several sustainable ways of disposing of e-waste, such as donating or selling to recycling firms. However, extant literature is not clear on how citizens in the global south deal with domestic e-waste, considering that they lack circular economy infrastructure. There is therefore the need to determine how individuals deal with e-waste material as a way of providing background information to mobile phone producers and environmental agencies that they will use to formulate sustainable e-waste disposal methods.

3. METHODOLOGY

Research Design

A qualitative research design was adopted for this study. The qualitative design used the phenomenological approach to understand the participants' lived experiences with regard to how they dispose of e-waste material generated from obsolete mobile phones. This approach is suitable for scrutinising what and how aspects of the phenomenon under study (Oluka, 2025; Brziak & Urban, 2025). This approach adopted the primary data collection method of semi-structured interviews. Semi-structured interviews provide a guided approach that is flexible enough to explore themes that arise from the discussion through probing (Maher et al., 2025; Buys et al., 2022).

Population and Sampling Procedures

This study targeted individuals with recently purchased mobile phones through using purposive sampling. Purposive sampling is a non-probability strategy of selecting participants for a qualitative study targeting those participants that possess the qualities that are relevant to the study (Ahmad & Wilkins, 2024; Andrade, 2021). These individuals were targeted on the reasoning that having a new mobile phone might be an indication that one had recently disposed of his old mobile phone. Thus, this purposive sampling helped the researcher to locate the relevant participants who had some experiences with disposing of obsolete mobile phones. This recruitment procedure excluded those who had disposed of their previous mobile phones through donations or selling to second-hand phone shops. It also excluded those who lost their previous cell phones.

The sample size was 30, which is generally the maximum standard for most qualitative studies and at which the study reached the saturation point. The saturation point indicates the stage at which data collection is terminated (Hennink & Kaiser, 2022; Sarfo et al., 2021). This sample size was distributed as follows: gender, males (13) and females (17). All the respondents were employed, with those in formal employment being 12 and those informally employed being 18. This reflects the status quo in the country, where the majority of the people are informally employed. All the respondents had reached at least the secondary level of education. The age groups were represented as follows: 18-30 (16), 31-39 (10), and above 40 (4). The age distribution indicates that generally it is the youth that frequently buy new mobile phones as a replacement for the old models.

Data Collection Procedures

Data was collected through semi-structured interviews. The interviewer guide sought some information related to how the participants had disposed of those previous mobile phones. There were different follow-up questions that emanated from the participants' responses. Each interview took approximately 20 minutes. The data collection period took slightly above 2 weeks.

Data Analysis Procedures

A thematic approach was adopted for data analysis. This approach involved several stages starting from data familiarisation. After familiarising with the data, the analysis process proceeded to data coding. Open coding was used in order to get all the aspects of the subject matter as presented by the participants (Rivas, 2012). Data coding was followed by theme development, which latterly culminated in results interpretation (Kabir et al., 2025; Naeem et al., 2023).

Validity and Reliability

Several strategies were employed to ensure the validity and reliability of the collected data. Member checking was conducted by inviting some of the participants to assess the preliminary findings with the objective of determining whether the results resonate with their lived experiences (Kullman & Chudyk, 2025; Schafer & Phillippi, 2025). Researcher triangulation was also used through inviting a colleague to independently analyse the data. His findings were compared with this researcher's findings, and the results did not significantly differ.

Ethical Considerations

Several tenets of ethical considerations were considered. Prior to data collection, verbal informed consent was secured from all the participants. Confidentiality of the data provided was also explicitly promised to the participants. The participants were also informed of their right to withdraw from the study at any given stage of the data collection process without incurring any liability. Finally, the anonymity of the respondents was preserved through eliminating all the identifying features in the data set, such as names.

4. RESULTS

The results are presented in each subsection related to themes that came out of the data analysis process. The major themes relate to how the participants indicated the ways they dispose of their e-waste from disused mobile phones. The themes that emerged from data analysis are toyification, sanitary disposal, incineration and landfilling and are presented in detail in subsections below.

Toyification

Toyification is the process of turning a general object into a toy (Ihamäki & Heljakka, 2019). Some of the respondents indicated that they normally give their obsolete mobile phones to their children to play with as toys. This was summarised by respondent number 22, who said:

"I rarely throw away any obsolete mobile phones. Rather, I give them to my children as toys. I understand that toys of that nature help the children to develop some technical knowledge of science through reverse engineering."

The other respondent (R24) added that they prefer disused mobile phones as toys for their ability to provide their children with experiences that are closer to the experiences with the real products.

"Toys from disused mobile phones excite our children, as they perform better than the real toys in terms of giving the children some exciting experiences."

The other respondent suggested that the toyification of disused phones has some significant cost savings.

"We rarely afford real toys for our children. As a result of that, we tend to prefer the use of disused mobile phones as toys for our children."

Sanitary disposal

The respondents also indicated that they dispose of their obsolete mobile phones by throwing them into sanitary bins. This was submitted by Respondent 11, who said:

"Everything that has turned into waste, we throw into the sanitary bins. Sanitary bins are the only waste material disposal method available to us. Hence, we do not have any specialised way of waste material disposal."

Respondent 13 added that, "*in our city there are no sanitary bins dedicated for segregated waste material. We therefore disposed of our obsolete e-waste together with the other household waste*."

Respondent 17 revealed that they use sanitary plastic bags for disposal of all their household waste, including discarded mobile phones. She said, "*We throw away all our household waste, including discarded mobile phones, into plastic sanitary bags which are then collected once every week by the municipal authorities*."

Incineration

Incineration was indicated as one of the methods that individuals use to dispose of domestic e-waste material. This involves backyard burning of e-waste material. Incineration was indicated as an e-waste disposal method by the residents in low-density areas where there is infrequent municipal domestic waste collection. This was captured from Respondent 12.

"We rarely have the municipality sanitary trucks plying in our neighbourhood, so we tend to burn all the waste at hand. The burning of such waste includes obsolete mobile phones and other electronic gadgets."

The other respondent (R27) added that they tend to use incineration due to the high magnitude of domestic waste that they generate and could not be accommodated by the available sanitary bins. She said, "*The sanitary bins which we have are not large enough to cater for all the household waste which we generate; as a result, we resort to incineration for excess waste, including e-waste material.*"

The use of incineration was also motivated by the delays in waste collection by municipal authorities. This was added by Respondent number 30. She said, "*The delays in municipal waste collection sometimes force us to dispose of e-waste together with the other general waste through incineration.*"

Landfilling

Landfilling was mentioned as the other e-waste material disposal method practised by most respondents. The landfills are usually small dumping pits in the respondents' backyards. The use of backyard landfills was mentioned by Respondent 14:

"The municipality sanitary trucks rarely collect garbage in our neighbourhood, so we tend to throw all our domestic waste material into the backyard pits. In those pits we also dump some of our obsolete electronic gadgets, such as mobile phones."

Respondent 19 added that they are motivated to use backyard landfilling due to the availability of large spaces of land in their backyards. "*Considering that our backyards are relatively spacious, we tend to use backyard landfilling as a quick means of getting rid of our e-waste.*"

5. DISCUSSION

The results of this study indicated several ways through which the participants dispose of their obsolete mobile phones. The most prominent revealed methods include toyification, sanitary disposal, incineration, and landfilling. However, a close examination of all these disposal methods indicates that none of them is sustainable and appropriate for handling e-waste.

Giving children disused electronic gadgets is a common practice in most developing countries where real toys are not affordable. These toys closely resemble real electronic toys, especially if they are decommissioned by the users whilst in good condition. Digital toys are more engaging and entertaining to the children. They also afford the children some amateurish practical reverse engineering lessons.

Moreover, children tend to regard such types of toys as more valuable than the conventional toys because they are sophisticated and tend to represent the reality associated with actual products (Heljakka, 2022). Tofication is considered an economically sustainable consumption practice, as it tends to extend the usefulness of a product that has otherwise technically reached its end of life. This reduces the need to spend money on buying toys. Furthermore, toyification resonates well with the circular economy, where products that had reached their end of life are reintroduced to other new uses.

There are, however, some challenges associated with the toyification of obsolete electronic gadgets such as mobile phones. Electronic gadgets are made of various harmful components such as acid and heavy metals such as lead, mercury, and cadmium (Kumar et al., 2024a), which, if children are exposed to, may result in them suffering from diseases such as cancer (Parida & Patel, 2023).

Discarding obsolete phones through sanitary bins was found to be the other method of disposing of mobile phones by the participants. The use of sanitary bins is part of the waste management systems in most municipalities. Regularly the sanitary collection trucks are supposed to move around the residential areas collecting waste in the sanitary bins for disposal at designated dumpsites (Tafadzwa & Steven, 2025). However, due to poor service delivery associated with many municipalities in developing countries, most sanitary bins go for months without being emptied (Takunda & Steven, 2023; Chikowore, 2021). This leads to attracting waste material pickers into scavenging through the sanitary bins, leading to spillage of the waste (Intauno & Poshai, 2023).

The other major challenge with the use of sanitary bins is that most domestic waste is rarely separated at source (Mukucha et al., 2025). Segregation at source involves grouping waste of similar nature together for the purposes of specialised disposal (Nawar et al., 2024). Waste separation at source is particularly important considering that most of the waste is not disposed of the same way. Some waste, such as plastics and metals, needs to be recycled (Schade et al., 2024), while bottle containers need to be reused (Mukucha et al., 2023). Harmful material such as e-waste needs to be landfilled (Sarkhoshkalat, 2024). Such variability in disposal methods depending on the nature of waste suggests that simply throwing it in a sanitary bin may not be appropriate.

Eventually, when such waste reaches the dumpsites, there is an inherent risk to the informal waste pickers. Informal waste pickers usually rummage through the dumpsites salvaging recyclable and reusable materials (Mukucha & Jaravaza, 2025). However, the presence of e-waste material tends to expose them to serious hazards that affect their health. Therefore, the disposing of e-waste material through the traditional municipality waste disposal systems, particularly in most developing countries where they are primitive, exposes the residents to severe health complications (Tafadzwa & Steven, 2025).

Backyard waste material incineration is one of the most common practices in developing countries, where municipality waste disposal systems are poorly managed (Shabani et al., 2024). This practice involves gathering all household waste and, on a designated day, burning it. However, unsanctioned burning is illegal in most local government areas, and it tends to attract heavy penalties (Zhang, 2023).

The practice of backyard incineration is highly dangerous (Jakhar et al., 2023). Most e-waste materials are highly inflammable (Mandadi et al., 2020). As such, they can produce uncontrollable flames that can spread to other residential properties, leading to severe loss of properties (Oswald, 2021). Moreover, some of the flames can have sudden bursts that can burn the people around. Thus, the burning of e-waste material poses a great risk of uncontrollable fire outbreaks.

Moreover, unsanctioned burning produces smoke that leads to air pollution (Onwudiegwu et al., 2024). The fumes from the burning materials affect the air quality by reducing oxygen and increasing carbon dioxide (Ćurić et al., 2022). This may cause breathing problems not only for the neighbours but also for the passing individuals. The smoke from the burning of e-waste material is linked to various respiratory diseases (Jerie et al., 2024).

Backyard landfilling is illegal in most municipalities in Zimbabwe. Moreover, landfilling e-waste material is a dangerous practice, as it tends to produce chemicals that contaminate underground water (Adu & Aneke, 2025). This is particularly the case where some residents have boreholes and wells at their residential places. A study by Matengo et al. (2025) in Harare has found that most of the boreholes located near dumpsites had contaminated water and therefore were unsafe for drinking. Drinking contaminated water has been attributed to terminal illnesses such as colon cancer (Espejo-Herrera et al., 2016).

Landfills are also a hazard to both animals and human beings (Mohajan, 2025; Podlasek et al., 2024). As the waste material decomposes, it may leave the firmness of the ground fragile. This weakness may cause any human being or animal that steps on it to fall. Falling into a landfill full of harmful waste material may lead to severe health consequences, such as coming into contact with toxic material. The use of landfills is also discouraged on the basis that over time landfills produce methane gas and other dangerous gases that compromise air quality (Ozbay et al., 2021). This tends to affect the residents in the vicinity of such landfills, leading to chest infections (Kiehbadroudinezhad et al., 2024).

6. CONCLUSIONS AND RECOMMENDATIONS

The study concluded that Zimbabwean consumers of mobile phones are not aware of the sustainable ways of disposing of e-waste material in general and their obsolete mobile phones in particular. The practised e-waste material disposal methods such as toyification, sanitary bins, incineration, and landfilling are all inappropriate for dealing with domestic e-waste. There is therefore a great need for providing some guidelines to the consumers on how they should deal with domestic e-waste. Individual consumers should be provided with manuals for handling obsolete mobile phones. Where possible, manufacturers should provide access to protective equipment for handling e-waste (Kumar et al., 2024b).

Local government authorities running various municipalities and the environmental management agency must enforce adherence to the prescribed waste material disposal methods. This is particularly important as most economies are transitioning from the linear approach to the circular mode (Mukucha et al., 2021). This suggests that waste material separation at source must be enforced at the household level. Waste material separation at source is necessitated by the fact that in a circular economy the fate of different types of waste material varies (Mukucha et al., 2023a). Some materials, such as e-waste, need to be recycled or refurbished, while others, like bottles, need to be reused. Therefore, mixing such materials may pose some challenges in successive stages in the waste material transformation supply chains.

Having mentioned the circular economy as a proper way of dealing with e-waste material, it is vital to note that none of the respondents in this study mentioned the recycling of e-waste material from mobile phones. This suggests that presently in Zimbabwe there are no recycling facilities for obsolete mobile phones. Therefore, this implies that the country should invest in recycling facilities for mobile phones, considering that almost every citizen in Zimbabwe has at least two mobile phones and that mobile phones go obsolete at an alarming rate.

REFERENCES

Abdirahman, A. A., Asif, M., & Mohsen, O. (2025). Circular economy in the renewable energy sector: A review of growth trends, gaps and future directions. *Energy Nexus*, *17*, 100395. DOI: 10.1016/j.nexus.2025.100395

Adu, J. T., & Aneke, F. I. (2025). Evaluation of heavy metal contamination in landfills from e-waste disposal and its potential as a pollution source for surface water bodies. *Results in Engineering*, *25*, 104431. DOI: 10.1016/j.rineng.2025.104431

Aga, N., Ani, P., Wijaningsih, D., & Kadek, S. (2024). From human-centered to ecofeminist environmental policies: Harnessing pancasila for global sustainability solutions. *Clío. Revista de Historia. Ciencias Humanas y Pensamiento Crítico.*, (9), 195–245.

Agrawal, S., Jain, V. K., & Verma, H. (2025). Circular economy! Sustainable growth and achieving zero net emissions-a way ahead for building sustainable tomorrow by optimising consumption. *International Journal of Sustainable Agricultural Management and Informatics*, *11*(1), 72–96. DOI: 10.1504/IJSAMI.2025.143112

Ahad, T., & Busch, P. (2024). Exploring the factors influencing Mobile-based Ubiquitous System adoption in the Bangladesh RMG sector: A view through DOI and TOE. *The Electronic Journal on Information Systems in Developing Countries*, *90*(1), e12291. DOI: 10.1002/isd2.12291

Ahmad, M., & Wilkins, S. (2024). Purposive sampling in qualitative research: A framework for the entire journey. *Quality & Quantity*, •••, 1–19.

Ajekiigbe, V. O., Agbo, C. E., Ogieuhi, I. J., Anthony, C. S., Onuigbo, C. S., Falayi, T. A., Oluwapelumi, O. Z., Amusa, O., Adeniran, G. O., Ogunleke, P. O., & Bakare, I. S. (2025). The increasing burden of global environmental threats: Role of antibiotic pollution from pharmaceutical wastes in the rise of antibiotic resistance. *Discover Public Health*, *22*(1), 1–10. DOI: 10.1186/s12982-025-00506-9

Alzaydi, A. (2024). Balancing creativity and longevity: The ambiguous role of obsolescence in product design. *Journal of Cleaner Production*, *445*, 141239. DOI: 10.1016/j.jclepro.2024.141239

Anbarasu, K., Thanigaivel, S., Sathishkumar, K., Alam, M. M., Al-Sehemi, A. G., & Devarajan, Y. (2025). Harnessing artificial intelligence for sustainable bioenergy: Revolutionizing optimization, waste reduction, and environmental sustainability. *Bioresource Technology*, *418*, 131893. DOI: 10.1016/j.biortech.2024.131893 PMID: 39608419

Andrade, C. (2021). The inconvenient truth about convenience and purposive samples. *Indian Journal of Psychological Medicine*, *43*(1), 86–88. DOI: 10.1177/0253717620977000 PMID: 34349313

Ateeq, K., Al Masaeid, T., Selim, H., Oswal, N., Alkubaiusy, A. A. A., Alami, R., & Ajdoobi, S. M. G. A. (2025). Harnessing AI for Faster Innovation: How AI Concept Generation Impacts Development Timelines and Market Agility. *Journal of Posthumanism*, *5*(2), 37–50. DOI: 10.63332/joph.v5i2.390

Bagwan, W. A. (2024). Electronic waste (E-waste) generation and management scenario of India, and ARIMA forecasting of E-waste processing capacity of Maharashtra state till 2030. *Waste Management Bulletin*, *1*(4), 41–51. DOI: 10.1016/j.wmb.2023.08.002

Bandura, A. (1982). Self-efficacy mechanism in human agency. *The American Psychologist*, *37*(2), 122–147. DOI: 10.1037/0003-066X.37.2.122

Bandura, A. (1986). Social foundations of thought and action. *Englewood Cliffs, NJ*, *1986*(23-28), 2.

Bandura, A. (2014). Social-cognitive theory. In *An introduction to theories of personality* (pp. 341–360). Psychology Press.

Baron, A., & Galizio, M. (2005). Positive and negative reinforcement: Should the distinction be preserved? *The Behavior Analyst*, *28*(2), 85–98. DOI: 10.1007/BF03392107 PMID: 22478443

Barros, M., & Dimla, E. (2021). From planned obsolescence to the circular economy in the smartphone industry: An evolution of strategies embodied in product features. *Proceedings of the Design Society*, *1*, 1607–1616. DOI: 10.1017/pds.2021.422

Bekele, N. A., Muffatto, M., & Ferrati, F. (2024). Desirability of consumer internet of things products: How emerging businesses address consumer desires to improve user experiences. *International Journal of Technology Marketing*, *18*(1), 34–61. DOI: 10.1504/IJTMKT.2024.135668

Benson, P. K. (2024). Concept testing and selection—where and why will it fail. In *Innovation Anesthesiology* (pp. 131–138). Academic Press. DOI: 10.1016/B978-0-12-818381-6.00013-9

Bhattacharjee, J., De, A., Kamila, B., & Mandal, A. (2025). Recent scenario of e-waste recycling: Chemical engineering. *International Journal of Chemical Reactor Engineering*, *23*(6), 675–700. DOI: 10.1515/ijcre-2024-0058

Bilici, F., & Özdemir, E. (2024). Consumer Perception of Planned Obsolescence: A Research on Smartphone Owners. *İşletme Araştırmaları Dergisi, 16*(1), 244-261.

Biyada, S., & Urbonavičius, J. (2025). Circularity in textile waste: Challenges and pathways to sustainability. *Cleaner Engineering and Technology, 24*, 100905. DOI: 10.1016/j.clet.2025.100905

Blahnik, V., & Schindelbeck, O. (2021). Smartphone imaging technology and its applications. *Advances in Optical Technologies, 10*(3), 145–232. DOI: 10.1515/aot-2021-0023

Bonifazi, G., Grosso, C., Palmieri, R., & Serranti, S. (2025). Current trends and challenges in construction and demolition waste recycling. *Current Opinion in Green and Sustainable Chemistry, 53*, 101032. DOI: 10.1016/j.cogsc.2025.101032

Bressanelli, G., Adrodegari, F., Pigosso, D. C., & Parida, V. (2022). Towards the smart circular economy paradigm: A definition, conceptualization, and research agenda. *Sustainability (Basel), 14*(9), 4960. DOI: 10.3390/su14094960

Brziak, M., & Urban, K. (2025). Phenomenology meets self-regulated learning: Exploring the lived experience of learning. *Review of Education, 13*(1), e70063. DOI: 10.1002/rev3.70063

Butturi, M. A., Neri, A., Mercalli, F., & Gamberini, R. (2025). Sustainability-oriented innovation in the textile manufacturing industry: Pre-consumer waste recovery and circular patterns. *Environments (Basel, Switzerland), 12*(3), 82. DOI: 10.3390/environments12030082

Buys, T., Casteleijn, D., Heyns, T., & Untiedt, H. (2022). A reflexive lens on preparing and conducting semi-structured interviews with academic colleagues. *Qualitative Health Research, 32*(13), 2030–2039. DOI: 10.1177/10497323221130832 PMID: 36194785

Cefis, E., Leoncini, R., & Marengo, L. (2024). Is innovation failure just a dead end? *Structural Change and Economic Dynamics, 70*, 481–493. DOI: 10.1016/j.strueco.2024.05.004

Chaudhari, M. M., & Shirude, S. B. (2025). Current E-library System: Challenges, Solutions and Open Issues. *Solutions and Open Issues (March 24, 2025).*

Chikowore, N. (2021). Factors influencing household waste management practices in Zimbabwe. *Journal of Material Cycles and Waste Management, 23*(1), 386–393. DOI: 10.1007/s10163-020-01129-9

Chioatto, E., & Sospiro, P. (2023). Transition from waste management to circular economy: The European Union roadmap. *Environment, Development and Sustainability, 25*(1), 249–276. DOI: 10.1007/s10668-021-02050-3

Christensen, D., Hjul-Nielsen, J., Moalem, R. M., & Johansen, B. (2021). Circular Economy in Denmark: Bornholm's Vision to Achieve 100 Percent Reuse and Recycling. In *Circular Economy: Recent Trends in Global Perspective* (pp. 385–424). Springer Nature Singapore. DOI: 10.1007/978-981-16-0913-8_13

Corrocher, N., & Paganuzzi, S. (2025). Planned obsolescence and smartphone replacement: Empirical evidence on the Italian market. *Telecommunications Policy, 49*(8), 103022. DOI: 10.1016/j.telpol.2025.103022

Costa, N., Oliveira, J., Martins, A., & Gomes, A. (2024). The planned obsolescence of knowledge: information with an expiry date. In *ICERI2024 Proceedings* (pp. 5407-5411). IATED. DOI: 10.21125/iceri.2024.1321

Ćurić, M., Zafirovski, O., Spiridonov, V., Ćurić, M., Zafirovski, O., & Spiridonov, V. (2022). Air quality and health. *Essentials of medical meteorology*, 143-182.

Danwitz, L., & von Helversen, B. (2025). Observational learning of exploration-exploitation strategies in bandit tasks. *Cognition, 259*, 106124. DOI: 10.1016/j.cognition.2025.106124 PMID: 40117983

Espejo-Herrera, N., Gràcia-Lavedan, E., Boldo, E., Aragonés, N., Pérez-Gómez, B., Pollán, M., & Villanueva, C. M. (2016). Colorectal cancer risk and nitrate exposure through drinking water and diet. *International Journal of Cancer, 139*(2), 334–346. DOI: 10.1002/ijc.30083 PMID: 26954527

Fazlollahtabar, H. (2024). Waste opportunity gap for energy productivity in new product design. *Energy Sources. Part B, Economics, Planning, and Policy, 19*(1), 2371407. DOI: 10.1080/15567249.2024.2371407

Freisinger, E., & McCarthy, I. P. (2024). What fails and when? A process view of innovation failure. *Technovation, 133*, 102995. DOI: 10.1016/j.technovation.2024.102995

Ghosh, S. K., Ghosh, S. K., & Baidya, R. (2021). Circular economy in India: Reduce, reuse, and recycle through policy framework. In *Circular Economy: Recent Trends in Global Perspective* (pp. 183–217). Springer Nature Singapore. DOI: 10.1007/978-981-16-0913-8_6

Gulivindala, A. K., Bahubalendruni, M. R., P, M. B., & Eswaran, M. (2023). Mechanical disassembly sequence planning for end-of-life products to maximize recyclability. *Sadhana, 48*(3), 109. DOI: 10.1007/s12046-023-02160-w

Guo, X., & Yan, K. (2017). Estimation of obsolete cellular phones generation: A case study of China. *The Science of the Total Environment, 575*, 321–329. DOI: 10.1016/j.scitotenv.2016.10.054 PMID: 27744159

Heljakka, K. (2022). On longevity and lost toys: sustainable approaches to toy design and contemporary play. In *Toys and Sustainability* (pp. 19–37). Springer Singapore. DOI: 10.1007/978-981-16-9673-2_2

Hennink, M., & Kaiser, B. N. (2022). Sample sizes for saturation in qualitative research: A systematic review of empirical tests. *Social Science & Medicine, 292*, 114523. DOI: 10.1016/j.socscimed.2021.114523 PMID: 34785096

Hudomiet, P., & Willis, R. J. (2022). Computerization, obsolescence and the length of working life. *Labour Economics, 77*, 102005. DOI: 10.1016/j.labeco.2021.102005 PMID: 37035254

Hunka, A. D., Linder, M., & Habibi, S. (2021). Determinants of consumer demand for circular economy products. A case for reuse and remanufacturing for sustainable development. *Business Strategy and the Environment, 30*(1), 535–550. DOI: 10.1002/bse.2636

Husin, N., Ibrahim, N. A., Rahim, N. A., Abd Manaf, K., & Shafie, I. S. M. (2024). Exploring the Impact of Authentic Personality and Social Media Posting on Self-Interested Charitable Behavior among Millennials Online Community through the Lens of Social Cognitive Theory. *Information Management and Business Review, 16*(1), 39–48. DOI: 10.22610/imbr.v16i1(I)S.3724

Ihamäki, P., & Heljakka, K. (2019, November). Out of the box, into the cubes: Envisioning user experiences through a tool for gamification, toyification and playification. In *International Conference on ArtsIT, Interactivity and Game Creation* (pp. 49-68). Cham: Springer International Publishing.

Intauno, K., & Poshai, L. (2023). Examining the occupational safety and health hazards encountered by municipal waste collectors in murewa rural district council, Zimbabwe. [IJLAPP]. *International Journal of Law and Public Policy, 5*(2), 59–71. DOI: 10.36079/lamintang.ijlapp-0502.532

Jakhar, R., Samek, L., & Styszko, K. (2023). A comprehensive study of the impact of waste fires on the environment and health. *Sustainability (Basel), 15*(19), 14241. DOI: 10.3390/su151914241

Jerie, S., Shabani, T., Mudyazhezha, O. C., & Shabani, T. (2024). A review towards developing a hierarchical model for sustainable hospital solid waste management in rural areas of Zimbabwe. *Environmental Monitoring and Assessment, 196*(3), 308. DOI: 10.1007/s10661-024-12488-3 PMID: 38407739

Kabir, S. A., Ali, F., Ahmed, R. L., & Sulaiman-Hill, R. (2025). Exploring the Use of AI in Qualitative Data Analysis: Comparing Manual Processing with Avidnote for Theme Generation. *International Journal of Qualitative Methods, 24*, 16094069251336810. DOI: 10.1177/16094069251336810

Kane, G. M., Bakker, C. A., & Balkenende, A. R. (2018). Towards design strategies for circular medical products. *Resources, Conservation and Recycling, 135*, 38–47. DOI: 10.1016/j.resconrec.2017.07.030

Kastanaki, E. (2025). Dynamic assessment of photovoltaic waste streams in the EU-27 countries under the circular economy principles of 'Reduce, Reuse and Recycle'. *Resources, Conservation and Recycling, 214*, 108033. DOI: 10.1016/j.resconrec.2024.108033

Kastanaki, E., & Giannis, A. (2022). Forecasting quantities of critical raw materials in obsolete feature and smart phones in Greece: A path to circular economy. *Journal of Environmental Management, 307*, 114566. DOI: 10.1016/j.jenvman.2022.114566 PMID: 35091243

Khajuria, A., Atienza, V. A., Chavanich, S., Henning, W., Islam, I., Kral, U., & Li, J. (2022). Accelerating circular economy solutions to achieve the 2030 agenda for sustainable development goals. *Circular Economy, 1*(1), 100001. DOI: 10.1016/j.cec.2022.100001

Kiehbadroudinezhad, M., Merabet, A., & Hosseinzadeh-Bandbafha, H. (2024). Landfill source of greenhouse gas emission. In *Advances and technology development in greenhouse gases: Emission, capture and conversion* (pp. 123–145). Elsevier. DOI: 10.1016/B978-0-443-19231-9.00023-5

Kullman, S. M., & Chudyk, A. M. (2025). Participatory member checking: A novel approach for engaging participants in co-creating qualitative findings. *International Journal of Qualitative Methods, 24*, 16094069251321211. DOI: 10.1177/16094069251321211

Kumar, A., Singh, P. P., & Khanna, M. K. (2024a). Sustainable e-waste management and its effect on environment and human health. In *Integrated Waste Management: A Sustainable Approach from Waste to Wealth* (pp. 349–373). Springer Nature Singapore. DOI: 10.1007/978-981-97-0823-9_17

Kumar, P., Singh, S., Gacem, A., Yadav, K. K., Bhutto, J. K., Alreshidi, M. A., & Alam, M. W. (2024b). A review on e-waste contamination, toxicity, and sustainable clean-up approaches for its management. *Toxicology, 508*, 153904. DOI: 10.1016/j.tox.2024.153904 PMID: 39106909

Kwatra, S., Pandey, S., & Sharma, S. (2014). Understanding public knowledge and awareness on e-waste in an urban setting in India: A case study for Delhi. *Management of Environmental Quality, 25*(6), 752–765. DOI: 10.1108/MEQ-12-2013-0139

Laeequddin, M., Kareem Abdul, W., Sahay, V., & Tiwari, A. K. (2022). Factors that influence the safe disposal behavior of E-waste by electronics consumers. *Sustainability (Basel), 14*(9), 4981. DOI: 10.3390/su14094981

Lechelt, S., Gorkovenko, K., & Speed, C. (2024). On Disused Connected Devices: Understanding Disuse,"Holding On," and Barriers to Circularity. *ACM Journal on Computing and Sustainable Societies, 2*(2), 1–27. DOI: 10.1145/3651171

Lev, N., Negev, M., & Ayalon, O. (2023). Sometimes littering is acceptable—Understanding and addressing littering perceptions in natural settings. *Sustainability (Basel), 15*(18), 13784. DOI: 10.3390/su151813784

Limna, P., Kraiwanit, T., & Siripipattanakul, S. (2023). The relationship between cyber security knowledge, awareness and behavioural choice protection among mobile banking users in Thailand. *International Journal of Computing Sciences Research, 7*, 1133–1151. DOI: 10.25147/ijcsr.2017.001.1.123

Lorang, S., Yang, Z., Zhang, H., Lü, F., & He, P. (2022). Achievements and policy trends of extended producer responsibility for plastic packaging waste in Europe. *Waste Disposal & Sustainable Energy, 4*(2), 91–103. DOI: 10.1007/s42768-022-00098-z

Ly, N. H. T., Hong, T. T. K., & Giao, N. T. (2025). Impact of Informal Electronic Waste Recycling and Dumping Activities on Soil Ecosystems: A review from 2000 to 2024. *Journal of Applied Science & Environmental Management, 29*(4), 1075–1085.

Maher, C., & Bedwei-Majdou, C. (2025). Semi-structured Qualitative Interview Guide: Process and Considerations for Doctoral Students. In *Qualitative Research Methods for Dissertation Research* (pp. 195-220). IGI Global Scientific Publishing.

Mairizal, A. Q., Sembada, A. Y., Tse, K. M., & Rhamdhani, M. A. (2021). Electronic waste generation, economic values, distribution map, and possible recycling system in Indonesia. *Journal of Cleaner Production, 293*, 126096. DOI: 10.1016/j.jclepro.2021.126096

Makov, T., & Fitzpatrick, C. (2021). Is repairability enough? big data insights into smartphone obsolescence and consumer interest in repair. *Journal of Cleaner Production, 313*, 127561. DOI: 10.1016/j.jclepro.2021.127561

Malinauskaite, J., & Erdem, F. B. (2021). Planned obsolescence in the context of a holistic legal sphere and the circular economy. *Oxford Journal of Legal Studies, 41*(3), 719–749. DOI: 10.1093/ojls/gqaa061 PMID: 34584495

Mandadi, G. K., Asmatulu, R., Khan, W. S., & Asmatulu, E. (2020). Fast and affordable recycling approach to electronic waste above the melting point using induction heat combined with centrifugal forces. *Asia-Pacific Journal of Chemical Engineering, 15*(4), e2483. DOI: 10.1002/apj.2483

Marciniuk-Kluska, A., & Kluska, M. (2025). Energy Recovery from Municipal Biodegradable Waste in a Circular Economy. *Energies, 18*(9), 2210. DOI: 10.3390/en18092210

Martín, H. R. (2024). *How do we learn? A scientific approach to learning and teaching (evidence-based education)*. John Wiley & Sons.

Martínez-Torres, J., Gonzálvez, C., & Antón, N. (2024). School refusal profiles maintained by negative reinforcement and their relationship with self-perceived health. *Frontiers in Psychology, 15*, 1340010. DOI: 10.3389/fpsyg.2024.1340010 PMID: 38562235

Matengo, T., Mwedzi, T., & Mbetu, J. (2025). Spatial and temporal patterns in water quality analysis near effluent dump-sites in Harare, Zimbabwe. *Journal of Water, Sanitation, and Hygiene for Development : a Journal of the International Water Association, 15*(3), 206–214. DOI: 10.2166/washdev.2025.270

Mohajan, H. K. (2025). Waste management strategy to save environment and improve safety of humanity. *Frontiers in Management Science, 4*(2), 74–81. DOI: 10.56397/FMS.2025.03.05

Mukucha, P., & Jaravaza, D. C. (2025). Waste material recycling: The horrors of informal waste material picking in Zimbabwe. *Journal of Responsible Production and Consumption, 2*(1), 235–253. DOI: 10.1108/JRPC-01-2024-0003

Mukucha, P., Jaravaza, D. C., & Nyengerai, S. (2023a). Circular economy of shopping bags in emerging markets: A demographic comparative analysis of propensity to reuse plastic bags versus cotton bags and paper bags. *Cogent Engineering, 10*(1), 2176582. DOI: 10.1080/23311916.2023.2176582

Mukucha, P., Katsande, T., Dube, T., Jaravaza, D. C., & Chisango, S. (2025a). Disseminating Sustainable End-of-life Product Disposal Information Through Social Media Nudging. In *Sustainable Green Marketing Strategies for a Circular Economy: African Realities and Imperatives* (pp. 239–264). Springer Nature Switzerland. DOI: 10.1007/978-3-031-81627-7_12

Mukucha, P., Mushanyuri, B. E., & Jaravaza, D. C. (2021). Green sourcing: effects on supplier performance metrics in fast food restaurants in frontier markets. *Logistics, supply chain, sustainability and global challenges, 12*(1), 47-59.

Mukucha, P., Mwapenya, D., & Jaravaza, D. C. (2025b). The efficacy of food waste reduction strategies in an African frontier market foodservices industry. *Tourism (Zagreb), 73*(2), 215–226. DOI: 10.37741/t.73.2.2

Mukucha, P., Sixpence, S., Muzanenhamo, L., Muskwe, N., & Ukpere, W. I. (2023b). Green supply chain management: The role of sustainable packaging on corporate image and patronage intentions. *Annals of Spiru Haret University. Economic Series, 23*(4), 553–572.

Munguiko, C., Masereka, E. M., & Sang, C. (2025). Applying Learning Theories to Clinical Teaching in Contemporary Settings: A Conceptual Analysis. *South Asian Res J Med Sci, 7*(2), 16–27. DOI: 10.36346/sarjms.2025.v07i02.001

Murthy, V., & Ramakrishna, S. (2022). A review on global E-waste management: Urban mining towards a sustainable future and circular economy. *Sustainability (Basel), 14*(2), 647. DOI: 10.3390/su14020647

Naeem, M., Ozuem, W., Howell, K., & Ranfagni, S. (2023). A step-by-step process of thematic analysis to develop a conceptual model in qualitative research. *International Journal of Qualitative Methods, 22*, 16094069231205789. DOI: 10.1177/16094069231205789

Nawar, K. N., Mahbub, T., Tashfiq, R. A., & Rashid, T. U. (2024). Municipal Solid Waste Collection, Transportation, and Segregation. In *Environmental Engineering and Waste Management: Recent Trends and Perspectives* (pp. 29–71). Springer Nature Switzerland. DOI: 10.1007/978-3-031-58441-1_2

Noliya, S., Gupta, A., & Kumar, V. (2025). E-waste management through Blockchain Technology: Supply chain led environmental sustainability. *EDPACS*, 1-11.

Nowakowski, P. (2019). Investigating the reasons for storage of WEEE by residents–A potential for removal from households. *Waste Management (New York, N.Y.), 87*, 192–203. DOI: 10.1016/j.wasman.2019.02.008 PMID: 31109518

Oluka, A. (2025). Phenomenological Research Strategy: Descriptive and Interpretive Approaches. *F1000 Research, 14*, 725. DOI: 10.12688/f1000research.166273.1

Onwudiegwu, C. A., Aigberua, A. O., & Elegbede, I. O. (2024). Uncontrolled Bush Burning and Air Quality. In *Air Pollutants in the Context of One Health: Fundamentals, Sources, and Impacts* (pp. 369–394). Springer Nature Switzerland. DOI: 10.1007/698_2024_1136

Onyeaka, H., Tamasiga, P., Nwauzoma, U. M., Miri, T., Juliet, U. C., Nwaiwu, O., & Akinsemolu, A. A. (2023). Using artificial intelligence to tackle food waste and enhance the circular economy: Maximising resource efficiency and minimising environmental impact: A review. *Sustainability (Basel), 15*(13), 10482. DOI: 10.3390/su151310482

Oraee, A., Pohl, L., Geurts, D., & Reichel, M. (2024). Overcoming premature smartphone obsolescence amongst young adults. *Cleaner and Responsible Consumption, 12*, 100174. DOI: 10.1016/j.clrc.2024.100174

Oswald, D. (2021). Homeowner vulnerability in residential buildings with flammable cladding. *Safety Science, 136*, 105185. DOI: 10.1016/j.ssci.2021.105185

Oyenuga, M. O., Singh, R., Apata, S. B., Khan, S., & Kumar, V. (2025). Circular Economy and Waste Reduction Technologies. In *Smart Technologies for Climate Change and Net Zero Policies: Practical Approaches Towards Sustainability* (pp. 117–129). Springer Nature Switzerland. DOI: 10.1007/978-3-031-92221-3_10

Ozbay, G., Jones, M., Gadde, M., Isah, S., & Attarwala, T. (2021). Design and operation of effective landfills with minimal effects on the environment and human health. *Journal of Environmental and Public Health, 2021*(1), 6921607. DOI: 10.1155/2021/6921607 PMID: 34531916

Özden, M. (2024). The Importance of Ecofeminism in Sustainable Development. *Reconstructing Feminism through Cyberfeminism, 275*, 88.

Parida, L., & Patel, T. N. (2023). Systemic impact of heavy metals and their role in cancer development: A review. *Environmental Monitoring and Assessment, 195*(6), 766. DOI: 10.1007/s10661-023-11399-z PMID: 37249740

Podlasek, A., Vaverková, M. D., Jakimiuk, A., & Koda, E. (2024). A comprehensive investigation of geoenvironmental pollution and health effects from municipal solid waste landfills. *Environmental Geochemistry and Health, 46*(3), 97. DOI: 10.1007/s10653-024-01852-4 PMID: 38393507

Poetz, M. K., & Schreier, M. (2012). The value of crowdsourcing: Can users really compete with professionals in generating new product ideas? *Journal of Product Innovation Management, 29*(2), 245–256. DOI: 10.1111/j.1540-5885.2011.00893.x

Power, J. R., Tanner, D., & Buckley, J. (2025). Self-efficacy development in undergraduate engineering education. *European Journal of Engineering Education, 50*(1), 1–25. DOI: 10.1080/03043797.2025.2468350

Prabhu, N. S., & Majhi, R. (2023). Disposal of obsolete mobile phones: A review on replacement, disposal methods, in-use lifespan, reuse and recycling. *Waste Management & Research, 41*(1), 18–36. DOI: 10.1177/0734242X221105429 PMID: 35722875

Proske, M., & Jaeger-Erben, M. (2019). Decreasing obsolescence with modular smartphones?–An interdisciplinary perspective on lifecycles. *Journal of Cleaner Production, 223*, 57–66. DOI: 10.1016/j.jclepro.2019.03.116

Pulido-Ocegueda, J. C., Santibáñez-Aguilar, J. E., Ramírez-Márquez, C., & Ponce-Ortega, J. M. (2025). Strategic development and economic impact of e-waste management in Mexico: A quantitative and perspective analysis on collection, recycling, and material recovery systems. *Journal of Material Cycles and Waste Management, 27*(5), 1–21. DOI: 10.1007/s10163-025-02344-y

Putra, A. N. (2024). *The Influence Of Observational Learning On Consumer Purchase Intention Is Mediated By Perceived Message Authenticity And Consumer Attitudes* (Doctoral dissertation, Universitas Muhammadiyah Surakarta).

Rashid, S., & Malik, S. H. (2023). Transition from a Linear to a Circular Economy. In *Renewable energy in circular economy* (pp. 1–20). Springer International Publishing. DOI: 10.1007/978-3-031-42220-1_1

Rauf, A., Puluhulawa, F. U., & Ahmad, A. (2024). Ideal Arrangements for Fines to Enhance Legal Awareness and Minimize Waste Effectively in Society. *Estudiante Law Journal, 6*(3), 593–606. DOI: 10.33756/eslaj.v6i3.28916

Reyes-Santias, F., Portela-Romero, M., Cinza-Sanjurjo, S., Cordova-Arevalo, O., & González-Juanatey, J. R. (2024). Estimated technological obsolescence of computed tomography equipment. *Academic Radiology, 31*(3), 951–955. DOI: 10.1016/j.acra.2023.07.006 PMID: 37541825

Rivas, C. (2012). Coding and analysing qualitative data. *Researching society and culture, 3*(2012), 367-392.

Sakthivelmurugan, E., Senthilkumar, G., & Karthick, K. N. (2022). Analysis of the impact of circular economy over linear economy in the paper processing industry. *Materials Today: Proceedings*, *66*, 1446–1452. DOI: 10.1016/j.matpr.2022.05.449

Sarfo, J. O., Debrah, T., Gbordzoe, N. I., Afful, W. T., & Obeng, P. (2021). Qualitative research designs, sample size and saturation: Is enough always enough. *Journal of Advocacy. Research in Education*, *8*(3), 60–65.

Sari, T. K., & Rahmani, P. H. (2024). The Role of Positive Reinforcement on Students in English Language Learning: A Skinnerian behaviorist: English. *JOEY: Journal of English Ibrahimy*, *3*(1), 1–5. DOI: 10.35316/joey.2024.v3i1.1-5

Sarkhoshkalat, M. M., Afkham, A., Bonyadi Manesh, M., & Sarkhosh, M. (2024). Circular Economy and the Recycling of E-Waste. In *New Technologies for Energy Transition Based on Sustainable Development Goals: Factors Contributing to Global Warming* (pp. 319–354). Springer Nature Singapore. DOI: 10.1007/978-981-97-2527-4_16

Schade, A., Melzer, M., Zimmermann, S., Schwarz, T., Stoewe, K., & Kuhn, H. (2024). Plastic waste recycling— A chemical recycling perspective. *ACS Sustainable Chemistry & Engineering*, *12*(33), 12270–12288. DOI: 10.1021/acssuschemeng.4c02551

Schafer, R., & Phillippi, J. C. (2025). Updating and advancing member-checking methods: Use of video and asynchronous technology to optimize participant engagement. *International Journal of Qualitative Methods*, *24*, 16094069251315395. DOI: 10.1177/16094069251315395 PMID: 40933861

Seif, R., Salem, F. Z., & Allam, N. K. (2023). E-waste recycled materials as efficient catalysts for renewable energy technologies and better environmental sustainability. *Environment, Development and Sustainability*, *26*(3), 1. DOI: 10.1007/s10668-023-02925-7 PMID: 36691418

Serpe, A., Purchase, D., Bisschop, L., Chatterjee, D., De Gioannis, G., Garelick, H., Kumar, A., Peijnenburg, W. J. G. M., Piro, V. M. I., Cera, M., Shevah, Y., & Verbeek, S. (2025). 2002–2022: 20 years of e-waste regulation in the European Union and the worldwide trends in legislation and innovation technologies for a circular economy. *RSC Sustainability*, *3*(3), 1039–1083. DOI: 10.1039/D4SU00548A

Shabani, T., Jerie, S., Mutekwa, T. V., & Shabani, T. (2024). Electronic waste: 21st century scenario in Zimbabwe—a review. *Circular Economy and Sustainability*, *4*(2), 1269–1284. DOI: 10.1007/s43615-024-00344-y

Shah, N. P.Nirali P Shah. (2024). Navigating challenges in new product development: Strategies for reducing failure rates in the medical device industry. *World Journal of Advanced Research and Reviews, 22*(3), 786–795. DOI: 10.30574/wjarr.2024.22.3.1740

Shrivastava, S. R., Bobhate, P. S., Singh, S., Quazi, S., & Jawade, S. (2025). Minimizing the dermatological consequences of exposure to e-waste. *Indian Dermatology Online Journal, 16*(5), 882–884. DOI: 10.4103/idoj.idoj_128_25 PMID: 40847670

Sonkusare, S., & Shastri, Y. (2025). A Systemic Evaluation of Reuse, Recycling, and Reduction Strategies in the Circular Economy for Global Sustainability Using an Integrated Planetary Model. *Circular Economy and Sustainability, 5*(3), 2009–2032. DOI: 10.1007/s43615-025-00513-7

Stephen, S. A. (2024). *Relationships between social identity, self-efficacy, other-efficacy, relation-inferred-self-efficacy and performance in sport* (Doctoral dissertation, Heriot-Watt University).

Tafadzwa, Z., & Steven, J. (2025). A critical review of innovative strategies for the sustainable management of solid waste generated in the health institutions of Zimbabwe. *Environmental Monitoring and Assessment, 197*(5), 611. DOI: 10.1007/s10661-025-14043-0 PMID: 40301144

Takunda, S., & Steven, J. (2023). Medical solid waste management status in Zimbabwe. *Journal of Material Cycles and Waste Management, 25*(2), 717–732. DOI: 10.1007/s10163-022-01578-4 PMID: 36686405

Tang, L., Murphree, M., & Breznitz, D. (2016). Structured uncertainty: A pilot study on innovation in China's mobile phone handset industry. *The Journal of Technology Transfer, 41*(5), 1168–1194. DOI: 10.1007/s10961-015-9432-9

Tao, X., & Ucbasaran, D. (2025). How does failure normalization foster product innovativeness in new product development? The role of passion and learning. *Journal of Product Innovation Management, 42*(2), 338–364. DOI: 10.1111/jpim.12755

Vail, J. P. (2022). The need for a sustainability pledge: Fighting planned obsolescence. *Geo. Wash. J. Energy & Env't L., 13*, 1.

van Baar, A., Knoote, F., & Consultancy, D. (2024). A fair price for human rights due diligence. *Bonn: Fairtrade International/Amsterdam: Vrije Universiteit Amsterdam. Retrieved, 30.*

Vishwakarma, A., Kanaujia, K., & Hait, S. (2023). Global scenario of E-waste generation: trends and future predictions. In *Global e-waste management strategies and future implications* (pp. 13–30). Elsevier. DOI: 10.1016/B978-0-323-99919-9.00013-1

Xu, C., & Yan, M. (2024). Product life cycle lengths and real activities management. *Journal of Accounting, Auditing & Finance*, *39*(4), 980–1007. DOI: 10.1177/0148558X221108970

Yin, J., Gao, Y., & Xu, H. (2014). Survey and analysis of consumers' behaviour of waste mobile phone recycling in China. *Journal of Cleaner Production*, *65*, 517–525. DOI: 10.1016/j.jclepro.2013.10.006

Zhang, L., Xuan, Y., Li, Z., Gao, P., & Zheng, Y. (2024). How to obtain customer requirements for each stage of the product life cycle from online reviews: Using mobile phones as an example. *Journal of Retailing and Consumer Services*, *80*, 103928. DOI: 10.1016/j.jretconser.2024.103928

Zhang, M. (2023). Burn to harvest, burn to sabotage: Between fire and water on a sugar plantation in Madagascar. *American Anthropologist*, *125*(1), 100–111. DOI: 10.1111/aman.13811

Zhilyaev, D., Cimpan, C., Cao, Z., Liu, G., Askegaard, S., & Wenzel, H. (2021). The living, the dead, and the obsolete: A characterization of lifetime and stock of ICT products in Denmark. *Resources, Conservation and Recycling*, *164*, 105117. DOI: 10.1016/j.resconrec.2020.105117

Zimba, A., Mukupa, G., & Chama, V. (2024). On emergent mobile phone-based social engineering cyberattacks in developing countries: The case of the Zambian ICT sector. *African Journal of Science, Technology, Innovation and Development*, *16*(6), 774-791. 1248.

Chapter 10
Prakriti Parv:
Nature-Friendly Festivals to Combat Environmental Pollution

Ginu George
http://orcid.org/0000-0001-5611-6174
Westford University College, UAE

K. Sudheesh
http://orcid.org/0000-0001-6277-2821
Christ (Deemed to be) University, India

ABSTRACT

Indian festivals are vibrant expressions of culture, tradition, and community that significantly contribute to the nation's social and economic fabric. They foster tourism, boost local economies, and strengthen cultural identity. However, these celebrations often generate environmental concerns, including air and water pollution, excessive waste, and resource depletion. This chapter examines the dual impact of festivals—celebrating their cultural and economic importance while addressing their ecological footprint. It explores sustainable solutions such as eco-friendly materials, waste management initiatives, policy guidelines, and public awareness campaigns. By integrating tradition with sustainability, the chapter emphasizes how communities can preserve cultural heritage while ensuring environmental protection, offering a roadmap for more responsible celebrations in the future.

DOI: 10.4018/979-8-2600-1960-3.ch010

INTRODUCTION: FESTIVALS AND ENVIRONMENTAL POLLUTION

India is one of the few countries with such rich cultural diversity. The country holds varied religions, customs, values, and dialects. It is a country with a rich, complex cultural tapestry. This diversity is one of the county's key strengths, reflecting a positive view of its religious landscape (Religion in India: Tolerance and Segregation, Pew Research, 2021). There are many festivals celebrated grandly across India. Among the most celebrated festivals are Holi, Durga Puja, Diwali, Ganesh Chaturthi, Christmas, and Ramadan. These festivals help preserve culture and various art forms. These are also the connecting lines between the present generation and ancient traditions. The festivals also exhibit considerable diversity in clothing, food, decorations, and festival-specific art forms. Every festival celebrated creates a distinct experience, atmosphere, and joyful mood, which in turn leads to exciting city programs. According to Chen, King, and Suntikul (2019), these festivals are times when locals and visitors unite, fostering festival vibrancy and enhancing a location's allure. Festival celebrations are also linked to numerous other benefits, as they help sustain local cultural expressions, foster social contact, reaffirm identities, and generate social capital (Marques, Borba, & Michael, 2020). We observe that numerous festival-related events are held across the country (Richards & Palmer, 2010), including the Kumbh Mela in Haridwar, Durga Puja in Kolkata, Navratri in Gujarat, the Pushkar Camel Fair in Rajasthan, and many others. According to Antchak et al. (2019), these events are significant elements of the event portfolio, as they provide economic, social, and cultural benefits.

In India, cities such as Mysore, Kolkata, Varanasi, Thrissur, Jaipur, and Madurai are among those that have branded themselves as "festival cities," celebrating specific festivals with grandeur and attracting large crowds during those periods. Cities that have branded themselves as "festive cities" (Gold & Gold, 2020) attract a special number of tourists during these periods. Table 1 provides a more comprehensive list of cities and festivals celebrated in India.

Table 1. Festival cities

FESTIVAL NAME	CITY/ STATE	VISITOR ESTIMATE	TOURISM IMPACT	SOURCES
Maha Kumbh Mela	Prayagraj, Uttar Pradesh	Approx. 400 million over 6 weeks (2025)	162% surge in flight bookings; significant hotel demand	Moneycontrol (2024); Curly Tales (2024); Firstpost. (2024)
Ratha Yatra	Puri, Odisha	Over 1.4 million	Significant tourism boost; 24 festivals annually in Puri	India Today (2022), Holidify (n.d.); Asian Confluence. (2022)
Thrissur Pooram	Thrissur, Kerala	1.8 to 2 million	Featured in the international tourism calendar	Kerala SDMA (2018), Travel and Tour World (2024)
Srivari Brahmotsavam	Tirupati, Andhra Pradesh	More than 1.5 million	High pilgrimage inflow; 9-day ritual attracts lakhs	Times of India (2023a), South Tourism (n.d.)
Durga Puja	Kolkata, West Bengal	400,000 flyers during festival week (2023)	10%+ increase in travel bookings and hospitality revenues	Times of India (2023b), Das (2023)

In 2023, India recorded 2,509 million domestic tourist visits, with religious tourism being a major driver (Press Information Bureau [PIB], 2024). The religious tourism market in India was valued at USD 10.8 billion in 2024 and is projected to reach USD 28.9 billion by 2030, with Hinduism contributing the largest share (Grand View Research, 2024). The Maha Kumbh Mela 2025 is estimated to have generated an economic impact of USD 30–35 billion (Reuters, 2025).

Branding a city as a "festive city", conducting special events during that period, attracting a large number of tourists, and local units, all these without a doubt provide a business opportunity and also attract a large amount of inflow of income. However, closer analysis reveals that these festival celebrations have a significant negative impact arising from failures in their handling and management. Most festivals in India are celebrated with large numbers of firecrackers, especially during the New Year, Diwali, Holi, and Ganesh Chaturti; firecrackers or sparklers are typically used during these festivals. These festival celebrations largely contribute to air pollution, and the use of firecrackers contains harmful metal-non-metal mixtures (Ambade, 2018), metalloids (Kulshrestha et al., 2004), and various chemical compounds (Gupta et al., 2020). The chemicals and pollutants emitted by these fireworks are highly harmful and toxic, and they increase atmospheric concentrations. Unfortunately, these toxic chemicals remain in the air for prolonged periods and travel long distances, affecting other parts of the city. Thus, even areas where no celebration had occurred are exposed to this polluted air (Klima et al., 2020; Liu et al., 2019; Camilleri & Vella, 2010).

Many researchers, including Sarkar et al. (2010), Singh et al. (2019), Beig et al. (2013), and Rastogi et al. (2018), have clearly demonstrated that the use of crackers as an integral part of celebrations can cause serious health issues. Various health problems, such as stress, cardiovascular disease, respiratory disease, and asthma, are increasing, particularly during the festive seasons. The data clearly show an increase in hospital cases during this period. However, the issues extend beyond health and also affect the environment, thereby harming all living creatures and humans. In a country known for grand, extensive festival celebrations often accompanied by fireworks, it is important to critically evaluate the extent to which these events affect air quality. One such measure is a detailed chemical characterization of pollutants released into the atmosphere. When cities conduct these analyses, they reveal specific facts, such as the types of chemicals present, their sources, and the concentration levels of airborne contaminants associated with the event. All of these have contributed to haze and fog pollution, which can significantly affect agricultural production. Because this haze and fog pollution hinders sunlight from reaching Earth, which is essential for all living things, this also affects the overall climatic conditions, which are not limited to a single city or state but extend globally; hence, this calls for an in-depth assessment (Tan et al., 2011). There are multiple ways to mitigate these kinds of problems. One such way is that, when festivals are celebrated, festival organizers can be provided with reports, such as ecological footprint analyses. These reports help the organizers and policymakers take appropriate measures, as they clearly indicate how festivals pollute and harm the environment. So, cautious decisions and measures can be taken to ensure that festival celebrations have minimal environmental impacts (Collins & Cooper, 2016). It is essential to continually raise awareness of the negative impacts, using all relevant reports and facts, and to highlight the large-scale consequences of these festivals.

In addition to sharing these reports with the relevant authorities and raising awareness, another approach to address this situation is to celebrate festivals more organically, ethically, and sustainably. These types of celebrations have gained considerable importance in recent years (Antchak, Ziakas, and Getz, 2019). More attention needs to be given to this, and efforts should be made to reduce or eliminate the use of firecrackers, as they substantially degrade air quality (Wang et al., 2019; Greven et al., 2019). Zhang et al., 2020). According to a study by Gupta et al. (2015), it was found that almost 70-80% of solid waste in India contributes to organic waste. This is due to a rapid shift in dietary habits and improper waste disposal after various celebrations at cultural, social, and regional levels.

Research on festivals and events has traditionally focused more on their economic effects than on their societal and cultural impacts. (Getz, 2010) Moreover, Gibson and Wong (2011) have highlighted the paucity of scholarly work on the environmental impacts of events and festivals, despite the rhetoric surrounding the need for sustainable development and the fight against climate change. However, in recent years, scholarly research has expanded to consider sociocultural and environmental effects (Mair & Whitford, 2013).

Researchers increasingly use the triple bottom line (TBL) framework to evaluate the economic, social, and environmental dimensions of festivals and events, thereby providing a more comprehensive and balanced assessment (Andersson & Lundberg, 2013). As a measure of the worldwide environmental impact, the Ecological Footprint (EF) has drawn much attention from scholars, practitioners, and policymakers (Collins & Flynn, 2015)

TYPES OF POLLUTION CAUSED BY FESTIVALS AND THEIR IMPACT ON THE ENVIRONMENT

Air Pollution and Its Impact

Any celebration harms the environment through air, water, and soil pollution, which persist despite efforts to implement specific measures. When we analyse the factors contributing to air pollution, we find that these include vehicle emissions, human activities, construction activities, and festival-related celebrations. This is also supported by a study by Ganguly et al. (2019), which identified significant sources of air pollution as industrial activities, power plant emissions, biomass burning, and vehicle emissions. It was also noted that two primary air pollutants are particulate matter and NOx (Mahtta, 2015). While looking at some of the significant sources that pollute the air, festivals are one such source, and one such festival is Diwali, and its contribution is significant (Ganguly et al., 2019; Business Standard, 2024; Ghosh et al., 2024). According to a detailed study published by AirVoice (2023), based on data from 14 states from 2019 to 2024, the researchers found alarming evidence of the negative impact of festivals such as Diwali on air pollution. AirVoice (2023) reports that air pollution levels across major Indian cities spike dramatically during Diwali, primarily due to the extensive use of firecrackers and increased vehicular emissions during the celebrations. The report highlights that PM2.5 and PM10 levels exceed the permissible limits, with Delhi, Lucknow, and Patna exhibiting the most hazardous air quality indices. The combination of firecracker smoke, low wind speed, and seasonal stubble burning intensifies pollution, creating severe respiratory health risks. Children and the elderly are especially vulnerable during this period. The report recommends urgent regulatory action and increased public awareness to reduce the use of fireworks and promote environmentally friendly alternatives.

Air pollution has a severe impact on the environment by altering natural processes and damaging ecosystems. Emissions of sulfur dioxide and nitrogen oxides lead to acid rain, which corrodes soil quality, weakens forest growth, and acidifies water bodies, endangering aquatic life (Scottish Environment Protection Agency, n.d.). Fine particulate matter and black carbon accumulate on glaciers and ice caps, particularly in the Himalayas, thereby accelerating snowmelt and disrupting river systems, such as the Ganges (Britannica, n.d.; The Hindu, 2020). Ground-level ozone, another air pollutant, impairs photosynthesis in crops and vegetation, reducing agricultural productivity and affecting plant biodiversity (WHO, n.d.). These pollutants also contribute to atmospheric warming, altered rainfall patterns, and increased frequency of droughts and floods, thereby destabilizing ecosystems across regions (NEXT IAS, n.d.).

Water Pollution and Its Impact

Water pollution is another issue of growing concern in our country. In addition to industrial-caused water pollution, festivals also contribute substantially to this, particularly those that include idol immersion or mass bathing as part of religious rituals. During Ganesh Chaturthi, thousands of plaster-of-Paris idols, coated with synthetic paint, are immersed in water bodies across states such as Maharashtra and Telangana. These paints often contain toxic chemicals such as mercury, cadmium, lead, and arsenic, which dissolve into rivers and lakes, contaminating aquatic ecosystems (PPS Thane, 2023). Festivals such as Durga Puja, which is celebrated grandly in Kolkata, are associated with increased pollution of the Ganges River during this period, as idols are immersed in the river, which contains many harmful chemicals. Alongside idol immersion, which is part of the religious process, they also throw plastic, clothing, and flowers (Zaman et al., 2018).

The Kumbh Mela, one of the largest human gatherings on Earth, entails mass bathing and ritual practices that discharge untreated sewage, soap residues, and substantial amounts of waste directly into the Ganga and Yamuna rivers (Civilsdaily, 2024; Nexteel, 2024). Festivals such as Bonalu, celebrated in Telangana, also contribute to water and soil degradation through the unregulated disposal of turmeric, vermilion, and other religious materials into local lakes (Ugandhar, 2024).

The environmental consequences of such practices are severe. The discharge of heavy metals and non-biodegradable materials during festivals reduces the water's oxygen-carrying capacity, thereby endangering fish and aquatic biodiversity (Zaman et al., 2018). This also causes bioaccumulation in aquatic species and may enter the human food chain. Eutrophication, caused by organic waste such as flowers and food offerings, promotes algal blooms, further degrading water quality and disrupting the natural aquatic balance (Ugandhar, 2024). These events not only disrupt local ecosystems but also strain municipal water treatment infrastructure. In the case of the Ganges, a river already overburdened by industrial discharge and urban waste, festival pollution accelerates ecological degradation and poses long-term risks to public health and biodiversity (Civilsdaily, 2024; Springer, 2024). While festivals are integral to India's socio-religious identity, these findings highlight the pressing need for eco-conscious celebration models that preserve cultural heritage without compromising environmental sustainability.

Soil Pollution and Its Impact

Specific festivals attract large-scale populations, and those with large gatherings and associated religious rituals are a significant source of soil pollution. Large-scale festivals are prevalent in India. For example, during festivals such as Holi and Diwali, large numbers of people generate non-biodegradable waste, including synthetic packaging, plastic utensils, and chemical-laden dyes. These are not disposed of properly, and they accumulate on the ground, contaminating the topsoil (Kulshrestha, 2018). Another festival is the Kumbh Mela. During this time, places such as Uttar Pradesh, Uttarakhand, Maharashtra, and Madhya Pradesh were where it was celebrated grandly, with extensive use of synthetic materials, and all of these were disposed of openly on the ground in and around the sites where these religious fairs take place. Along with these other wastes, like food items, flowers, and ritual remains, thereby degrading soil health (Singh, 2024). Additionally, fireworks during Diwali release heavy metals such as lead, cadmium, and barium, which accumulate in the soil and cause long-term toxicity (Khade & Sawant, 2023).

This soil contamination has a detrimental impact on plant growth, disrupts the microbial balance of soil ecosystems, and leaches harmful chemicals into nearby water sources through surface runoff, particularly during the monsoon season (Saini, 2024). Polluted soil reduces agricultural productivity and disrupts ecological balance by affecting insects and other organisms that depend on healthy soil systems. Moreover, prolonged exposure to heavy metals through the food chain poses significant health risks to both humans and animals, underscoring the need for eco-conscious celebrations and policy-level interventions to mitigate festival-induced environmental degradation (Kulshrestha, 2018; Saini, 2024).

The festival's organizers have focused on waste management, procurement, and venue selection to address the event's direct environmental impacts. To date, the following have been accomplished: 35% less resources (including printed materials); 20% less diesel; 15% less electricity; 100% renewable energy on-site; 75% of on-site waste is recycled; 4.8 tonnes of food waste from on-site catering outlets is composted; and a solar water heating system is being tested (Hay Festival Wales. (2013). *Hay on Earth Festival Report.*)

ECO-FRIENDLY ALTERNATIVES TO REDUCE POLLUTION AND PROTECT PLANTS

Indian festivals maintain their cultural value, but their celebrations now harm the environment more each year. The discharge of firecrackers on Diwali results in elevated PM2.5 and PM10 concentrations that exceed safety standards, thereby worsening health outcomes and degrading air quality (Ghei & Sane, 2018). The practice of immersing Plaster of Paris idols in Durga Puja and Ganesh Chaturthi celebrations causes water pollution and toxic effects on aquatic life due to the content of Plaster of Paris and chemical-based paints, as confirmed by the Central Pollution Control Board (2020). The practices create ecosystem disturbances, damaging biodiversity and degrading soil and water quality. The adoption of clay idols, along with natural dyes and decorations featuring embedded seeds, helps reduce pollution. Sustainable practices can be successfully implemented in any community, as exemplified by Pune through their eco-Ganesh movement (Kumar & Tripathi, 2021). Community-based celebrations are promoted to reduce idol use and waste generation. Decorations made from biodegradable materials, such as banana leaves and flowers, and the minimal use of plastic help reduce solid waste. Immersing idols in artificial tanks or designated water containers, rather than in rivers or lakes, also mitigates water contamination (De Koning, 2023). Cultural traditions require environmental ethics to protect native wildlife and flora while fostering ecological resilience. Establishing eco-friendly cultural events requires joint efforts by politicians, teachers, and civil society members to achieve conservation and cultural celebration goals.

Despite their cultural importance, firework displays remain significant contributors to air pollution because they generate hazardous levels of particulate matter, toxic gases, and heavy metals into the atmosphere. Pollutants emitted from these sources degrade air quality and pose significant health risks to individuals with respiratory conditions. Green crackers emerged as sustainable substitutes after scientists developed this environmentally friendly alternative with a reduction potential of up to 30%. This type of cracker technology produces lower noise levels and trace particles while satisfying environmental safety standards and maintaining traditional celebratory practices. Technical innovations, such as laser light shows and synchronized drone displays, provide environmentally friendly visual displays that replace traditional, polluting pyrotechnics. The use of these methods reduces the environmental impact of celebration activities.

During festive seasons, the initiation of plantation drives is an effective means of mitigating carbon emissions and improving air quality. These greening initiatives offer two benefits: they both boost urban biodiversity and help clean the atmosphere in the future. When incorporated into public awareness programs and policy strategies, these practices can transform conventional celebrations into environmentally conscious events. The adoption of green alternatives and ecological restoration programs during festivals represents an essential approach to achieving environmental sustainability and public health benefits.

The immersion of synthetic and chemically painted idols during religious festivals significantly contributes to water pollution, affecting aquatic ecosystems and public health (CPCB, 2020). A sustainable alternative involves promoting eco-friendly idols made of natural clay and plant-based, biodegradable colors, which dissolve harmlessly in water without releasing toxins (MoEFCC, 2019). Several cities, including Mumbai and Hyderabad, have pioneered the use of artificial immersion tanks and mobile immersion ponds to mitigate direct discharge into natural water bodies (The Hindu, 2020). These controlled environments enable safe immersion while facilitating the recovery and reuse of materials. For instance, the Pune Municipal Corporation introduced mobile immersion tanks in 2022, resulting in a 30% reduction in idol immersion in natural water bodies (PMC Report, 2022). Additionally, awareness campaigns by NGOs, such as Jal Shakti Abhiyan, have been instrumental in promoting community adoption of sustainable practices. Integrating such models at scale can significantly reduce pollutant loads in urban water systems, aligning with the objectives of the National Water Mission. Institutionalizing these methods through local governance and community participation is imperative for ensuring ecological and cultural sustainability.

Using eco-friendly decorations for festivals offers a sustainable alternative to traditional synthetic ornaments. Natural materials, such as flowers, banana leaves, mango leaves, and coconut fronds, help reduce the waste output of non-biodegradable substances (Kumar et al., 2020). Organic decorations such as thermocol artifacts and plastic buntings break down naturally into beneficial soil content when customers return them to the compost (Gupta & Yadav, 2019). The South Indian cultural traditions of Onam and Ugadi emphasize pookkalam and toran floral displays, showing that environmentally conscious customs derive from cultural practices. Such events become more sustainable through composting organic waste, as this enhances the waste value chain while reducing landfilling and promoting a circular bioeconomy (Patel et al., 2021). Pune and Indore are examples of municipalities that have launched community composting activities during festive periods, which have proven effective in distributing waste management across urban areas (Swachh Bharat Mission, 2022). The implementation of bio-based decorative elements serves dual purposes by aligning with national environmental policies and restoring traditional practices that protect ecological balance. By running educational campaigns, organizations can teach people appropriate, sustainable celebration habits.

CASE STUDIES OF ECO-FRIENDLY FESTIVALS

This section will provide real-world examples of festivals that have successfully adopted sustainable practices to reduce pollution.

Case Study 1: Ganesh Chaturthi in Pune

Ganesh Chaturthi is one of the Hindu festivals celebrated nationwide. It marked the birth of the deity Ganesha through the installation of Ganesha idols, followed by their immersion in rivers. This celebration has raised environmental concerns due to the use of toxic materials in the manufacture of idols.

Although many steps have been taken to reduce the environmental impacts of this practice, the Pune Municipal Corporation's (PMC) efforts have yielded results. The PMC initiated a campaign in 2020, titled "Punaravartan," to promote sustainable environmental practices and reduce water pollution.

Punaravartan is a collaborative effort involving more than 200 volunteers from 20 NGOs in Pune city. Punaravartan encouraged the use of environmentally friendly materials, particularly natural clay, for idol-making. In 2023, they set up 619 artificial tanks to immerse the idols, 60 more than the previous year, thereby significantly reducing water pollution.

Similarly, the initiative encouraged the reuse of clay for future idol making. The corporation further initiated various awareness programmes as a part of this initiative. In 2022, PMC collected more than 23,000 kg of clay from various sites and returned it to artisans at no cost for reuse. In 2024, the volunteers collected over 50 tonnes of clay for recycling. Additionally, a portion of the collected clay sludges is made available to educational institutions for experimentation. These practices help to replace the POP (Plaster of Paris) idols and make Ganesha Chaturthi a green festival.

Case Study 2: Diwali Green Crackers Campaign

Diwali is the Hindu festival of light, which is celebrated with great enthusiasm in India and abroad. The people celebrate it by lighting oil lamps (diyas) and offering prayers. Bursting firecrackers is a popular activity that adds loud bangs and colours to the celebrations. However, celebrating Diwali has an environmental impact on the social sphere, contributing to pollution. Cities in India have been showing higher levels of air pollution during Diwali. In 2024, the Air Quality Index (AQI) in some areas of Delhi exceeded 800 (the highest in Karkarduma, at 806) during Diwali. Celebrations have a dark side; it is not possible to ban such celebrations. However, authorities have initiated many restrictions.

The Government of India has banned the import of Chinese-made firecrackers, citing concerns about their environmental impact. Similarly, in 2018, the Supreme Court banned the use of traditional firecrackers. The movement was a significant intervention by the apex court of India, which mandated the use of green crackers. Similarly, in Karnataka, the Public Health Action Team (PACT), in collaboration with the State Pollution Control Board, has conducted many awareness campaigns to educate the public on the benefits of green crackers. Another movement is the Delhi government's "10-Point Action Plan, which imposes various restrictions during Diwali, such as a ban on construction and the use of water sprinklers and anti-smog guns to reduce dust.

Figure 1. Diwali Air Quality in Delhi 2024

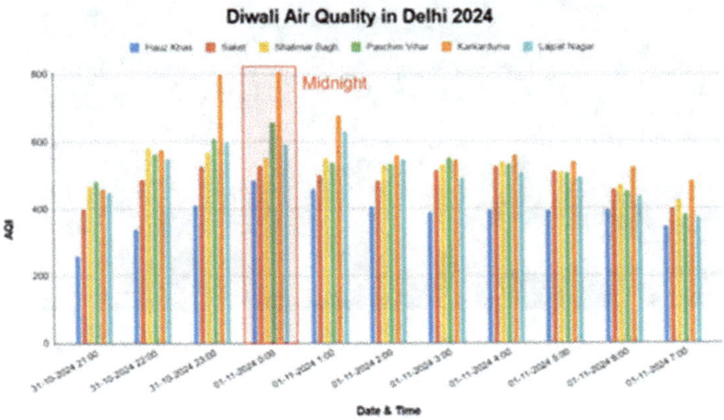

Source: *Diwali Air Quality Report 2024 by AQI | AQI India*

Many schools have Eco Clubs to raise students' awareness about a greener Diwali. In Kolkata, the Police Department has initiated a program to visit schools and ensure students' sensitization. Similarly, various resident associations and NGOs conduct door-to-door campaigns and eco-friendly Diwali fairs. The "#GlowGreenThisDiwali" is a social media campaign launched by Ecofy in 2024 to promote eco-friendly celebrations during Diwali and encourage sustainability.

Figure 2. #GlowGreenThisDiwali

Case Study 3: Green Christmas in Europe

Like any celebration, Christmas also generates significant environmental hazards. In 2020, the European Environment Agency (EEA) reported that during the holiday season, plastic waste increased by 30% compared to earlier periods [Plastics, the circular economy and Europe's environment | European Environment Agency's home page]. To address these, numerous measures have been introduced in Europe. In 2022, the EU introduced the Circular Economy Action Plan to reduce plastic waste and promote recycling and reuse (The Circular Economy for Plastics – A European Analysis 2024 • Plastics Europe).

Recently, European countries have introduced eco-friendly Christmas celebrations by promoting the use of recyclable decorations, LED lights, and locally grown real Christmas trees. During the 2024 Christmas season, the European Commission encouraged people to use reusable decorations made of wood and fabric. The EEA report for 2024 states that there has been a significant rise in eco-friendly practices, resulting in greener and more sustainable celebrations, thanks to initiatives such as public awareness campaigns.

Government policies across Europe have played a substantial role in encouraging environmentally friendly Christmas celebrations. The European Union's "Fit for 55" initiative has successfully reduced the environmental impact of celebrations. The initiative emphasized energy-efficient lighting, waste reduction, and the promotion of local products. Similarly, European cities have shown their model of positive environmental impact. The Christmas market in Cologne, Germany, has switched to 100% renewable electricity, resulting in an estimated 152 tons of CO2 savings.

Case Study 4: Kumbh Mela – Managing Environmental Challenges at Scale

The Kumbh Mela in India, one of the world's largest religious gatherings, poses significant environmental challenges due to the influx of millions of devotees. The primary threat is water pollution resulting from ritual baths at the Triveni Sangam, along with the substantial volume of waste generated in the local ecosystem.

To address these threats, various eco-friendly practices have been initiated. During the 2019 Prayagraj Kumbh Mela, Biodegradable waste management systems were established to control the immense amount of waste generated. To limit water pollution, authorities have established Portable bio-toilets and sewage treatment plants. The government has invested significant effort in conducting awareness campaigns to encourage pilgrims to avoid plastic and littering. Over 15,000 sanitation workers and 2,000 "Ganga Sewa Doots" actively volunteered to maintain cleanliness. Furthermore, the fairgrounds are designated as plastic-free zones and have banned single-use plastics. To prepare for the 2025 Maha Kumbh Mela, the Prayagraj Municipal Corporation has, over the past two years, initiated the development of a 56,000-square-meter dense forest using the Miyawaki Technique, which has improved air quality and enhanced the area's green appearance.

The impact of these initiatives was significant. There was a notable reduction in plastic waste compared to previous events. The efforts were acclaimed as a model for protecting eco-friendliness. The Maha Kumbh Mela 2025 saw over 66 crore pilgrims taking a holy bath at the Triveni Sangam. The government provided the necessary security, sanitation, and medical facilities through a collaborative effort with the community. The Mela serves as a model for large-scale events, demonstrating the success of planning and execution aligned with environmental preservation.

POLICIES AND STRATEGIES TO ENCOURAGE NATURE-FRIENDLY FESTIVALS

Although many initiatives have been implemented to celebrate events and festivals in an environmentally friendly manner, we still lack policies and strategies to mitigate their environmental impacts. It is not only the role of government authorities to take the initiative; other stakeholders, such as NGOs, associations, educational institutions, local vendors, business units, residents, volunteers, and the media, must also be involved. Organizing nature-friendly festivals is crucial for promoting environmental awareness, enhancing community engagement, and promoting sustainable practices. Some of the policies and strategies to enhance eco-friendly festivals are as follows:

Technology-supported festivals

The authorities can implement regulations to ensure the use of solar energy to power festival stages and lighting. Similarly, digital entry tickets for places of worship can be introduced to reduce paper use. Introducing recycling apps can enhance devotees' awareness of waste segregation and recycling processes in places of worship. Similarly, more environmentally friendly transportation options can be introduced, along with electric vehicle charging stations. Furthermore, the technology can be utilized to mitigate water pollution. In large public gatherings, we can utilize sensors, drones, and AI technologies to monitor, detect, and address contaminants, thereby ensuring cleaner water sources.

Subsidies and government support for promoting eco-friendly festivals:

Economic incentives play a significant role in encouraging eco-friendly practices. Authorities can provide subsidies for the procurement of environmentally friendly materials and substances for the production of decorative items, food containers, cutlery, and other products. The government can reduce the GST for bags and clothing made from sustainable materials. Governments can further introduce financial assistance schemes to promote green businesses for festival celebrations.

Mandatory guidelines: Authorities can implement festival-specific regulations and protocols, such as mandating the use of environmentally friendly materials for celebrations. Similarly, encourage the use of natural colours by charging a high GST on artificially made colours and other substances. Another possibility is banning single-use plastics. The government and municipal authorities can impose heavy fines for the use of plastics and other toxic materials in the production of various festival essentials.

The government can impose strict guidelines, such as designating pilgrimage sites as zero-waste areas and imposing fines on worship places, including temples, mosques, and churches, for generating excess waste during festive seasons. The government can also implement laws requiring mandatory waste management plans in every place of worship to segregate waste at the source and recycle it for reuse. All these efforts are essential to minimize environmental damage.

Environmentally friendly gifting: Individuals can choose to exchange only meaningful, sustainable, and eco-friendly gifts during festivals. Authorities should promote items like organic candles, natural products, and terracotta plants as gifts to reduce waste. Additionally, there is a need to change the pattern of gifting. It is recommended to opt for edible gifts, such as homemade sweets, organic chocolates, and gift baskets featuring locally made sustainable products. Similarly, individuals can present gifts such as bamboo-made items, wooden utensils, reusable cloth bags, and seed packets. Furthermore, to promote sustainable living, gifting indoor plants and saplings is an effective practice. Similarly, the selection of wrapping materials for gifts should be aligned with environmental consciousness. The use of biodegradable paper and reusable clothes will show a gesture and help make the festivals joyous while also caring for the environment.

Collaborative efforts for nature restoration:

Governments can collaborate with various environmental organisations and communities to undertake nature restoration projects. Participants can engage in activities such as planting trees, habitat restoration, and beach cleaning.

In addition, other strategies can be implemented, such as eco-friendly, sustainable accommodations for festival attendees. Eco-friendly interactive installations made from recycled materials, as well as virtual reality shows and workshops that explain sustainable living practices, can help increase environmental consciousness among festivalgoers. Finally, educational institutions can honour students, teachers, and families by presenting green awards to those who demonstrate eco-friendly festival practices.

ROLE OF PLANTS IN SUSTAINABLE CELEBRATIONS

Plant-based dyes for Holi.

Natural colours derived from flowers, leaves, and fruits are integral to the vibrant Holi festivities. They offer Ayurvedic benefits in addition to aesthetic appeal, as many contain turmeric and neem (Basumatary et al., 2021). However, the recent commercialisation of festivals and celebrations has also affected the observance of Holi. This is evident in the high-level use of synthetic colors. Such alternatives are considered harmful to the skin and pose environmental threats. Basumatary et al. (2021) highlighted the life-threatening nature of Holi powders, which are made from various chemicals, such as mercury, lead, copper sulfate, and chromium. Similarly, other hazardous materials, such as asbestos and silica, used in these powders cause ecological imbalances, including soil and water pollution (Indian Express, 2024).

The adverse effects of an unsustainable approach to festivities opened the public's eyes to a moderate extent and increased interest in plant-based Holi colours. In India, herbal powders made from natural ingredients are increasingly popular.

These natural items mainly include spinach for green, turmeric and yellow marigold for yellow, and orange marigold for orange shades. Such natural colors added to very eco-friendly powders made of tapioca and rice flour. Tapioca flour, in particular, appears ideal due to its low cost, higher return, and smooth texture (Basumatary et al., 2021). The makers of Holi powders are more dependent on natural fragrances made from orange peels, which enhance consumer pleasure while ensuring non-toxicity.

"From gardens to gulal", a movement supported by the Indian non-profit organisation Kalpavriksh. The initiative promotes the use of environmentally friendly dyes for Holi celebrations. Educational campaigns by biodiversity conservation groups and initiatives in India are increasing, reinforcing traditional methods of natural dye extraction (CockColours, 2024). Many families and communities are returning to backyard plants, such as hibiscus, rose, and mint, to form safe Holi powders at home. Such initiatives help create and expand livelihood opportunities. India opens its rich botanical diversity and traditional knowledge systems to entrepreneurs, especially women and other rural communities, who find new opportunities by creating Holi colours from natural products.

Recent studies indicate a shift among consumers toward more natural and sustainable products. Customers' awareness of product composition informs their choices. General studies have found that most informed participants chose plant-based Holi dyes over chemical alternatives because of perceived advantages, such as safety, fragrance, and texture. However, 60% still prefer the appearance and feel of eco-friendly powders, even though they are unaware of their exact composition (Basumatary et al., 2021).

It has been observed that the Plant-based Holi not only promotes safe and healthy celebrations but also ensures cultural authenticity and commitment. It helps to protect environmental sustainability, maintain indigenous knowledge, and foster conscious consumerism. With greater awareness of these natural alternatives for festivities, society can mould the Holi celebration into a more eco-friendly and sustainable one.

Flower decorations instead of synthetic ones

In India, natural flowers are generally used for various festivals. This is more consistent with the view that festivals are associated with temples, where non-native flowers are typically not used. However, plastic and synthetic flowers are available everywhere in the market during festival seasons. Using natural flowers not only adds color and fragrance to the festivals but also reduces the ecological hazards. During festivities such as Diwali, natural flowers invite more ambience than synthetic flowers and garlands. This also helps keep home environments safe and creates biodegradable waste from celebrations (Times of India, 2024).

Although it is cheaper to use synthetic decorations made from plastics and glitter, in the long run, such non-recyclable leftovers create uncontrollable pollution. Thus, promoting sustainable sources not only helps protect the environment but also supports local farmers and florists. Use of flowers holds cultural meanings that symbolise prosperity, purity, and joy (Smaller Footprint, 2024). Also, fosters a deeper connection with nature. Similarly, the large quantities of flowers used in festivities can be used to produce organic dyes, incense, and handmade paper, offering value-added products. As awareness increases, households shift their preferences for celebrations toward more choices that honour tradition and sustainability.

Tree planting initiatives as part of festival celebrations

India has a wide culture of planting trees during festivals and celebrations. This culture has made the festivities more sustainable and connected to nature. People nowadays consider festivals an opportunity to give back to nature by planting new trees and even gifting plants, which symbolise renewal and gratitude. However, the celebrations have deviated from this culture. However, we can see many organisations, such as Grow Billion Trees and Nelda, promoting tree-planting drives during Indian festivals such as Diwali, Holi, Ganesh Chaturthi, and Independence Day, thereby inspiring people and communities to celebrate with purpose (Grow Billion Trees, 2024).

Such awareness initiatives impart a sense of responsibility on individuals and the community towards the environment. Recently, due to frequent environmental awareness campaigns by government and non-government organisations, people are increasingly embracing green festivals by planting saplings of native trees. Such initiatives yield many benefits, such as improved air quality, enhanced biodiversity, and a cleaner environment for future generations (Nelda, 2024).

Such green drives are implemented by schoolchildren, residents, and corporate volunteers. It has been observed that schoolchildren have felt such drives as educational experiences, while others have shown volunteerism toward creating a sustainable future. Since they are aligned with cultural and religious events, an emotional bond between traditions and environmental stewardship is naturally fostered. Gradually, such purposeful gestures can transform the way festivals are celebrated, shaping moments of joy into enduring, sustainable impact.

Recommendations and Roadmap for Change

Towards Sustainable Celebrations: A Roadmap for Community-Level Implementation of Eco-Friendly Festival Practices

A transition in the celebration of festivals is necessary to protect nature for future generations. It is a necessity of the time and considered a cultural and social need. Such green practices require strategic planning and community-wide volunteerism to engage with deeply rooted cultural festivities. This can be achieved through a multifaceted plan of action, including raising awareness among municipal leadership, public behaviour education, and measures to sustain traditional practices, to initiate green festivities and celebrations.

1. Municipal Governance and Sustainable Waste Management Local government bodies, such as municipal authorities, play a significant role in promoting sustainable festival management. They can implement crucial waste-management policies during festival seasons. Local bodies can provide designated waste collection points alongside door-to-door domestic waste collection services. Similarly, establishing decomposting units can streamline and expedite waste management. Such initiatives are essential, especially during festivals such as Ganesh Chaturthi, Durga Puja, and Diwali. As noted in many studies, the Pune Municipal Corporation (PMC) is a good example of its waste management strategies. The PMC has adopted a system in which it deploys self-help groups (SHGs) to collect and decompose waste during Ganesh Visarjan (Joshi & Khanvilkar, 2018). Similarly, the Bhubaneswar Municipal Corporation (BMC) regulates idol immersion by restricting it to artificial ponds, thereby reducing pollution of natural water bodies (BMC Circular No. 3823, 2020). It is the local government bodies that need to issue special orders requiring the use of only sustainable products to make idols during Ganesha Chaturthi. Additionally, strict monitoring of plastic use and waste-disposal protocols should be implemented. The incorporation of Extended Producer Responsibility frameworks into festival planning, as outlined by MoEF&CC (2021), is expected to reduce packaging and decorative waste.

2. Behavioral Change through Education and Awareness Campaigns To ensure that festivals are celebrated in accordance with the green protocol, the public must adopt environmentally friendly practices. To this end, authorities can rely more on awareness campaigns, such as community-wide educational programs. This is evident in Kolkata, where the West Bengal Pollution Control Board (WBPCB) and civil society organisations have initiated a campaign "Green Durga Puja". The campaign offered incentives to community clubs to initiate sustainable festivities. Additionally, they promoted awareness through various channels, including visual media, street dramas, and engaging local community leaders to educate people about environmental issues (Mukherjee & Chakraborty, 2022). The study has found that awareness levels are more effectively achieved when involvement is facilitated through cultural elements. Well-established learning institutions and houses of worship continue to serve as vital educational platforms for teaching ecological principles to young populations. The ripple effect can be achieved by conducting sustainable idol-making workshops, combining them with sustainability quiz contests, and sharing traditional environmental narratives. Community engagement is an essential practice for fostering ecological awareness among youth. Similarly, the wider distribution of eco-friendly practices requires the use of digital and social media platforms. Supporting this perspective, TERI (2020) reported that various environmental non-governmental organisations (ENGOs) utilise #GreenGanesh and #SustainableDiwali campaigns to demonstrate how technological platforms facilitate behavioural transformation and engagement within local communities

3. Integrating Sustainability into Festival Traditions The eco-friendly festivities are a blend of traditional customs and sensitivity to nature. Such a transformation of cultural expression comprises a deliberate adaptation of traditional rituals to respect environmental values. One example of this is the use of Shadu clay (a natural clay) to make idols instead of Plaster of Paris (PoP), which demonstrates the authenticity of Indian culture. It is evident in practice that the government of Maharashtra has initiated training programs, market integration, and financial incentives (Government of Maharashtra, 2022).

Initiatives like collaborations with traditional scholars and ecologists are inevitable during community festivals. They can inspire and lead the community through programmes such as 'Green Ritual Demonstrations' by promoting the use of natural pigments derived from turmeric and beetroot. Such practices can redefine ritual offerings, illumination arrangements, and aesthetic embellishments through sustainable methods.

Under the 'Seed Ganesh' initiative, religious devotees can include plant seeds within the revered statues, which subsequently transform the sacred water immersion ritual into a tree-planting celebration. By incorporating these symbols into religious rituals, organisations can maintain their spiritual integrity while promoting environmental awareness.

Singh (2021) has identified that leaders and role models in religious and community settings can serve as cultural protectors, promoting green practices during festivals. Such a green partnership will enhance support for cultural heritage and lead to improved sustainability outcomes.

The fundamental need of the time regarding the conduct of festivals in India involves navigating the boundaries between governmental regulations, policies, educational practices, and historical customs. This will help to make the festivities eco-friendly. Similarly, proper cultural scheduling is essential for municipal waste management systems. Furthermore, authorities need to ensure that educational programmes are aligned with the development of an environmentally mindful population.

A mindful celebration of festivals enables a shift from ecological harm to multifunctional ecological and community engagement initiatives. The proposed roadmap outlines a practical approach to cultural adaptation for achieving this transformational shift.

CONCLUSION

Making festivals nature-friendly is no longer just a choice; it is an urgent necessity. As this study reveals, while festivals offer immense cultural, economic, and spiritual value, they also exert tremendous pressure on our natural environment. Air, water, and soil pollution from firecrackers, synthetic decorations, chemical-laden idols, and plastic waste not only damage ecosystems but also pose significant threats to public health and biodiversity. Transitioning to sustainable celebration models is essential to safeguard these fragile natural systems and ensure that the joy of festivals does not come at the cost of ecological harm.

Adopting eco-conscious practices, such as using plant-based dyes during Holi, biodegradable floral decorations during Diwali, and clay idols during Ganesh Chaturthi, demonstrates that it is entirely possible to uphold traditional rituals while reducing pollution. Similarly, initiatives such as tree-planting drives during festivals not only restore green cover but also instill environmental awareness among participants, particularly the younger generation. These practices support the broader goals of biodiversity conservation, carbon reduction, and responsible consumption, all while enhancing the spiritual and communal essence of the celebrations.

Moreover, sustainable celebrations align cultural heritage with ecological responsibility. By using natural materials, supporting local artisans, minimizing waste, and promoting community-led green initiatives, festivals can serve as platforms for environmental education and civic engagement. Successful case studies from Pune, Hyderabad, and European cities demonstrate that, with appropriate policies, awareness, and community collaboration, festivals can serve as models for ecological restoration and cultural preservation.

Ultimately, the future of cultural festivities must be grounded in sustainability. Preserving time-honored traditions should go hand in hand with protecting the very air, water, and soil that sustain life. When communities embrace eco-friendly practices, they not only celebrate their heritage but also nurture a healthier planet for generations to come. In this way, nature-friendly festivals can serve as symbols of harmony among people, their beliefs, and the environment they cherish. It is through such conscious celebration that we can truly honor the spirit of our festivals while ensuring the well-being of Earth's ecosystems and future inhabitants.

REFERENCES

AirVoice. (2023, November 6). AirVoice releases comprehensive report on Diwali air pollution in India. https://airvoice.global/tpost/8bhy30c0c1-airvoice-releases-comprehensive-report-o

Ambade, B. (2018). The air pollution during Diwali festival by the burning of fireworks in Jamshedpur city, India. *Urban Climate*, *26*, 149–160. DOI: 10.1016/j.uclim.2018.08.009

Andersson, T. D., & Lundberg, E. (2013). Commensurability and sustainability: Triple impact assessments of a tourism event. *Tourism Management*, *37*, 99–109. DOI: 10.1016/j.tourman.2012.12.015

Antchak, V., Ziakas, V., & Getz, D. (2019). *Event portfolio management: Theory and methods for event management and tourism.* https://derby.openrepository.com/handle/10545/624187

Asian Confluence. (2022). *India's soft power: The case of Ratha Yatra.* https://www.asianconfluence.org/publication-details-full/india-s-soft-power-the-case-of-ratha-yatra

Basumatary, P., Bhuyan, M. K., Sarmah, P., & Saikia, D. (2021). Eco-friendly Holi using natural dye and low-cost filler base. *International Journal of Current Microbiology and Applied Sciences*, *10*(2), 1067–1075. DOI: 10.20546/ijcmas.2021.1002.126

Beig, G., Chate, D. M., Ghude, S. D., Mahajan, A. S., Srinivas, R., Ali, K., Sahu, S. K., Parkhi, N., Surendran, D. E., Trimbake, H. K., & Tiwari, S. (2013). Evaluating population exposure to environmental pollutants during Deepavali fireworks displays using air quality measurements of the SAFAR network. *Chemosphere*, *92*(1), 116–124. DOI: 10.1016/j.chemosphere.2013.02.043 PMID: 23561573

Bhubaneswar Municipal Corporation. (2020). *Guidelines for eco-friendly festival practices* (BMC Circular No. 3823).

Business Standard. (2024, October 25). Diwali fireworks and air pollution: How to protect your respiratory health. https://www.business-standard.com/india-news/diwali-fireworks-and-air-pollution-how-to-protect-your-respiratory-health-124102500784_1.html

Camilleri, R., & Vella, A. J. (2010). Effect of fireworks on ambient air quality in Malta. *Atmospheric Environment*, *44*(35), 4521–4527. DOI: 10.1016/j.atmosenv.2010.07.057

Central Pollution Control Board. (2020). *Guidelines for idol immersion.*

Chen, Z., King, B., & Suntikul, W. (2019). Festivalscapes and the visitor experience: An application of the stimulus–organism–response approach. *International Journal of Tourism Research, 21*(6), 758–771. DOI: 10.1002/jtr.2302

Civilsdaily. (2024). *The Maha Kumbh and river pollution: Are we learning from past mistakes?* https://www.civilsdaily.com/the-maha-kumbh-and-river-pollution-are-we-learning-from-past-mistakes/

CockColours. (2024). *How to prepare natural colours for the festivities of Holi.* https://cockcolours.com/article/How-to-Prepare-Natural-Colours-for-the-Festivities-of-Holi

Collins, A., & Cooper, C. (2016). Measuring and managing the environmental impact of festivals: The contribution of the ecological footprint. *Journal of Sustainable Tourism, 25*(1), 148–162. DOI: 10.1080/09669582.2016.1189922

Collins, A. J., & Flynn, A. (2015). *The ecological footprint: New developments in policy and practice.* http://ci.nii.ac.jp/ncid/BB18739155

Curly Tales. (2024, April 12). Maha Kumbh 2025 to boost spiritual tourism in Prayagraj; 162% YOY increase in flight, 187% surge in train bookings reported. https://curlytales.com/maha-kumbh-2025-boost-spiritual-tourism-in-prayagraj-162-y-o-y-increase-in-flight-187-surge-in-train-bookings-reported/

Das, D. (2023, October 23). From culture to economy: How Kolkata's Durga Puja fuels economic growth. *LinkedIn.* https://www.linkedin.com/pulse/from-culture-economy-how-kolkatas-durga-puja-fuels-economic-dona-das

De Koning, D. (2023). Green Ganesha Chaturthi: The ritualising and materialising of a green Hindu identity and the emerging of an alternative representation of Ganesha. *Religions, 14*(1), 22. DOI: 10.3390/rel14010022

Firstpost. (2024, April 13). Maha Kumbh 2025: Soaring prices, hotels in Prayagraj & flight bookings explained. https://www.firstpost.com/explainers/maha-kumbh-2025-soaring-prices-hotels-in-prayagraj-flight-train-bookings-13852657.html

Ganguly, N. D., Tzanis, C. G., Philippopoulos, K., & Deligiorgi, D. (2019). Analysis of a severe air pollution episode in India during Diwali festival – A nationwide approach. *Atmósfera (México), 32*(3), 225–236. DOI: 10.20937/ATM.2019.32.03.05

Getz, D. (2010). The nature and scope of festival studies. *International Journal of Event Management Research, 5*(1), 1–47. https://ijemr.org/wp-content/uploads/2014/10/Getz.pdf

Ghei, D., & Sane, R. (2018). *Air pollution and fireworks: Evidence from Diwali.* Indian Statistical Institute.

Ghosh, B., Barman, H. C., & Padhy, P. K. (2024). Trends of air pollution variations during pre-Diwali, Diwali and post-Diwali periods and health risk assessment using HAQI in India. *Discover Environment, 2*(1), 81. DOI: 10.1007/s44274-024-00082-w

Gibson, C., & Wong, C. (2011). Chapter 6. Greening rural festivals: Ecology, sustainability and human–nature relations. In *Multilingual Matters eBooks* (pp. 92–106). https://doi.org/DOI: 10.21832/9781845411688-009

Gold, J. R., & Gold, M. M. (2020). *Festival cities.* Routledge., DOI: 10.4324/9781003130802

Government of Maharashtra. (2022). *Sustainable Ganesh Idol Making Scheme – Department of Environment and Climate Change.*

Grand View Research. (2024). *India religious tourism market size, share & trends analysis report, 2024–2030.* https://www.grandviewresearch.com/industry-analysis/india-religious-tourism-market-report

Greven, F. E., Vonk, J. M., Fischer, P., Duijm, F., Vink, N. M., & Brunekreef, B. (2019). Air pollution during New Year's fireworks and daily mortality in the Netherlands. *Scientific Reports, 9*(1), 5735. Advance online publication. DOI: 10.1038/s41598-019-42080-6 PMID: 30952902

Grow Billion Trees. (2024). *Tree plantation on festivals.* https://growbilliontrees.com/collections/tree-plantation-on-festivals

Gupta, A., Bherwani, H., Gautam, S., Anjum, S., Musugu, K., Kumar, N., Anshul, A., & Kumar, R. (2020). Air pollution aggravating COVID-19 lethality? Exploration in Asian cities using statistical models. *Environment, Development and Sustainability, 23*(4), 6408–6417. DOI: 10.1007/s10668-020-00878-9 PMID: 32837279

Gupta, N., & Yadav, K. K. (2019). Sustainable solid waste management in India. *Environmental Science and Pollution Research International, 26*(3), 2164–2175.

Gupta, N., Yadav, K. K., & Kumar, V. (2015). A review on current status of municipal solid waste management in India. *Journal of Environmental Sciences (China), 37*, 206–217. DOI: 10.1016/j.jes.2015.01.034 PMID: 26574106

Holidify. (n.d.). *Jagannath Rath Yatra, Puri overview.* https://www.holidify.com/places/puri/jagannath-rath-yatra-sightseeing-3321.html

India Today. (2022, June 29). After two Covid-curtailed years, Puri Rath Yatra preparations in full swing. https://www.indiatoday.in/india-today-insight/story/after-two-covid-curtailed-years-puri-rath-yatra-preparations-in-full-swing-1968178-2022-06-29

Joshi, M., & Khanvilkar, V. (2018). Urban waste management during religious festivals: A study of Pune city. *Journal of Environmental Planning and Management, 61*(11), 1934–1950.

Khade, S., & Sawant, M. (2023). Assessment of festival-related pollution impacts on urban soil. *CMR Journal, 11*, 43–54. https://cmrj.in/011-May-23-Khade-Savant.pdf

Klima, V., Chadyšienė, R., Ivanec-Goranina, R., Jasaitis, D., & Vasiliauskienė, V. (2020). Assessment of air pollution with polychlorinated dibenzodioxins (PCDDs) and polychlorinated dibenzofuranes (PCDFs) in Lithuania. *Atmosphere (Basel), 11*(7), 759. DOI: 10.3390/atmos11070759

Kulshrestha, U. C. (2018). The social rituals and environmental degradation: How to tackle? *Current World Environment, 13*(3), 285–287. DOI: 10.12944/CWE.13.3.01

Kulshrestha, U. C., Nageswara Rao, T., Azhaguvel, S., & Kulshrestha, M. J. (2004). Emissions and accumulation of metals in the atmosphere due to crackers and sparklers during Diwali festival in India. *Atmospheric Environment, 38*(27), 4421–4425. DOI: 10.1016/j.atmosenv.2004.05.044

Kumar, R.. (2020). Eco-friendly festive practices in India. *Journal of Environmental Management, 260*, 110143.

Kumar, R., & Tripathi, S. (2021). Sustainable celebrations: A case study of eco-friendly Ganeshotsav in Pune. *Journal of Environmental Management.*

Liu, J., Chen, Y., Chao, S., Cao, H., & Zhang, A. (2019). Levels and health risks of PM2.5-bound toxic metals from firework/firecracker burning during festival periods in response to management strategies. *Ecotoxicology and Environmental Safety, 171*, 406–413. DOI: 10.1016/j.ecoenv.2018.12.104 PMID: 30634092

Mahtta, R. (2015). *Green growth and air pollution in India* (18 pp.). The Energy and Resources Institute (TERI).

Mair, J., & Whitford, M. (2013). An exploration of events research: Event topics, themes and emerging trends. *International Journal of Event and Festival Management, 4*(1), 6–30. DOI: 10.1108/17582951311307485

Marques, L., Borba, C., & Michael, J. (2020). Grasping the social dimensions of event experiences: Introducing the Event Social Interaction Scale (ESIS). *Event Management, 25*(1), 9–26. DOI: 10.3727/152599520X15894679115448

Ministry of Environment, Forest and Climate Change. (MoEFCC) (2021). *Guidelines for implementation of extended producer responsibility (EPR).*

Moneycontrol. (2024, April 11). Oh my God! Maha Kumbh drives 162% jump in flight bookings; hotels brimming with tourists. https://www.moneycontrol.com/travel/oh-my-god-maha-kumbh-drives-162-jump-in-flight-bookings-hotels-brimming-with-tourists-article-12908601.html

Mukherjee, R., & Chakraborty, A. (2022). Cultural campaigns and environmental awareness: A study of Green Durga Puja. *International Journal of Cultural Studies, 25*(3), 255–270.

Nelda. (2024). *Festivals plantation.* https://nelda.org.in/festivals-plantation/

NEXT IAS. (n.d.). *Environmental pollution.* https://www.nextias.com/blog/environmental-pollution

Nexteel. (2024). *Balancing tradition and environmental sustainability: Water challenges at the Kumbh Mela.* https://nexteel.in/balancing-tradition-and-environmental-sustainability-water-challenges-at-the-kumbh-mela/

Patel, M.. (2021). Circular economy in organic waste management. *Resources, Conservation and Recycling, 168,* 105291. DOI: 10.1016/j.resconrec.2021.105291

Press Information Bureau. (2024, February 10). *India records 2,509 million domestic tourist visits in 2023.* https://pib.gov.in/PressReleasePage.aspx?PRID=2024XXXX

Rastogi, N., Singh, A., & Satish, R. (2018). Characteristics of submicron particles coming from a big firecracker burning event: Implications to atmospheric pollution. *Atmospheric Pollution Research, 10*(2), 629–634. DOI: 10.1016/j.apr.2018.11.002

Reuters. (2025, January 15). *Kumbh Mela 2025 projected to boost Indian economy by up to $35 billion.* https://www.reuters.com/article/kumbh-mela-2025-economic-impact-idUSKBNXXXX

Richards, G., & Palmer, R. (2010). *Eventful cities: Cultural management and urban revitalization.* http://ci.nii.ac.jp/ncid/BB04121970

Saini, H. (2024). Impact of solid waste dumping on soil quality and its potential risk on human health and environment. *Environmental Monitoring and Assessment, 196*(2). https://www.researchgate.net/publication/382532973 PMID: 38183498

Sarkar, S., Khillare, P. S., Jyethi, D. S., Hasan, A., & Parween, M. (2010). Chemical speciation of respirable suspended particulate matter during a major firework festival in India. *Journal of Hazardous Materials, 184*(1–3), 321–330. DOI: 10.1016/j.jhazmat.2010.08.039 PMID: 20817345

Scottish Environment Protection Agency. (n.d.). *Air pollution impacts on the environment.* https://www.environment.gov.scot/our-environment/air/air-pollution-impacts-on-the-environment/

Sharpley, R., & Stone, P. (2014). *Contemporary tourist experience: Concepts and consequences.* Routledge. DOI: 10.4324/9780203139110

Singh, A., Pant, P., & Pope, F. D. (2019). Air quality during and after festivals: Aerosol concentrations, composition and health effects. *Atmospheric Research, 227*, 220–232. DOI: 10.1016/j.atmosres.2019.05.012

Singh, N. (2021). *Religion, ecology, and sustainability: Engaging faith leaders for environmental transformation.* Sage Publications.

Singh, R. (2024). Ecosystem dynamics of Prayagraj. In *Environmental impact of ritual practices in India* (pp. 79–88). ResearchGate., https://www.researchgate.net/publication/392315941

Smaller Footprint. (2024). *Festivals and flowers.* https://smallerfootprint.co/blogs/good-steps/festivals-and-flowers

South Tourism. (n.d.). *Brahmotsavam Festival – Tirupati.* https://www.southtourism.in/andrapradesh/festivals/brahmotsavam.php

Springer. (2024). *Festival-related pollution in the River Ganges.* https://link.springer.com/content/pdf/10.1007/s13201-024-02227-x.pdf

Swachh Bharat Mission. (2022). *Best practices compendium.* Ministry of Housing and Urban Affairs, Government of India.

Tan, J., Guo, S., Ma, Y., Duan, J., Cheng, Y., He, K., & Yang, F. (2011). Characteristics of particulate PAHs during a typical haze episode in Guangzhou, China. *Atmospheric Research, 102*(1–2), 91–98. DOI: 10.1016/j.atmosres.2011.06.012

TERI. (2020). *Sustainable celebrations: A guidebook for eco-friendly festivals.* The Energy and Resources Institute.

Thane, P. P. S. (2023). *Effects of Ganapati festival.* https://www.ppsthane.com/blog/effects-of-ganapati-festival

The Hindu. (2020, February 15). *Why India needs to clean its air.* https://www.thehindu.com/sci-tech/energy-and-environment/why-india-needs-to-clean-its-air/article69400021.ece

The Indian Express. (2024). *From gardens to gulal: Celebrating Holi with sustainable herbal colours.* https://indianexpress.com/article/lifestyle/art-and-culture/from-gardens-to-gulal-celebrating-holi-with-sustainable-herbal-colours-9885601/

Times of India. (2023a, October 2). *Six lakh devotees attend grand Brahmotsavams at Tirumala.* https://timesofindia.indiatimes.com/city/vijayawada/six-lakh-devotees-attend-grand-brahmotsavams-at-tirumala/articleshow/114195251.cms

Times of India. (2023b, October 25). *Kolkata airport sees record footfall during Durga Puja 2023.* https://timesofindia.indiatimes.com/city/kolkata/kolkata-airport-sees-record-footfall-during-durga-puja-2023/articleshow/114229931.cms

Times of India. (2024). *Unique Diwali festive decor ideas with sustainable materials.* https://timesofindia.indiatimes.com/life-style/home-garden/unique-diwali-festive-decor-ideas-with-sustainable-materials/articleshow/114726870.cms

Travel and Tour World. (2024, April 17). *India tourism rolls out free pavilion access for international visitors at Kerala's iconic Thrissur Pooram.* https://www.travelandtourworld.com/news/article/india-tourism-rolls-out-free-pavilion-access-for-international-visitors-at-keralas-iconic-thrissur-pooram/

Ugandhar, T. (2024). Eco-conscious celebrations: Bonalu festival's impact on environmental sustainability and public health in Telangana, India. *ResearchGate.* https://www.researchgate.net/publication/383467369

Wang, S., Yu, R., Shen, H., Wang, S., Hu, Q., Cui, J., Yan, Y., Huang, H., & Hu, G. (2019). Chemical characteristics, sources, and formation mechanisms of PM2.5 before and during the Spring Festival in a coastal city in Southeast China. *Environmental Pollution, 251*, 442–452. DOI: 10.1016/j.envpol.2019.04.050 PMID: 31103004

World Health Organization. (n.d.). *Air pollution.* https://www.who.int/india/health-topics/air-pollution

Zaman, S.. (2018). Water quality of the River Ganga in and around the city of Kolkata during and after Goddess Durga immersion. *Parana Journal of Science and Education, 4*(9), 1–7.

Zhang, X., Shen, H., Li, T., & Zhang, L. (2020). The effects of fireworks discharge on atmospheric PM2.5 concentration in the Chinese Lunar New Year. *International Journal of Environmental Research and Public Health, 17*(24), 9333. DOI: 10.3390/ijerph17249333 PMID: 33322228

Chapter 11
Environmental Applications of Disinfection Technologies Against COVID-19

Muhammad Rizwan Ullah
The University of Faisalabad, Pakistan

Mamoona Javed
The University of Faisalabad, Pakistan

Muhammad Jamil Khan
http://orcid.org/0009-0007-2388-1970
The University of Faisalabad, Pakistan

ABSTRACT

World are facing the manifestation of a respiratory disease that belongs to a "coronaviral family". The first case of this virus appeared in China. The corona virus (COVID-19) entered in Pakistan in February 26, 2020. Currently, the number of confirmed cases in Pakistan reached 1,581,936 by April 2024. Pakistan's national institute of health offered an instantaneous authorize action and management plan to tackle this situation. This action plan is comprised of different disinfection technologies and techniques with the efforts of different agencies. The study aims to discuss the technologies, techniques and strategies to disinfect COVID-19 from different environmental settings including hospitals, healthcare centers and other public places. Therefore, the study presents disinfection technologies and techniques to disinfect COVID-19. Moreover, a SWOT analysis and the application of various

DOI: 10.4018/979-8-2600-1960-3.ch011

disinfection technologies and strategies are discussed. This chapter is crucial to the development of a strategy for controlling, preventing and disinfecting the similar pandemic in future.

1. INTRODUCTION

The emergence of infectious diseases with pandemic potential represents one of the most significant global public health challenges of the twenty-first century. Rapid urbanization, increased human mobility, climate change, and globalization have collectively intensified the frequency and scale of disease outbreaks, placing unprecedented strain on healthcare systems and governance structures worldwide. Among these challenges, the COVID-19 pandemic stands out as a defining global crisis, exposing systemic vulnerabilities in public health preparedness, environmental management, and infection control practices. While pharmaceutical interventions such as vaccines and antiviral treatments have played a crucial role in mitigating disease severity, non-pharmaceutical interventions—particularly environmental disinfection—have proven indispensable in controlling transmission, especially during the early and peak phases of the pandemic.

Environmental disinfection has long been recognized as a foundational pillar of infection prevention and control. Historically, surface sanitation, waste management, and hygiene practices have been instrumental in limiting the spread of infectious diseases such as influenza, severe acute respiratory syndrome (SARS), and Middle East respiratory syndrome (MERS). The COVID-19 pandemic reaffirmed the importance of these measures by demonstrating that pathogens can persist outside the human body and remain infectious on surfaces, in wastewater, and in shared environments for extended periods. As a result, environmental settings—once considered secondary in disease transmission—have emerged as critical battlegrounds in the fight against SARS-CoV-2.

The role of environmental transmission became particularly evident during the early stages of the COVID-19 outbreak, when scientific understanding of the virus was still evolving and vaccines were unavailable. Healthcare facilities, public transport systems, workplaces, educational institutions, and recreational spaces quickly transformed into high-risk environments due to frequent human interaction and shared surface contact. In these settings, environmental disinfection was not merely a preventive measure but a necessity to protect healthcare workers, essential service providers, and the general population. The widespread adoption of disinfectants, sanitization protocols, and waste management strategies underscored the centrality of environmental control in pandemic response.

Despite its importance, environmental disinfection presents complex scientific, operational, and policy challenges. The effectiveness of disinfection technologies depends on multiple interacting factors, including the physicochemical properties of disinfectants, surface characteristics, environmental conditions, and human behavior. Furthermore, improper or excessive use of disinfectants can result in unintended consequences, such as chemical exposure risks, antimicrobial resistance, and environmental pollution. These challenges highlight the need for balanced, evidence-based approaches that consider both efficacy and sustainability.

The COVID-19 pandemic also revealed stark disparities between countries in their capacity to implement effective disinfection strategies. High-income countries were often able to deploy advanced technologies such as automated disinfection systems, ultraviolet irradiation, and sophisticated waste treatment facilities. In contrast, low- and middle-income countries faced significant constraints related to limited resources, inadequate infrastructure, and weak regulatory enforcement. These disparities underscore the importance of context-specific research that accounts for socio-economic realities and institutional capacities when evaluating disinfection interventions.

Pakistan provides a particularly relevant case for examining environmental disinfection strategies during the COVID-19 pandemic. As a densely populated developing country with diverse urban and rural settings, Pakistan faced considerable challenges in managing the rapid spread of the virus. Healthcare facilities experienced increased pressure due to surges in patient numbers, while public spaces remained vulnerable due to overcrowding and limited sanitation infrastructure. In response, national and local authorities implemented a range of disinfection measures, including chemical spraying, hand hygiene campaigns, waste management protocols, and the installation of disinfection units in public areas. These interventions were deployed under significant time and resource constraints, making their evaluation both timely and necessary.

While a growing body of international literature has examined individual disinfection technologies, much of this research remains fragmented and context-neutral. Many studies focus on laboratory-based efficacy testing or isolated interventions without considering real-world implementation challenges. Moreover, few studies offer integrated assessments that simultaneously address healthcare waste management, environmental surface disinfection, and public-space interventions within a single analytical framework. This fragmentation limits the practical utility of existing research for policymakers and practitioners, particularly in developing countries where strategic prioritization is essential.

Against this backdrop, there is a pressing need for comprehensive and context-sensitive evaluations of environmental disinfection technologies. Such evaluations must move beyond technical descriptions to address strategic questions related to feasibility, risks, opportunities, and long-term sustainability. The present chapter responds to this need by synthesizing existing scientific evidence and policy guidance while situating disinfection technologies within the specific institutional and environmental context of Pakistan. By adopting a strategic analytical lens, the chapter aims to bridge the gap between theoretical knowledge and practical application.

In doing so, this study positions environmental disinfection not as an isolated emergency response but as a core component of pandemic preparedness and resilience. The lessons learned from COVID-19 have implications far beyond the current crisis, offering valuable insights for managing future outbreaks of infectious diseases. Understanding which disinfection technologies work, under what conditions, and at what cost is essential for building robust public health systems capable of responding effectively to emerging threats.

From the last few months, all the countries in the world are facing the manifestation of a respiratory disease that belongs to a "coronaviral family". The first outbreak of this virus appeared in Wuhan; an emergent business hub of China. During the first six days of this virus, more than 70,000 people of Wuhan were diseased. The international committee on Taxonomy virus (ICTV) called this virus as "SARS-CoV-2", and the infection caused by this virus is termed as "Corona virus disease 2019 (COVID-19)". Corona virus is microscopic in size i.e., *"65-125 nm in diameter"*, and comprised of a single marooned "*ribonucleic acid*" (RNA) as nuclide substantial mass, ranges from *"26 to 32 kbs"* in length.

According to CDC (Centers for Disease Control and Prevention), its gestation period falls between 2-14 days. The symptoms of this virus include fatigue, diarrhea, loss of sense of smell, cough and breathlessness and fever. The virus is highly infectious and transmits through the droplets released during coughing and sneezing or while contacting with infected person. With the passage of time this virus started spreading across the world and exaggerated 222 nations. This virus diseased around 158.3 million people around the globe and caused more than 3,295,982 deaths (as on 09 May, 2021) all over the world (WHO, 2020). Alike other nations, the virus also affected many people in Pakistan. This virus entered into Pakistan on 26 February, 2020 when first patient was infected from this virus in Karachi and speedily spread all over the country. The virus has affected about 854,240 people and about 18,797 deaths have been recorded in Pakistan till 09 May, 2021. The health centers (i.e., hospitals and pharmacies) are facing number of challenges while handling the corona patients. The expulsion of health centers, specifically those without the proper management would depict the public in *"danger of infection,* and increases the health risk. For instance, many types of restorative and dangerous waste i.e., gloves, diseased masks, used injections and other defensive equipment increase the risk of spread of this novel virus from person to person (United Nations Environment Program (UNEP), 2020). Therefore, it is of significant importance to reduce these health risks, particularly in the context of Pakistan, which is possible by using different disinfecting technologies and techniques. A number of academic researchers have conducted their studies on the disinfection techniques of corona virus and concluded that disinfection techniques are crucial in controlling the spread of COVID-19 (Alwan et al., 2023).

In addition to the disinfection of corona virus in health centers, it is also of significant importance to disinfect the virus from the environmental surfaces like furniture, medical equipment, toilets, floor, wall, sinks and electronics etc. This is, because when an infected person touches the wall, walks on floor, uses toilets, furniture etc., it increases the probability of spread of virus (Rai, 2020). Numerous researchers have worked on the relationship between environmental surfaces and spread of corona virus and concluded that environmental surfaces have positive contributions to the spread of corona virus (Qasim et al., 2020; Sharma, 2020). The recent past studies concluded that environmental surfaces are more probable to be polluted with the corona virus in the health centers (where definite medical techniques are performed on patients), and public places (where people gathered to have some fun i.e., shopping malls, restaurants and food points etc.) (Ye et al., 2020). In public places, people come from different corners of the country, and the authority do not have an idea that which person is infectious. Therefore, environmental surfaces of both places (health care centers and public places) should be disinfected to reduce the spread of this COVID-19 (Sarada et al., 2020). Present study, therefore, is an attempt to discuss the disinfectant technologies and techniques to managing the wastes of health care centers and to clean the environmental surfaces of health care centers and public places in context of Pakistan. To the best of author's knowledge, until now, no study on this issue has been conducted in Pakistan. Present study will not only enhance the existing debate of COVID-19 in the context of Pakistan, but also help the policy makers to disinfect the virus from different environmental surfaces of health centers and other public places.

Despite the growing body of international research on COVID-19 disinfection technologies, most existing studies focus on individual techniques—such as chemical disinfectants, hand hygiene, or waste management—without offering an integrated and context-specific evaluation of their effectiveness across different environmental settings. Moreover, the majority of published research is concentrated in high-income or technologically advanced countries, with limited attention paid to developing countries where institutional capacity, resource availability, and governance structures differ significantly. In the context of Pakistan, empirical and review-based studies addressing environmental disinfection strategies remain fragmented and largely descriptive, offering little guidance for strategic decision-making. This chapter addresses this gap by providing a comprehensive and context-sensitive synthesis of disinfection technologies used in Pakistan, evaluated through a SWOT framework.

By integrating healthcare waste management, environmental surface disinfection, and public-space interventions, the study moves beyond isolated assessments and offers a strategic perspective that is currently absent from the literature.

1.1. Scientific Contribution

This chapter makes three key scientific contributions to the existing literature. First, it provides one of the few comprehensive, Pakistan-focused syntheses of COVID-19 disinfection technologies, bridging a notable geographical and contextual gap in pandemic-related environmental health research. Second, by applying a SWOT analytical framework, the study offers a strategic evaluation of disinfection methods, highlighting not only their effectiveness but also operational challenges, risks, and opportunities—an approach rarely adopted in prior studies. Third, the chapter integrates healthcare waste management and environmental surface disinfection within a unified framework, thereby contributing to a more holistic understanding of infection control in both clinical and non-clinical settings. Collectively, these contributions enhance the theoretical and practical relevance of disinfection research for managing current and future pandemics.

1.2. COVID-19 in Pakistan and Research Questions

As stated earlier, the corona virus entered in Pakistan in February 26, 2020 when the first case was confirmed in Karachi. On the same day, another case was confirmed in Pakistan. Currently, the number of confirmed cases reached 1,581,936. Gradually, Pakistan faced rapid increase in corona virus cases. Figure 1 shows the total corona virus cases, total deaths, daily new cases and daily new deaths in Pakistan due to corona infection since February 26, 2020 to May 09, 2021.

Figure 1.

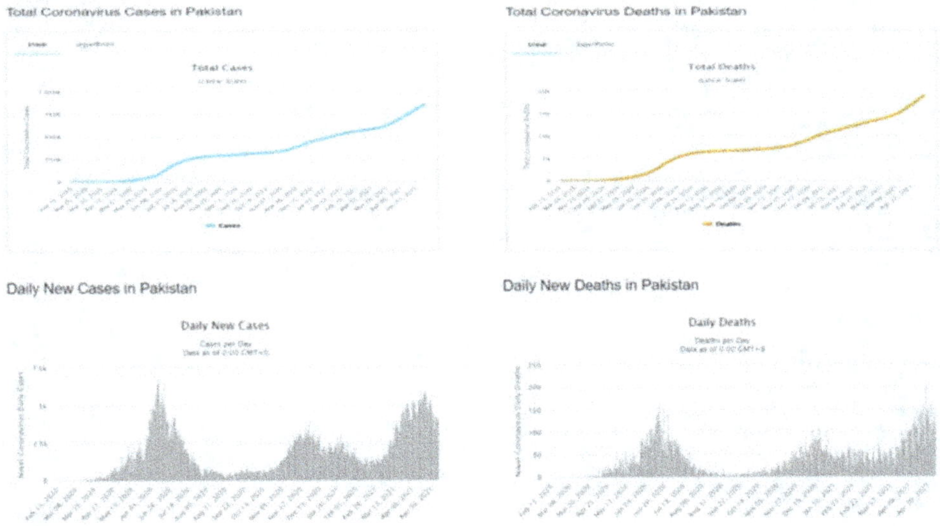

Bearing in mind this threatening situation of corona virus, Pakistan's national institute of health (NIH) offered an instantaneous authorize action and management plan "National Action Plan for Preparedness & Response to Corona Virus Disease (Covid-19) Pakistan" to tackle this situation (Ren et al., 2020). This action plan is comprised of different combined techniques with the dual efforts of national, international, regional and local agencies. These agencies indicate that there is a huge need to disinfect COVID-19 from different environmental settings to reduce the spread of virus. Now the question arises that what technologies/techniques and strategies are adopted by Pakistan to disinfect the virus? What are these technologies? How these strategies can be used and where can they be used? (see Figure 2).

Figure 2.

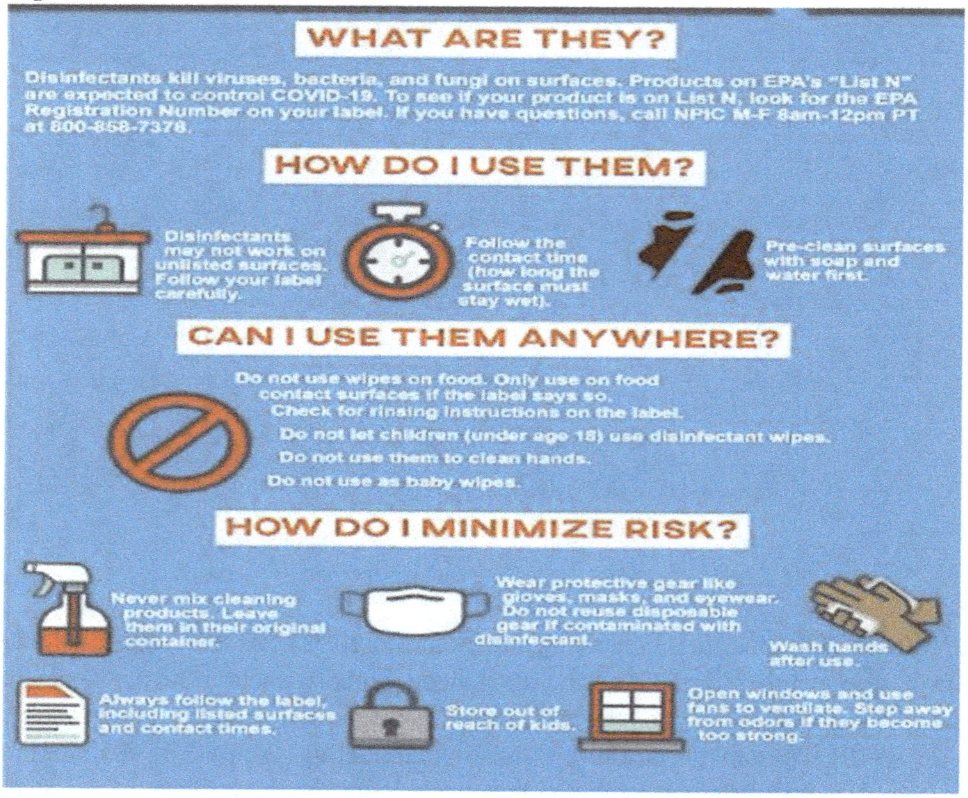

2. LITERATURE REVIEW

The literature on COVID-19 disinfection can be broadly categorized into three thematic streams: (i) environmental persistence and transmission of SARS-CoV-2, (ii) effectiveness of disinfection technologies and techniques, and (iii) institutional and policy-level implementation challenges. Studies within the first stream establish that SARS-CoV-2 can persist on various surfaces for extended periods, thereby necessitating systematic disinfection (Kramer et al., 2006; Ye et al., 2020). The second stream evaluates specific technologies such as chemical disinfectants, ultraviolet irradiation, and hand hygiene practices, often under controlled conditions (Wang et al., 2020; Rutala et al., 2020). The third stream highlights operational constraints, particularly in developing countries, including inadequate waste management systems and limited regulatory enforcement (Kulkarni & Anantharama, 2020).

However, these streams remain largely disconnected in the literature, underscoring the need for integrative analyses that link environmental risk, technological effectiveness, and contextual feasibility—an approach adopted in the present study. The concept of disinfection is initiated from the perception that a virus can persist on different surfaces (Kramer et al., 2006). For example, the virus can survive outer the mass body on non-living surfaces for flexible times and transferred consequently from person to person when the individuals touch these places. Therefore, it is essential to disinfect the virus from different environmental surfaces to stop the spread of COVID-19. Disinfection is a process of cleaning or purifying something (specifically from certain chemicals) in order to destroy or finish the bacteria or virus. The center of disease control and prevention (CDC) of USA has issued all-inclusive disinfecting techniques of corona virus from different public and private places to get rid of the virus (Gina et al., 2025). Similarly, there are following technologies, techniques and strategies that are being used in Pakistan to disinfect the corona virus from different environmental surfaces, health care centers and other public places.

2.1. Disinfection Technologies and Strategies for Managing the Waste of Health Care Centers

Dispose the waste material of health centers

The first and foremost strategy to disinfect the health centers from COVID-19 is to *"dispose the hospitals' or pharmacies' waste materials"*. The process of disposing the particular waste materials is shown in Figure 3. Classification of the wastes of health centers is the initial step to dispose the wastage of health care centers (Ren et al., 2020). It is the preeminent exercise that the corona waste should be disposed-off at its origin. This strategy is not only time effectual but also evades the probability of virus spread (Luo et al., 2020). The second step is to manage the corona waste into the plastic bags and put the wasted material into the bins. Beware, the wasted material should be sealed into *"double-layered plastic bags"* (Capoor & Parida, 2021).In study by (Sahoo et al., 2024) suggested that the waste material should be classified into different plastic bags (i.e., chemical wastes should be sealed in one bag and other infectious wastes such as used masks should be sealed in another bag, similarly, bio medical wastes, municipal waste etc. should be put in some other bags and so on). The third step is the collection of this waste by a municipal authority. For instance, a person from the waste management department of Pakistan should come and collect this waste material from the temporary storage places and dispose them by using *"high temperature treatment facility"*. This process disinfects the particular places from novel corona virus (Kulkarni and Anantharama, 2020).

Figure 3.

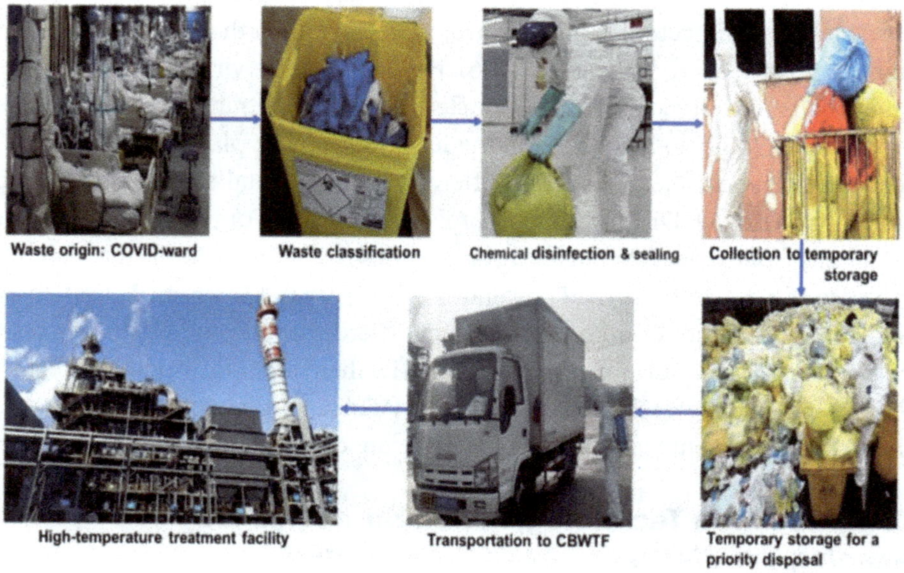

Disinfection of health center's waste water

The second strategy to disinfect the COVID-19 from the health centers is the *"disinfection of health center's waste water"*. This is necessary as waste water is one of the most prominent reasons of the spread of virus. Thus, it is crucial that waste water of health care centers should be managed (Wang et al., 2020) accordingly. Paleologos et al. (2020) conclude that liquid chlorine, sodium hypochlorite, chlorine dioxide, ultraviolet irradiation and ozone are commonly used equipment to disinfect the water waste management of health care centers. The process of water waste management in the health care centers is shown in Figure 4.

Figure 4.

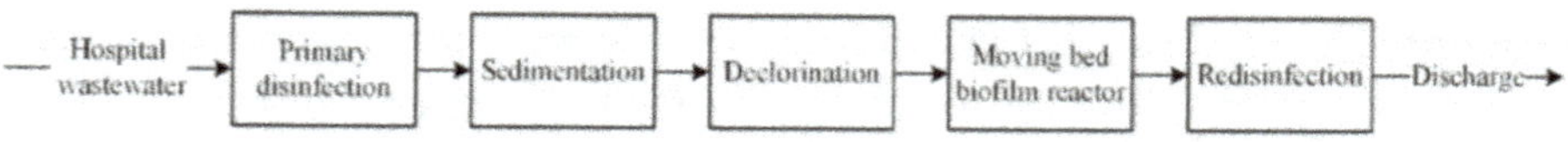

Sterilizing the medical instruments

Sterilizing the medical instruments (such as syringes and surgical equipment) is another important strategy to disinfect the virus from health care centers. There are following three steps to sterilize the medical equipment from the virus (Allison and McDaniel, 2020). *Cleaning:* First step is to clean the instrument. *Disinfecting:* The second step is to use liquid chemicals to kill the bacteria from instruments. *Sterilize:* The final step is to apply most suitable method to kill the disease-causing viruses. The instruments can be sterilized using four methods suggested by (Zhuang et al., 2023). *a. Steam sterilization:* In the steam sterilization, a suitable purifier is used to hygiene the upper surfaces of medical instruments (Rutala et al., 2020). *b. Dry heat sterilization:* Dry heat sterilization uses air of approximately *"340-degree F"* to kill infectious material from the medical instruments (Alavi et al., 2020). *c. Chemical sterilization:* In this method of sterilization, different chemicals (such as bleach, ozone, hydrogen peroxide and ethylene) are used to purify the medical instruments (McDonnell, 2020). These chemicals have supremacy to kill the expansive range of *"pathogens properties"* that could be dangerous for the people. In order to kill the virus, a definite amount of these chemicals is required. After the chemical sterilization the equipment should be washed with fresh water and leave it to get dry by itself. *d. Plasma gas sterilization:* This method of sterilization is applicable to the dental equipment that uses "peroxide-based gas plasma" to kill the virus from instrument (Starikovskiy and Usmanova, 2020).

2.2. Disinfection Technologies, Techniques and Strategies for Environmental Surfaces of Health Care Centers and Public Places

Understanding the environmental persistence and transmission mechanisms of SARS-CoV-2 is fundamental to the development of effective disinfection strategies. Unlike earlier coronaviruses, SARS-CoV-2 has demonstrated a remarkable ability to survive on a wide range of surfaces and environmental media, thereby increasing the risk of indirect transmission. Numerous studies confirm that viral particles can remain viable on non-living surfaces such as metal, plastic, glass, cardboard, and fabric for hours or even days, depending on environmental conditions (Kramer et al., 2006; Ye et al., 2020). This persistence has elevated environmental surfaces from passive elements to active vectors in the transmission chain of COVID-19.

Transmission of SARS-CoV-2 primarily occurs through respiratory droplets and aerosols; however, indirect contact transmission through contaminated surfaces (fomites) has been widely documented. When infected individuals cough, sneeze, talk, or even breathe, viral droplets settle on nearby surfaces, increasing the likelihood of transmission when these surfaces are touched by others. Studies conducted in hospital wards and quarantine facilities have revealed widespread contamination of frequently touched surfaces such as bed rails, door handles, medical equipment, toilets, and floors (Ye et al., 2020). These findings underscore the importance of routine and terminal disinfection, particularly in high-risk environments such as healthcare facilities.

Environmental conditions play a critical role in determining viral survival. Temperature, humidity, ultraviolet radiation, and surface material composition significantly influence the persistence of SARS-CoV-2. Research suggests that lower temperatures and low humidity levels prolong viral viability, whereas exposure to sunlight and higher temperatures accelerates viral inactivation (Han et al., 2020). In densely populated regions and tropical climates such as Pakistan, these environmental factors interact with overcrowding and limited ventilation, further complicating infection control efforts.

Healthcare environments represent a particularly vulnerable setting for environmental transmission. The concentration of infected individuals, combined with frequent medical procedures and waste generation, increases the probability of surface contamination. Wang et al. (2020) emphasize that improper handling of contaminated waste, wastewater, and reusable medical instruments can significantly amplify viral spread. This is especially concerning in resource-constrained healthcare systems where standardized infection control protocols may not be consistently enforced. Consequently, environmental disinfection emerges as a non-pharmaceutical intervention of critical importance in reducing transmission risks.

Beyond healthcare facilities, public spaces such as shopping malls, public transport systems, educational institutions, and recreational areas have also been identified as high-risk environments. In such locations, large numbers of individuals interact with shared surfaces, often without awareness of prior contamination. Studies conducted during the early phases of the pandemic indicate that surface contamination in public spaces played a contributory role in community transmission, particularly in settings with limited hand hygiene facilities (Sharma, 2020). These findings reinforce the necessity of implementing disinfection strategies beyond clinical environments.

Importantly, the persistence of SARS-CoV-2 on environmental surfaces has implications not only for immediate transmission control but also for long-term public health preparedness. The COVID-19 pandemic has demonstrated that environmental transmission pathways must be systematically addressed in pandemic response planning. Disinfection technologies, therefore, should not be viewed as temporary emergency measures but as integral components of sustainable infection prevention systems.

In the context of developing countries, including Pakistan, environmental transmission risks are exacerbated by infrastructural limitations, inadequate waste management systems, and high population density. Informal settlements, overcrowded hospitals, and insufficient sanitation facilities increase reliance on surface disinfection as a primary barrier against viral spread. However, the effectiveness of such measures depends heavily on proper implementation, public awareness, and institutional capacity. This highlights the need for context-specific evaluations of disinfection technologies, rather than adopting strategies developed for high-income settings without adaptation.

Overall, the literature clearly establishes that environmental persistence and surface-mediated transmission of SARS-CoV-2 are significant contributors to the spread of COVID-19. These findings provide the theoretical foundation for examining disinfection technologies and justify the emphasis placed on environmental cleaning, waste management, and surface sterilization in the present study. By synthesizing insights from environmental virology and infection control research, this chapter situates disinfection technologies as essential tools in mitigating both current and future pandemic threats.

Environmental surfaces of different health care centers (hospitals and pharmacies) and non-health care centers (schools, offices, parks, restaurants, gyms etc.) usually suffer from the high chances of COVID-19 infection. This is, because people normally touch the environmental surfaces, and if the person suffering from corona, touches the environmental surfaces, these surfaces become infectious too, it causes to increase the people being affected by COVID-19. Therefore, it is crucial that the environmental surfaces of different public places and health care centers should be disinfected.

Disinfection technologies developed to control SARS-CoV-2 can be broadly categorized into chemical, physical, and hybrid approaches. Each category operates through distinct mechanisms, offers varying levels of effectiveness, and presents unique advantages and limitations depending on the environmental context in which it is applied. The COVID-19 pandemic has accelerated the deployment of these technologies across healthcare facilities and public spaces, underscoring the importance of understanding their functional principles and operational constraints.

Chemical disinfection remains the most widely used approach due to its relative affordability, availability, and rapid virucidal action. Common chemical disinfectants include alcohol-based solutions, chlorine compounds, hydrogen peroxide, quaternary ammonium compounds, and iodine-based formulations. Alcohol-based disinfectants, particularly those containing ethanol or isopropanol at concentrations between 60% and 80%, have been shown to effectively inactivate SARS-CoV-2 by disrupting the viral lipid envelope (McDonnell, 2020). These disinfectants are especially effective for hand hygiene and small surface applications, making them indispensable in both healthcare and community settings.

Chlorine-based disinfectants, such as sodium hypochlorite and calcium hypochlorite, are extensively used for surface disinfection, wastewater treatment, and medical waste management. Their strong oxidizing properties enable rapid inactivation of viral particles, even in the presence of organic matter (Wang et al., 2020). However, improper concentration levels or prolonged exposure may pose risks to human health and the environment, including respiratory irritation and the formation of harmful by-products. Consequently, standardized guidelines and training are essential to ensure their safe and effective application.

Physical disinfection technologies rely on non-chemical mechanisms to inactivate pathogens. Among these, ultraviolet (UV) irradiation—particularly UV-C light—has gained prominence during the COVID-19 pandemic. UV-C radiation damages viral nucleic acids, thereby preventing replication and rendering the virus non-infectious. UV-based systems have been deployed in hospital rooms, laboratories, and public transportation systems to supplement conventional cleaning methods (Rutala et al., 2020). Despite their effectiveness, physical disinfection methods often require specialized equipment and controlled exposure conditions, limiting their widespread use in resource-constrained environments.

Thermal disinfection is another physical approach commonly applied to sterilize medical instruments and manage infectious waste. Steam sterilization (autoclaving) and dry heat treatment are effective in eliminating SARS-CoV-2 and other pathogens by denaturing viral proteins and nucleic acids. These methods are widely used in healthcare facilities due to their reliability and broad-spectrum efficacy (Alavi et al., 2020). However, their application is generally limited to heat-resistant materials, and they require significant energy inputs and infrastructural support.

Hybrid disinfection approaches combine chemical and physical mechanisms to enhance effectiveness and overcome individual limitations. For example, plasma gas sterilization integrates chemical agents such as hydrogen peroxide with low-temperature plasma to achieve rapid sterilization of heat-sensitive medical equipment. This method has demonstrated effectiveness against a wide range of pathogens, including enveloped viruses like SARS-CoV-2 (Starikovskiy & Usmanova, 2020). Hybrid approaches are particularly valuable in healthcare settings where precision and safety are paramount.

Another emerging hybrid approach involves automated disinfection systems, such as fogging, misting, and spray-based technologies. These systems disperse disinfectant solutions uniformly across large areas, reducing manual labor and improving coverage. During the COVID-19 pandemic, such technologies were widely used in public spaces, transportation hubs, and quarantine centers (Fatimah et al., 2020). While effective on exposed surfaces, their limitations include uneven penetration, potential chemical exposure, and reduced efficacy on porous materials.

The selection of an appropriate disinfection technology depends on several factors, including the type of surface, level of contamination, available resources, and environmental conditions. Studies emphasize that no single technology is universally optimal; instead, layered strategies that combine multiple approaches yield the most reliable results (Sarada et al., 2020). This integrated perspective aligns with infection control frameworks that prioritize redundancy and adaptability.

In developing countries, the adoption of advanced disinfection technologies is often constrained by financial and infrastructural limitations. Consequently, low-cost chemical disinfectants and manual cleaning remain dominant, particularly in rural and semi-urban areas. However, overreliance on chemical agents without proper training and regulation may lead to misuse, health risks, and environmental degradation (Gharpure et al., 2020). These challenges highlight the importance of context-sensitive evaluations that balance effectiveness with feasibility.

Overall, the literature demonstrates that chemical, physical, and hybrid disinfection technologies each play critical roles in controlling SARS-CoV-2 transmission. Their effectiveness is maximized when deployed as part of an integrated infection prevention strategy tailored to specific environmental and institutional contexts. This understanding provides a strong conceptual foundation for the SWOT-based evaluation presented in this chapter and supports the strategic assessment of disinfection practices implemented in Pakistan.

Routine environmental cleaning

The simplest and easy way to disinfect the environmental surfaces of different public and private places is through routine environmental cleaning as cleaning is the crucial part of disinfection (Gharpure et al., 2020). It helps in the eradication of the "pathogens" and significantly reduce their burden on polluted surfaces. Cleaning the different environmental places (such as walls, floors, doors, furniture) with detergent, water, or soap etc. reduce the freight of soil which in turn allow the disinfection to work. However, the length of time of cleaning causes the virus to persist on environmental surfaces will differ with different factors (Han et al., 2020; Hines et al., 2020). For instance, it depends on the amount of unhygienic body liquid (i.e., respirational condensations), humidity and environmental temperature. Consequently, it is necessary to make some routine environmental cleaning to disinfect the environmental surfaces. However, the steps and recommendations are presented in Figure 5.

Chemical disinfection

Many researchers argue that the routine environmental cleaning with soap, water, detergent, scrubbing, or brushing is effective to confiscate the dust from any place, but not an effective technique as it does not kill the virus. Thus, there is a need to use some chemical antiseptics such like "alcohol or chlorine" to kill the corona virus from environmental surfaces.

Terminal cleaning

Terminal cleaning (also known as infectious cleaning) is another important disinfection technique that should be used to purify the environmental surface in different health care centers. It is a method used to disinfect the discharged corona patient in an effective way. Terminal cleaning is accomplished in two steps:

The first step of terminal cleaning is to prepare the room i.e., *"remove the medical equipment, and patient used items"*. After that following steps are taken (as shown in Figure 6)

- Wear masks, gloves and protecting eyewear.
- Change bedsheet and curtains of the room.
- Moist the dust from all the environmental surfaces (tables, chairs, doors, etc.) of the room.

Figure 5.

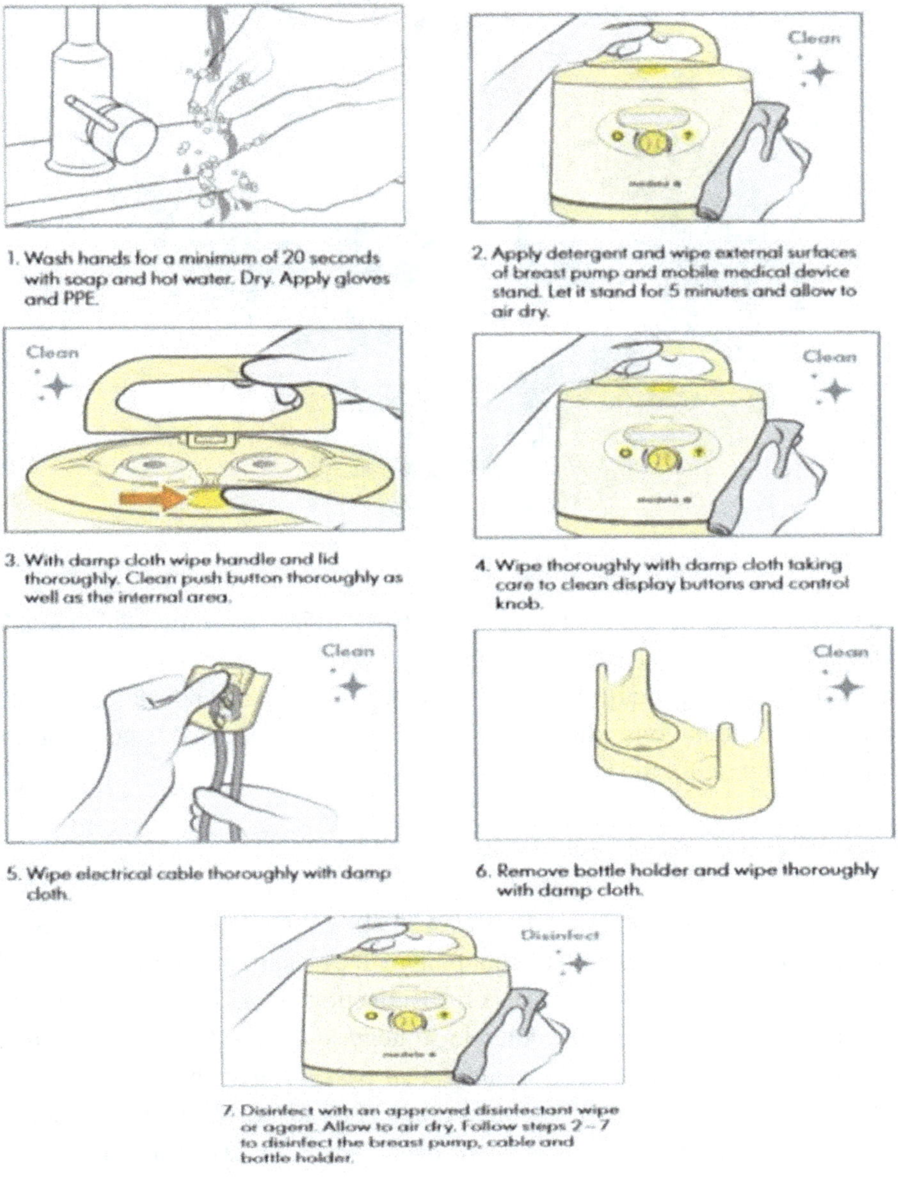

- Clean the windows and ceiling fans.
- Clean the floor with the detergent.
- Remove the mask, glove and protecting eyewear and discard.
- Use sanitizer or hand hygiene to clean hands.

Figure 6.

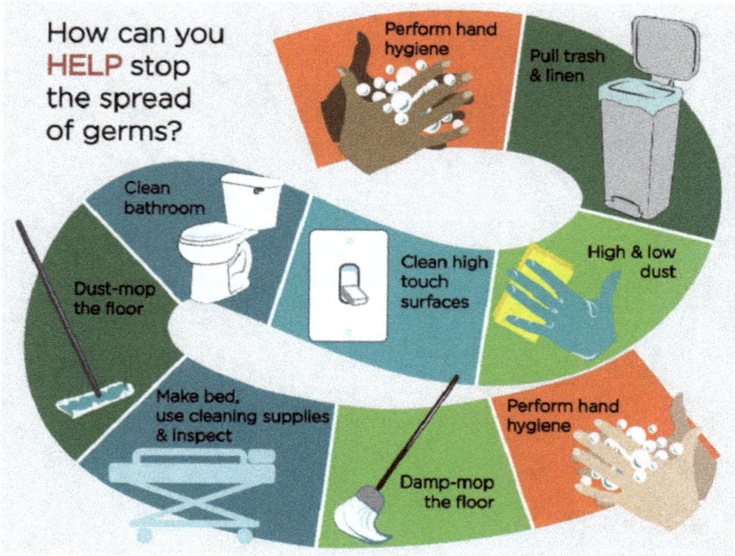

The second step of terminal cleaning is the physical cleaning. This includes disinfecting the virus using *"chlorine-based products"* like sodium hypochlorite. Such products contain "liquid (sodium hypochlorite), solid and powdered (calcium hypochlorite) formulations which liquefy in water and produce a watery aqueous chlorine solution to kill the virus.

Anti COVID-19 Disinfection Tuner

Another important strategy to disinfect COVID-19 is to place the disinfection tuners on the entrances of hospitals, shopping malls, restaurants and various public places such as schools, offices and parks. Disinfection tuner is also known as auto sanitizing gate which is the most suitable strategy to disinfect the virus (Gardezi et al., 2020). For instance, when an individual pass through the tuner, the tuner automatically sprays some purifiers on him/her, and the individual disinfects from the virus before entering in the particular place. Anti COVID disinfection tuners are shown in Figure 7.

Figure 7.

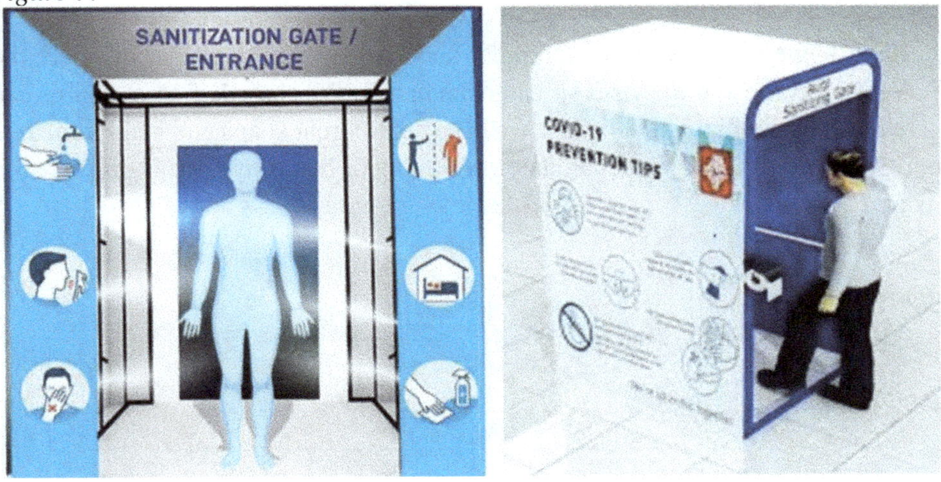

Spraying Disinfectants

Spraying is another effective technique to disinfect the virus from public places such as roads, schools, universities and parks etc. This technique completely cleans, disinfects and disposes the virus from the environmental surfaces (Fatimah et al., 2020). Disinfecting spray is done using an *"electric spray machine"*. This spray is comprised of "accelerated Hydrogen Peroxide 0.5%, Benzalkonium Chloride 0.05%, Chloroxylenol 0.12%, Ethyl Alcohol 62-71%, Iodine in Iodophor 50 ppm, Isopropanol 50%, Pine Oil 0.23%, Povidone-iodine 1%, Sodium Hypochlorite 0.05-0.5%, Sodium Chlorite 0.23%, and Sodium dischloroisocyanurate 0.1-0.5%" (Fatimah et al., 2020). All these chemicals have been liquefied using water which helps to kill the virus from public places. The spraying disinfectants are shown in Figure 8.

Figure 8.

Hand Hygiene

Hand hygiene is the most prominent and effective technique to disinfect the COVID-19 as most of the virus spread from hands. The spread of corona virus can be reduced by sanitizing the hands. We can use different soaps and hand sanitizers to hygiene the hands. The technique of sanitizing or hands hygiene is presented in Figure 9.

Figure 9.

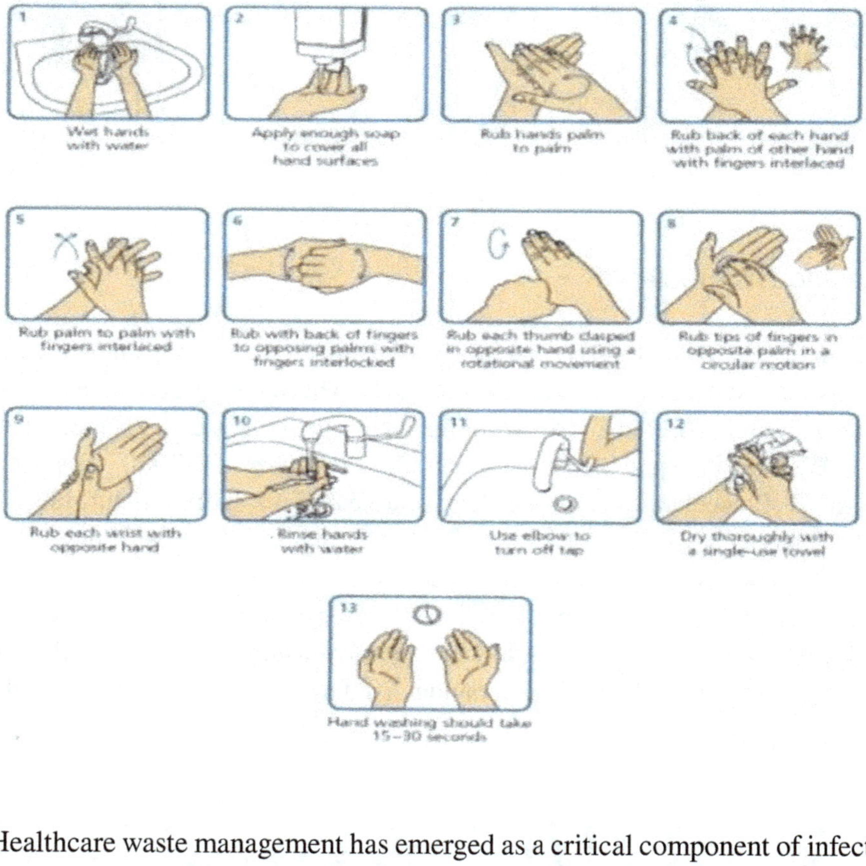

Healthcare waste management has emerged as a critical component of infection prevention and control during pandemics, particularly in the context of COVID-19. The unprecedented volume of infectious waste generated during the pandemic—including used personal protective equipment (PPE), disposable medical supplies, testing materials, and contaminated packaging—posed significant challenges to healthcare systems worldwide. Improper handling and disposal of such waste can facilitate secondary transmission of SARS-CoV-2, endangering healthcare workers, waste handlers, and the general public (UNEP, 2020).

Biomedical waste generated in healthcare facilities is broadly categorized into infectious, pathological, chemical, pharmaceutical, and general waste. During COVID-19, the proportion of infectious waste increased dramatically due to the widespread use of disposable masks, gloves, gowns, and face shields. Studies report that the volume of healthcare waste in some hospitals increased by three to five times during peak infection periods (Datta et al., 2018; Kulkarni & Anantharama, 2020). This surge overwhelmed existing waste management infrastructures, particularly in developing countries where segregation, storage, and treatment capacities were already limited.

Effective healthcare waste management during pandemics requires a multi-stage approach involving waste segregation at source, safe collection, temporary storage, transportation, treatment, and final disposal. Segregation at the point of generation is widely recognized as the most critical step, as it minimizes cross-contamination and reduces the volume of waste requiring specialized treatment. Color-coded containers and double-layered bags have been recommended for COVID-19 waste to ensure containment and reduce leakage risks (Capoor & Parida, 2021). Failure to implement proper segregation practices has been linked to increased occupational exposure and environmental contamination.

Treatment and disposal technologies play a central role in mitigating the risks associated with infectious waste. Incineration, autoclaving, chemical disinfection, and microwave treatment are among the most commonly used methods. High-temperature incineration is effective in destroying pathogens and reducing waste volume; however, it requires strict emission controls to prevent air pollution and the release of toxic by-products. Autoclaving, which employs steam sterilization, is considered a safer alternative for treating infectious waste, although it does not significantly reduce waste volume (McDonnell, 2020).

Wastewater management is another often overlooked but crucial aspect of healthcare waste management. SARS-CoV-2 RNA has been detected in hospital wastewater, raising concerns about potential environmental dissemination (Wang et al., 2020). Disinfection of wastewater using chlorine-based compounds, ultraviolet irradiation, or ozone treatment is therefore essential to prevent viral spread through sewage systems. Inadequate wastewater treatment infrastructure in many regions increases the risk of contamination of surface and groundwater resources, posing long-term public health and environmental threats.

In developing countries, including Pakistan, healthcare waste management faces additional structural and governance challenges. Limited funding, lack of trained personnel, outdated equipment, and weak regulatory enforcement hinder effective waste handling. Informal waste recycling practices further exacerbate risks, as contaminated materials may re-enter the market without proper decontamination. Studies emphasize that waste handlers and sanitation workers are among the most vulnerable occupational groups during pandemics, underscoring the need for targeted training and protective measures (Sahoo et al., 2024).

Policy frameworks and institutional coordination are essential for strengthening healthcare waste management during public health emergencies. International organizations such as the World Health Organization (WHO) and the United Nations Environment Programme (UNEP) have issued guidelines to support countries in managing pandemic-related waste. However, successful implementation depends on local adaptation, inter-agency collaboration, and public-private partnerships. In Pakistan, municipal authorities and healthcare institutions collaborated to establish temporary waste treatment facilities and specialized collection teams during the COVID-19 crisis, highlighting the importance of coordinated response mechanisms.

Overall, the literature underscores that healthcare waste management is not merely a logistical concern but a critical public health intervention during pandemics. Effective waste management reduces environmental transmission risks, protects frontline workers, and enhances the resilience of healthcare systems. These insights justify the inclusion of waste management strategies as a core component of the present study and support the subsequent SWOT-based evaluation of disinfection technologies in the Pakistani context.

The reviewed literature collectively demonstrates that environmental disinfection is a multidimensional intervention encompassing technological effectiveness, institutional capacity, and human behavior. While laboratory-based studies provide strong evidence regarding the virucidal efficacy of chemical and physical disinfection methods, their real-world performance is often shaped by contextual factors such as resource availability, regulatory oversight, and user compliance. This gap between controlled efficacy and practical effectiveness has been particularly evident during the COVID-19 pandemic, where rapid implementation frequently outpaced systematic evaluation.

Moreover, existing studies tend to examine disinfection technologies in isolation, focusing on individual methods such as chemical disinfectants, hand hygiene, or waste treatment systems. Few investigations adopt an integrated perspective that assesses how multiple interventions interact within complex healthcare and public environments. This limitation constrains the development of holistic disinfection strategies capable of addressing diverse transmission pathways simultaneously.

Consequently, there is a clear need for strategic analytical frameworks that synthesize technological, environmental, and organizational dimensions of disinfection practices. By consolidating insights from prior research and evaluating them through a structured SWOT approach, the present study contributes to a more comprehensive understanding of environmental disinfection and provides a practical bridge between academic research and policy-oriented decision-making.

3. MATERIAL AND METHODS

This study employs a qualitative approach based on government publications, scientific literature, and international health guidelines from agencies like the WHO, CDC, and Pakistan's National Institute of Health (NIH), this study takes a qualitative approach. The effectiveness of several disinfection technologies against SARS-CoV-2 was identified and assessed. These technologies included chemical sterilization, waste management, routine cleaning, air purification, spray disinfection, and hand hygiene. Each method's practical applications and limitations were evaluated through a SWOT (Strengths, Weaknesses, Opportunities, and Threats) analysis. Incorporating local data and institutional strategies employed in hospitals and quarantine centers, the study also took into account Pakistan's national response. Peer-reviewed publications, public health records, and professional opinions were among the sources. To gain important insights and guide strategic recommendations, all data was subjected to a thematic analysis. This method made it possible to fully comprehend disinfection procedures and how important they are for controlling COVID-19 and other infectious disease outbreaks in the future. Furthermore, only publicly accessible and validated sources were cited in order to adhere to ethical standards. The study excluded pharmaceutical treatments and vaccine interventions and concentrated only on disinfection technologies.

4. ANALYSIS AND RESULTS

Table 1 provides a SWOT analysis of different disinfection technologies and strategies used in Pakistan.

Table 1. SWOT Analysis of Different Disinfection Technologies

Technology	Strategy	Strength (S)	Weakness (W)	Opportunity (O)	Threat (T)
Waste disposition disinfection technique	Dispose of waste material Dispose of waste water	The involvement of expert waste management team	An organizing team is essential to the process of waste management	Public inducements to environment related initiatives such as cleaner production	Fondness argument of contract in which every contractor is accountable for handling the waste
Sterilization/ chemical disinfection technique	Steam, heat, chemical and plasma gas sterilization Terminal cleaning Chemical disinfection	Stable & rapid performance and wide spectrum of sterilization	Do not lessen the size of waste	On spot application of disinfection technology, destroy spores of virus and controls the spread of virus	Anthropogenic vaporizers shaped can pierce alveoli upon breathing, absorbance of atomized disinfectant
Cleaning disinfection techniques	Routine cleaning of environment	Economical, simple and easy	Beneficial for removing dust instead of killing virus	Can be utilized at any surface of environment	Risky method as it is not confirmed whether it kills the virus or not
Technology disinfection	Air purifiers Disinfection tuner	Demand is based on need	Prices are high	Reduce air pollution	Enhances disposable income in the affected areas
Spray disinfection	Spray disinfection	absolutely cleans, disposes and disinfects the virus from environmental surfaces	Futile in eradicating contaminants exterior of direct spray zones	Economical and easy to use	spraying is risky to eyes, respiratory or skin irritation and injurious to health
Hand hygiene	Hand sanitizer	Reduces the virus spread	Poor market inundation in rural areas	Reduces pollution	Dangerous for face and eyes

A brief SWOT analysis of several disinfection technologies is provided in Table 1, emphasizing their usefulness in infection control. Solid and wastewater disposal are two examples of waste disposal techniques that work well under professional supervision, but they also present contractual accountability issues and call for coordinated management. Although chemical sterilization methods that use heat, steam, or plasma are quick and efficient against viruses and spores, they do not lower waste volume and may release harmful vapors.

4.1. Discussion and Implications

The SWOT analysis reveals that while Pakistan has adopted a wide range of disinfection technologies, their effectiveness is highly contingent upon institutional capacity, public compliance, and environmental context. For instance, waste disposal and sterilization techniques demonstrate high effectiveness when managed by trained personnel; however, their reliance on centralized infrastructure limits scalability in rural areas. Similarly, routine environmental cleaning, though economically viable, offers limited protection against viral persistence unless supplemented with chemical disinfection. From a theoretical perspective, these findings support risk management frameworks that emphasize layered and context-specific interventions rather than reliance on single solutions. Practically, the results suggest that policymakers should prioritize integrated disinfection strategies that balance cost, safety, and operational feasibility. The findings are particularly relevant for low- and middle-income countries facing similar infrastructural constraints, thereby extending the relevance of this study beyond Pakistan.

Although routine cleaning is inexpensive and simple to do, it mostly eliminates dust rather than infections, and it is still unknown how effective it is against viruses. Although they are costly and usually used only, when necessary, air purifiers, a type of technological disinfection, aid in lowering airborne pollutants. Although spray disinfection is economical and effective on contact surfaces, its limited reach and chemical exposure present health risks. Hand washing, particularly with hand sanitizers, effectively prevents the spread of viruses and lowers pollution in the environment. It can irritate delicate areas like the face or eyes, though, and its market penetration is low in rural areas. Overall, the table emphasizes that although each technique has advantages, combining several strategies according to the environment and degree of risk guarantees more thorough and dependable disinfection.

The findings of this study provide important insights into the effectiveness, feasibility, and limitations of various disinfection technologies employed during the COVID-19 pandemic in Pakistan. The SWOT analysis reveals that while multiple disinfection strategies were implemented across healthcare facilities and public spaces, their success was uneven and highly dependent on institutional capacity, public compliance, and environmental context. These results highlight the necessity of adopting a multi-layered and adaptive approach to infection control rather than relying on isolated interventions.

One of the most significant findings relates to healthcare waste management and disposal practices. The SWOT analysis demonstrates that professionally managed waste segregation and disposal systems significantly reduce the risk of secondary viral transmission. However, weaknesses such as inadequate coordination, limited infrastructure, and reliance on contractual accountability mechanisms often undermine their effectiveness. These findings are consistent with prior studies emphasizing that waste management failures can transform healthcare facilities into sources of environmental contamination rather than sites of disease control (UNEP, 2020). From a theoretical standpoint, this underscores the importance of systems-based approaches in infection prevention, where waste management is treated as an integral component of public health strategy rather than a peripheral service.

Chemical disinfection and sterilization techniques, including chlorine-based disinfectants, alcohol formulations, and thermal sterilization, emerged as highly effective in inactivating SARS-CoV-2 when applied correctly. The strengths identified in the SWOT analysis—such as rapid action and broad-spectrum efficacy—align with existing literature on virucidal performance (McDonnell, 2020; Wang et al., 2020). However, the analysis also reveals critical weaknesses, including potential health hazards from chemical exposure and the inability of certain methods to reduce waste volume. These limitations highlight the need for standardized protocols, worker training, and monitoring mechanisms to mitigate unintended consequences. The findings support risk management theories that emphasize balancing effectiveness with safety and sustainability.

Routine environmental cleaning was identified as an economically viable and widely adopted disinfection strategy, particularly in resource-constrained settings. Its primary strength lies in reducing surface contamination and facilitating subsequent chemical disinfection. However, the SWOT analysis confirms that routine cleaning alone is insufficient to eliminate viral pathogens, a finding echoed in multiple empirical studies (Gharpure et al., 2020). This result has important practical implications, suggesting that policymakers and facility managers should avoid overreliance on basic cleaning measures and instead integrate them into comprehensive disinfection frameworks.

Technological disinfection interventions, such as air purifiers, disinfection tunnels, and automated spray systems, present a mixed picture. While these technologies offer opportunities for rapid and large-scale disinfection, their effectiveness is often constrained by high costs, maintenance requirements, and uncertain long-term health impacts. For example, disinfection tunnels gained popularity during the pandemic; however, subsequent evidence questioned their efficacy and raised concerns regarding chemical inhalation risks (Gardezi et al., 2020). The SWOT analysis highlights these trade-offs, reinforcing the importance of evidence-based decision-making in adopting emerging technologies.

Hand hygiene consistently emerged as one of the most effective and accessible disinfection strategies. Its strengths include low cost, ease of implementation, and strong empirical support for reducing viral transmission. Nevertheless, weaknesses such as limited access to sanitizers in rural areas and potential skin irritation were also identified. These findings underscore the importance of equitable distribution and public education to maximize the benefits of hand hygiene interventions. From a public health perspective, hand hygiene exemplifies a low-technology, high-impact intervention that should remain central to pandemic preparedness strategies.

The results also carry significant policy implications. First, the findings suggest that national disinfection strategies should prioritize integrated frameworks that combine waste management, surface disinfection, technological interventions, and behavioral measures. Second, investments in training, infrastructure, and regulatory enforcement are critical to enhancing the effectiveness of disinfection technologies. Third, the study highlights the need for context-specific guidelines tailored to the socio-economic realities of developing countries. Policies designed for high-income settings may not be directly transferable without adaptation.

Theoretical implications of this study extend to infection control and environmental health literature. By applying a SWOT framework, this chapter contributes to a more strategic understanding of disinfection technologies, moving beyond efficacy-focused evaluations to consider feasibility, risks, and opportunities. This aligns with contemporary public health models that emphasize resilience, adaptability, and system integration. The findings support the argument that effective pandemic response requires not only technological solutions but also institutional coordination and public engagement.

Overall, the discussion demonstrates that disinfection technologies are most effective when deployed as part of a coordinated and context-aware strategy. The COVID-19 pandemic has underscored the limitations of fragmented interventions and highlighted the value of comprehensive approaches that address environmental, institutional, and behavioral dimensions of infection control. These insights are particularly relevant for developing countries seeking to strengthen their preparedness for future public health emergencies.

4.2. Application of Disinfection Technologies and Strategies in Pakistan

- Hand hygiene is most common and effective technique that has been used by the management of various public and private hospitals, health centers, pharmacies, restaurants, shopping malls, and so on. The management has placed sanitizers on the entrance of these places so that the people first sanitize their hands and then enter into the particular place.
- Anti COVID disinfecting units has also been seen in the front of different public and private hospitals, shopping malls, universities and restaurants to disinfect the virus.
- Pakistan's public and private medical and dental authorities are also using sterilizing techniques to purify the medical instruments.

- The country has also been adapted the strategy of spray disinfection to disinfect the roads, parks and other public places.
- Pakistan's municipal committee department has arranged a team (named as Tiger Force) to manage or dispose the waste materials of different public and private places.
- Pakistan's water and sanitation agency has also made a team to manage the waste water of hospitals and other public and private health centers.
- Routine environmental cleaning has also been done in different hospitals, and other public places to disinfect the corona virus from the non-living organisms.

5. CONCLUSION

This chapter examined environmental disinfection technologies employed during the COVID-19 pandemic, with a particular focus on healthcare facilities and public spaces in Pakistan. The analysis demonstrates that no single disinfection method is sufficient to mitigate viral transmission; rather, a combination of complementary strategies tailored to specific environments is essential. The SWOT analysis highlights both the strengths and limitations of commonly used technologies, underscoring the importance of strategic planning and institutional coordination.

Despite its contributions, this study has certain limitations. It relies primarily on secondary data and qualitative analysis, which may limit the generalizability of findings. Additionally, the effectiveness of specific technologies was not empirically tested under controlled conditions. Future research should incorporate quantitative assessments, comparative cross-country analyses, and longitudinal studies to evaluate the long-term sustainability and health impacts of disinfection strategies. Further investigation into community-level compliance and cost–benefit trade-offs would also enhance policy relevance.

In recent years, the COVID-19 pandemic has become one of the biggest threats to world health. In order to reduce its spread, both public and private areas must be thoroughly cleaned. This study examined several disinfection techniques and technologies and provided a thorough SWOT analysis of each. The findings highlight that no single method is universally sufficient; rather, the effectiveness of disinfection depends on context, accessibility, and proper implementation. A combined approach tailored to specific environments is necessary to ensure thorough and reliable virus mitigation. Overall, this chapter provides a strategic foundation for improving environmental disinfection practices and offers valuable insights for managing future public health emergencies.

Recommendation

The study makes a number of important recommendations to improve disinfection procedures and stop the spread of COVID-19 based on the analysis. To guarantee appropriate waste management, medical and contaminated waste must first be disposed of in double-sealed, specially marked bags or bins that must be available in all isolation wards, patient rooms, and quarantine centers. Regarding personal hygiene, people are advised to stay away from high-touch surfaces and always carry hand sanitizers. Hand sanitization must be done right away if contact happens. Routine environmental cleaning procedures might work well in homes, but they are insufficient in public areas and healthcare facilities where there is a much greater chance of viral transmission. The application of cutting-edge disinfection technologies is highly advised in such settings. To reduce external contamination, anti-COVID disinfection units should also be placed at the entrances of pharmacies, hospitals, shopping centers, gyms, schools, colleges, and universities. To further lower the risk of transmission, the general public should be advised to wear protective gear like masks, gloves, and eyewear. Lastly, it is crucial to make sure that all environmental surfaces and medical equipment that come into contact with COVID-19 patients are quickly cleaned using the proper sterilization ways described in this study.

The COVID-19 pandemic has underscored the critical importance of environmental disinfection as a core component of public health preparedness and response. This chapter examined a wide range of disinfection technologies and strategies employed in healthcare facilities and public spaces in Pakistan, providing a comprehensive assessment of their effectiveness, feasibility, and associated risks. By integrating findings from scientific literature, policy documents, and practical implementations, the study highlights that effective infection control requires a coordinated and multilayered approach rather than reliance on any single intervention.

The SWOT-based analysis demonstrates that while Pakistan adopted several effective disinfection measures—such as hand hygiene, chemical sterilization, waste management protocols, and routine environmental cleaning—their success varied considerably across contexts. Strengths such as cost-effectiveness, accessibility, and rapid virucidal action were often offset by weaknesses including infrastructural constraints, health risks associated with chemical exposure, and uneven implementation. These findings reinforce the need for context-sensitive strategies that align technological solutions with institutional capacity and public behavior.

Despite its contributions, this study has several limitations that must be acknowledged. First, the analysis is based primarily on qualitative methods and secondary data sources, which may limit the generalizability of the findings. While the SWOT framework provides valuable strategic insights, it does not quantify the relative effectiveness of individual disinfection technologies under controlled conditions.

Second, the study focuses on institutional and environmental disinfection practices and does not assess individual behavioral compliance in detail, which is a critical determinant of intervention success. Third, data availability constraints limited the ability to conduct region-specific comparisons within Pakistan, particularly between urban and rural settings.

These limitations present important opportunities for future research. Quantitative studies assessing the comparative effectiveness and cost–benefit profiles of disinfection technologies would significantly enhance evidence-based decision-making. Longitudinal research examining the sustainability and environmental impacts of large-scale chemical disinfection is also needed, particularly in low- and middle-income countries. Future studies should further explore behavioral dimensions, including public perceptions, compliance patterns, and risk communication strategies related to disinfection practices.

Additionally, comparative cross-country analyses could provide valuable insights into how different governance structures and resource levels influence the effectiveness of disinfection interventions. Research focusing on rural and underserved communities is especially important, as these areas often face heightened vulnerabilities due to limited infrastructure and healthcare access. Finally, integrating disinfection strategies into broader pandemic preparedness and resilience frameworks would help policymakers transition from reactive responses to proactive risk management.

In conclusion, this chapter contributes to the growing body of literature on environmental disinfection by offering a strategic, context-specific evaluation of technologies used during the COVID-19 pandemic in Pakistan. The findings emphasize that disinfection is not merely a technical process but a complex public health intervention requiring coordination, regulation, and public engagement. By addressing existing gaps and outlining clear directions for future research, this study provides a valuable foundation for strengthening infection control systems and improving preparedness for future global health emergencies.

REFERENCES

Alavi, S., Asadi, F., Raji, S. A. H., & Samie, S. (2020). Effect of steam and dry heat sterilization on the insertion and fracture torque of orthodontic miniscrews. *Dental Research Journal (Isfahan)*, *17*(3), 219–224. DOI: 10.4103/1735-3327.284729 PMID: 32774800

Allison, D. L., & McDaniel, J. A. (2020). Cleaning and sterilization of dental instruments. In *Infection control in the dental office* (pp. 193–203). Springer. DOI: 10.1007/978-3-030-30085-2_13

Alwan, N., Almazrouei, S., Almazrouei, M., Aldhaheri, J., Alismaili, F., & Ghach, W. (2023). Evaluation of public awareness and performance toward the safe use of household disinfectants-cleaners to prevent COVID-19 in the Emirate of Abu Dhabi. *Frontiers in Public Health*, *11*, 1214240. DOI: 10.3389/fpubh.2023.1214240 PMID: 37448658

Capoor, M. R., & Parida, A. (2021). Current perspectives of biomedical waste management in context of COVID-19. *Indian Journal of Medical Microbiology*, *39*(2), 171–178. DOI: 10.1016/j.ijmmb.2021.03.003 PMID: 33766404

Datta, P., Mohi, G. K., & Chander, J. (2018). Biomedical waste management in India: Critical appraisal. *Journal of Laboratory Physicians*, *10*(1), 6–14. DOI: 10.4103/JLP.JLP_89_17 PMID: 29403196

Fatimah, H., & Hafsah, U. M. (2020). Social Work's Program By Universitas Negeri Malang Spraying Desinfectant of Distribution of Covid-19 Trafficking Kunti District, Bungkal Ponorogo. *International Research-Based Education Journal*, *2*(2), 63–68. DOI: 10.17977/um043v2i2p63-68

Gardezi, A., Rathore, F. A., & Farooq, F. (2020). Sanitization Walk-Through Gates During COVID-19 Pandemic: Effective or A False Sense of Protection? *Annals of King Edward Medical University*, *26*(Special Issue), 259–261.

Gharpure, R., Hunter, C. M., Schnall, A. H., Barrett, C. E., Kirby, A. E., Kunz, J., & Garcia-Williams, A. G. (2020). Knowledge and practices regarding safe household cleaning and disinfection for COVID-19 prevention—United States, May 2020. *Morbidity and Mortality Weekly Report*, *69*(23), 705709. PMID: 32525852

Gina, M., Ofenloch, R., Schwebke, I., Hübner, N. O., Brüning, T., & Fartasch, M. (2025). The Effect of Alcohol-Based Virucidal Hand Sanitizers on Skin Barrier Function—A Randomised Experimental Study. *Contact Dermatitis*, *93*(2), 119–130. DOI: 10.1111/cod.14808 PMID: 40356051

Han, Z., Pappas, E., Simmons, A., Fox, J., Donskey, C. J., & Deshpande, A. (2020). Environmental cleaning and disinfection of hospital rooms: A nationwide survey. *American Journal of Infection Control*, *49*(1), 34–39. DOI: 10.1016/j.ajic.2020.08.008 PMID: 32798634

Hines, S. E., Brown, C. H., Oliver, M., Gucer, P., Frisch, M., Hogan, R., Roth, T., Chang, J., & McDiarmid, M. (2020). Cleaning and Disinfection Perceptions and Use Practices Among Elastomeric Respirator Users in Health care. *Workplace Health & Safety*, *68*(12), 572–582. DOI: 10.1177/2165079920938618 PMID: 32812845

Kramer, A., Galabov, A. S., Sattar, S. A., Döhner, L., Pivert, A., Payan, C., Wolff, M. H., Yilmaz, A., & Steinmann, J. (2006). Virucidal activity of a new hand disinfectant with reduced ethanol content: Comparison with other alcohol-based formulations. *The Journal of Hospital Infection*, *62*(1), 98–106. DOI: 10.1016/j.jhin.2005.06.020 PMID: 16307826

Kulkarni, B. N., & Anantharama, V. (2020). Repercussions of COVID-19 pandemic on municipal solid waste management: Challenges and opportunities. *The Science of the Total Environment*, *743*, 140693. DOI: 10.1016/j.scitotenv.2020.140693 PMID: 32663690

Luo, H., Fu, M., Chen, X., Chen, Y., Li, Y., Liu, X., Su, X., & Chen, Y. (2020). Exploration of Infection Prevention and Control Practice of Designated Hospitals in During Corona Virus Disease 2019 Epidemic Period. *Journal of Clinical and Nursing Research*, *4*(4), 80–90. DOI: 10.26689/jcnr.v4i4.1347

McDonnell, G. E. (2020). *Antisepsis, disinfection, and sterilization: types, action, and resistance*. John Wiley & Sons.

Paleologos, E. K., O'Kelly, B. C., Tang, C. S., Cornell, K., Rodríguez-Chueca, J., Abuel-Naga, H., & Goli, V. S. N. S. (2020). Post COVID-19 Water and Wastewater Management to Protect Public Health and Geoenvironment. *Environmental Geotechnics*, *40*, 1–15. DOI: 10.1680/jenge.20.00067

Qasim, M., Ahmad, W., Yoshida, M., Gould, M., & Yasir, M. (2020). Analysis of the worldwide corona virus (COVID-19) pandemic trend; a modelling study to predict its spread. *medRxiv*. https://doi.org/DOI: 10.1101/2020.03.30.20048215

Rai, P. (2020). Disinfection of Endoscopy and Reusability of Accessories. *Journal of Digestive Endoscopy*, *11*(1), 61–66. DOI: 10.1055/s-0040-1712238 PMID: 40477028

Ren, Y., Liu, S., Yang, L., Shan, J., Liu, Y., Li, H., Chen, L.-H., & Chen, H. (2020). Practice and exploration of infection prevention and control measures based on risk management of surgical patients during the epidemic of corona virus disease 2019 (COVID-19). *American Journal of Infection Control, 49*(2), 151–157. DOI: 10.1016/j.ajic.2020.07.023 PMID: 32702389

Rutala, W. A., Gergen, M. F., Sickbert-Bennett, E. E., & Weber, D. J. (2020). Comparative evaluation of the microbicidal activity of low-temperature sterilization technologies to steam sterilization. *Infection Control and Hospital Epidemiology, 41*(4), 391–395. DOI: 10.1017/ice.2020.2 PMID: 32098638

Sahoo, S., Rathod, W., Vardikar, H., Biswal, M., Mohanty, S., & Nayak, S. (2024). Biomedical waste plastic: Bacteria, disinfection and recycling technologies—a comprehensive review. *International Journal of Environmental Science and Technology, 21*(1), 1141–1158. DOI: 10.1007/s13762-023-04975-w PMID: 37360566

Sarada, B. V., Vijay, R., Johnson, R., Rao, T. N., & Padmanabham, G. (2020). Fight against COVID-19: ARCI's technologies for disinfection. *Transactions of the Indian National Academy of Engineering : an International Journal of Engineering and Technology, 5*(2), 349–354. DOI: 10.1007/s41403-020-00153-3 PMID: 38624431

Sharma, S. (2020). How To Reduce Corona Spread. *Journal of Medical Case Reports and Reviews, 3*(5), 658–662.

Starikovskiy, A. Y., & Usmanova, D. R. (2020). Medical mask with plasma sterilizing layer. *arXiv preprint arXiv:2004.00807*. Available at: https://arxiv.org/abs/2004.00807

UNEP. (2020). BASEL: Waste management an essential public service in the fight to beat [The United Nations Environment Programme] [UNEP] [and The Basel Convention.]. *COVID*, •••, 19.

United Nations Environment Programme (UNEP). (2020). Waste management: An essential public service in the fight against. *COVID*, •••, 19.

Wang, J., Shen, J., Ye, D., Yan, X., Zhang, Y., Yang, W., Li, X., Wang, J., Zhang, L., & Pan, L. (2020). Disinfection technology of hospital wastes and wastewater: Suggestions for disinfection strategy during coronavirus Disease 2019 (COVID-19) pandemic in China. *Environmental Pollution, 262*, 114665. DOI: 10.1016/j.envpol.2020.114665 PMID: 32443202

World Health Organization (WHO). (2020). *Coronavirus disease (COVID-19) outbreak situation.*

Ye, G., Lin, H., Chen, L., Wang, S., Zeng, Z., Wang, W., & Yang, Z. (2020). Environmental contamination of the SARS-CoV-2 in healthcare premises: an urgent call for protection for healthcare workers. *medRxiv*. https://doi.org/DOI: 10.1101/2020.03.11.20034546

Zhuang, E., Chen, H. H., Kolesnik, O., & Hines, S. E. (2023). Tolerability, user acceptance and preference for a novel reusable respirator among health care workers. *American Journal of Infection Control*, *51*(7), 821–826. DOI: 10.1016/j.ajic.2022.09.006 PMID: 36122632

Compilation of References

Aazam, M., St-Hilaire, M., Lung, C. H., & Lambadaris, I. (2016, October). Cloud-based smart waste management for smart cities. In 2016 IEEE 21st international workshop on computer aided modelling and design of communication links and networks (CAMAD) (pp. 188-193). IEEE. DOI: 10.1109/CAMAD.2016.7790356

Abbas, U., Akhai, S., Abbass, M., Abass, S., Chouksey, A., & Wadhwa, A. S. (2026). Sustainable Growth by Navigating Complexity Through Innovation and Adaptive Industrial Policies. In Tzouvelekas, M., Zarotiadis, G., & Varsakelis, N. (Eds.), *Industrial Policy, Innovation, and Complexity: Comparative Analysis* (pp. 29–82). Trends, and Prospects., DOI: 10.4018/979-8-3373-1767-0.ch002

Abdirahman, A. A., Asif, M., & Mohsen, O. (2025). Circular economy in the renewable energy sector: A review of growth trends, gaps and future directions. *Energy Nexus*, *17*, 100395. DOI: 10.1016/j.nexus.2025.100395

Abubakar, I. R. (2017). Household Response to Inadequate Sewerage and Garbage Collection Services in Abuja, Nigeria. *Journal of Environmental and Public Health*, *5314840*. Advance online publication. DOI: 10.1155/2017/5314840 PMID: 28634496

Access, O. (2016).. . *International Journal of Science and Engineering*.

Addas, A., Khan, M. N., & Naseer, F. (2024). Waste management 2.0 leveraging internet of things for an efficient and eco-friendly smart city solution. *PLoS One*, *19*(7), e0307608. DOI: 10.1371/journal.pone.0307608 PMID: 39083562

Adu, J. T., & Aneke, F. I. (2025). Evaluation of heavy metal contamination in landfills from e-waste disposal and its potential as a pollution source for surface water bodies. *Results in Engineering*, *25*, 104431. DOI: 10.1016/j.rineng.2025.104431

Affordofe, M., Udofia, E. A., Akyeampong, E., Koranteng, F. O., Tettey, P., Botwe, P. K., Andoh, T., Peprah, E. K., Attah, H. Y., Tabong, P., Boateng, G., & Quansah, R. (2025). Knowledge and practice of solid healthcare waste management among waste handlers in hospitals in Southern Ghana: A qualitative study. *BMC Public Health, 25*(1), 702. DOI: 10.1186/s12889-025-21874-6 PMID: 39979880

Aga, N., Ani, P., Wijaningsih, D., & Kadek, S. (2024). From human-centered to ecofeminist environmental policies: Harnessing pancasila for global sustainability solutions. *Clío. Revista de Historia. Ciencias Humanas y Pensamiento Crítico.*, (9), 195–245.

Aggarwal, P., Rana, M., & Akhai, S. (2022). Briefings on e-waste hazard until COVID era in India. *Materials Today: Proceedings, 71*(2), 389–393. DOI: 10.1016/j.matpr.2022.09.507

Agrawal, S., Jain, V. K., & Verma, H. (2025). Circular economy! Sustainable growth and achieving zero net emissions-a way ahead for building sustainable tomorrow by optimising consumption. *International Journal of Sustainable Agricultural Management and Informatics, 11*(1), 72–96. DOI: 10.1504/IJSAMI.2025.143112

Ahad, T., & Busch, P. (2024). Exploring the factors influencing Mobile-based Ubiquitous System adoption in the Bangladesh RMG sector: A view through DOI and TOE. *The Electronic Journal on Information Systems in Developing Countries, 90*(1), e12291. DOI: 10.1002/isd2.12291

Ahamad, A., Kumar, P., & Kumar, B. (2025). Treatment and Recycling of Plastic Waste: A Dire Need for a Sustainable Future. In Kumar, P., & Kumar, B. (Eds.), *Global Waste Management* (pp. 359–381). John Wiley & Sons., DOI: 10.1002/9781394318414.ch14

Ahmad, M., & Wilkins, S. (2024). Purposive sampling in qualitative research: A framework for the entire journey. *Quality & Quantity*, •••, 1–19.

Ahmed, K., Dubey, M. K., Kumar, A., & Dubey, S. (2024). Artificial intelligence and IoT driven system architecture for municipality waste management in smart cities: A review. *Measurement. Sensors, 36*, 101395. DOI: 10.1016/j.measen.2024.101395

Aiello, G., Muriana, C., Quaranta, S., & Abusohyon, I. A. S. (2025). A sustainable inventory management model for closed loop supply chain involving waste reduction and treatment. *Cleaner Logistics and Supply Chain, 16*, 100244. Advance online publication. DOI: 10.1016/j.clscn.2025.100244

AirVoice. (2023, November 6). AirVoice releases comprehensive report on Diwali air pollution in India. https://airvoice.global/tpost/8bhy30c0c1-airvoice-releases-comprehensive-report-o

Ajekiigbe, V. O., Agbo, C. E., Ogieuhi, I. J., Anthony, C. S., Onuigbo, C. S., Falayi, T. A., Oluwapelumi, O. Z., Amusa, O., Adeniran, G. O., Ogunleke, P. O., & Bakare, I. S. (2025). The increasing burden of global environmental threats: Role of antibiotic pollution from pharmaceutical wastes in the rise of antibiotic resistance. *Discover Public Health, 22*(1), 1–10. DOI: 10.1186/s12982-025-00506-9

Akhai, S. (2023). Biogas plants for sustainable municipal waste management. In Akhai, S., Pant, P., & Bandara, J. M. S. J. (Eds.), *Optimizing sustainable wastewater treatment in developing nations* (pp. 162–179). IGI Global., DOI: 10.4018/978-1-6684-8117-2.ch012

Akubia, G. K., Gaffar, V., Sultan, M. A., & Andriana, D. (2025). The Impact of Green Logistics Management Practices on Manufacturing Firms' Sustainability Performance in Ghana and Indonesia. *International Journal of Supply and Operations Management, 12*(2), 215–235. DOI: 10.22034/IJSOM.2024.110506.3185

Al, A., Ahmed, A., & Asadullah, A. (2020). Artificial Intelligence and Machine Learning in Waste Management and Recycling. Engineering International, 8(1).

Alavi, S., Asadi, F., Raji, S. A. H., & Samie, S. (2020). Effect of steam and dry heat sterilization on the insertion and fracture torque of orthodontic miniscrews. *Dental Research Journal (Isfahan), 17*(3), 219–224. DOI: 10.4103/1735-3327.284729 PMID: 32774800

Al-Hassan, R. A., Alyousuf, A. N., Jasim, Z. A., Kamil, K., & Yaseen, A. (2026a). Fashion Waste Management in the Age of Artificial Intelligence: Strategies and Challenges. In Mouloudj, K., & Arantes, L. (Eds.), *Ethical Consumption and Responsible Innovation in Fashion through AI*. IGI Global., DOI: 10.4018/979-8-3373-7200-6.ch009

Al-Hassan, R. A., Dirawi, W. M., Almosafri, W. H., Alabduladheem, M. F., Jahloul, I. H., & Alyousuf, A. N. (2026b). Green HRM in Hospitals as a Strategic Approach to Promoting Employees' Pro-Environmental Behavior and Achieving Organizational Sustainability. In Rodrigues, R., Mohanty, V., & Figueiredo, P. (Eds.), *Organizational Culture and Employee Experience in the AI-Driven Organization* (pp. 269–308). IGI Global., DOI: 10.4018/979-8-2600-0610-8.ch010

Alighardashi, M., Mousavi, S. A., Almasi, A., & Mohammadi, P. (2025). A study of indoor air pollution caused by disinfection equipment as a consequence of infectious waste management. *International Journal of Environmental Health Research, 35*(3), 634–644. DOI: 10.1080/09603123.2024.2368123 PMID: 38899861

Ali, Q., & Mahmood, M. A. (2025). Transforming Leadership: The Role of Artificial Intelligence in Modern Organizational Management. In Baig, S., Mohsin, M., Lodhi, R., & Batool, G. (Eds.), *Reshaping Organizational Management and Workplace Culture with AI* (pp. 223–262). IGI Global., DOI: 10.4018/979-8-3693-8960-7.ch009

Allison, D. L., & McDaniel, J. A. (2020). Cleaning and sterilization of dental instruments. In *Infection control in the dental office* (pp. 193–203). Springer. DOI: 10.1007/978-3-030-30085-2_13

Almokdad, E., & Mouloudj, K. (2026). Exploring Actual Sustainable Behavior of Tourists: Qualitative Analysis. In Mouloudj, K., & Bouarar, A. (Eds.), *Insights on Consumer Psychology in the Digital Landscape* (pp. 185–210). IGI Global., DOI: 10.4018/979-8-3373-2424-1.ch007

Altassan, M. A. (2024). The nexus between green HR practices and firm sustainable performance in Saudi Arabia manufacturing industry: The role of green innovation and green transformation leadership. *Uncertain Supply Chain Management, 12*(4), 2207–2220. DOI: 10.5267/j.uscm.2024.6.010

Alwan, N., Almazrouei, S., Almazrouei, M., Aldhaheri, J., Alismaili, F., & Ghach, W. (2023). Evaluation of public awareness and performance toward the safe use of household disinfectants-cleaners to prevent COVID-19 in the Emirate of Abu Dhabi. *Frontiers in Public Health, 11*, 1214240. DOI: 10.3389/fpubh.2023.1214240 PMID: 37448658

Alzaydi, A. (2024). Balancing creativity and longevity: The ambiguous role of obsolescence in product design. *Journal of Cleaner Production, 445*, 141239. DOI: 10.1016/j.jclepro.2024.141239

Ambade, B. (2018). The air pollution during Diwali festival by the burning of fireworks in Jamshedpur city, India. *Urban Climate, 26*, 149–160. DOI: 10.1016/j.uclim.2018.08.009

Anbarasu, K., Thanigaivel, S., Sathishkumar, K., Alam, M. M., Al-Sehemi, A. G., & Devarajan, Y. (2025). Harnessing artificial intelligence for sustainable bioenergy: Revolutionizing optimization, waste reduction, and environmental sustainability. *Bioresource Technology, 418*, 131893. DOI: 10.1016/j.biortech.2024.131893 PMID: 39608419

Andersson, T. D., & Lundberg, E. (2013). Commensurability and sustainability: Triple impact assessments of a tourism event. *Tourism Management*, *37*, 99–109. DOI: 10.1016/j.tourman.2012.12.015

Andrade, C. (2021). The inconvenient truth about convenience and purposive samples. *Indian Journal of Psychological Medicine*, *43*(1), 86–88. DOI: 10.1177/0253717620977000 PMID: 34349313

Antchak, V., Ziakas, V., & Getz, D. (2019). *Event portfolio management: Theory and methods for event management and tourism.* https://derby.openrepository.com/handle/10545/624187

Aparcana, S., & Salhofer, S. (2013). Development of a social impact assessment methodology for recycling systems in low-income countries. *The International Journal of Life Cycle Assessment*, *18*(5), 1106–1115. DOI: 10.1007/s11367-013-0546-8

Arous, J. A., ElRabbat, S. A., ElAwad, I. H., Hima, N. A., & El Hamdi, R. (2025). Assessing the impact of environmental training programs on awareness and job performance in institutions in the Kingdom of Saudi Arabia. *International Area Studies Review*, *28*(1), 1–23. DOI: 10.69473/iasr.2025.28.1.1

Aseweh Abor, P. (2013). Managing healthcare waste in Ghana: A comparative study of public and private hospitals. *International Journal of Health Care Quality Assurance*, *26*(4), 375–386. DOI: 10.1108/09526861311319591 PMID: 23795428

Ashuro, T., Kant, S., & Adula, M. (2025). Implementing reverse logistics for recycling, reusing, and reducing waste at the source by mediation of waste reduction initiatives in the horn of Africa. In *Multi-Stakeholder Collaboration for Sustainable Supply Chain*. IGI Global., DOI: 10.4018/979-8-3693-8925-6.ch008

Asian Confluence. (2022). *India's soft power: The case of Ratha Yatra.* https://www.asianconfluence.org/publication-details-full/india-s-soft-power-the-case-of-ratha-yatra

Ateeq, K., Al Masaeid, T., Selim, H., Oswal, N., Alkubaiusy, A. A. A., Alami, R., & Ajdoobi, S. M. G. A. (2025). Harnessing AI for Faster Innovation: How AI Concept Generation Impacts Development Timelines and Market Agility. *Journal of Posthumanism*, *5*(2), 37–50. DOI: 10.63332/joph.v5i2.390

Attia, A. (2025). The effect of commitment to sustainable supply chain management and reverse logistics on performance in context of UAE food industry. *Sustainable Futures : An Applied Journal of Technology, Environment and Society*, *9*, 100442. Advance online publication. DOI: 10.1016/j.sftr.2025.100442

Bagwan, W. A. (2024). Electronic waste (E-waste) generation and management scenario of India, and ARIMA forecasting of E-waste processing capacity of Maharashtra state till 2030. *Waste Management Bulletin, 1*(4), 41–51. DOI: 10.1016/j.wmb.2023.08.002

Baldé, C. P., Forti, V., Gray, V., Kuehr, R., & Stegmann, P. (2024). *The global e-waste monitor 2024*. United Nations University, United Nations Institute for Training and Research, International Telecommunication Union, & International Solid Waste Association.

Bandura, A. (1986). Social foundations of thought and action. *Englewood Cliffs, NJ, 1986*(23-28), 2.

Bandura, A. (1982). Self-efficacy mechanism in human agency. *The American Psychologist, 37*(2), 122–147. DOI: 10.1037/0003-066X.37.2.122

Bandura, A. (2014). Social-cognitive theory. In *An introduction to theories of personality* (pp. 341–360). Psychology Press.

Baron, A., & Galizio, M. (2005). Positive and negative reinforcement: Should the distinction be preserved? *The Behavior Analyst, 28*(2), 85–98. DOI: 10.1007/BF03392107 PMID: 22478443

Barros, M., & Dimla, E. (2021). From planned obsolescence to the circular economy in the smartphone industry: An evolution of strategies embodied in product features. *Proceedings of the Design Society, 1*, 1607–1616. DOI: 10.1017/pds.2021.422

Basumatary, P., Bhuyan, M. K., Sarmah, P., & Saikia, D. (2021). Eco-friendly Holi using natural dye and low-cost filler base. *International Journal of Current Microbiology and Applied Sciences, 10*(2), 1067–1075. DOI: 10.20546/ijcmas.2021.1002.126

Behal, A., Islam, S., & Tan, C. S. L. (2025). Exploring Circular Economy Practices in School Uniforms: A Study on Parental Perspectives from Australia. *Sustainability (Basel), 17*(11), 5158. Advance online publication. DOI: 10.3390/su17115158

Beig, G., Chate, D. M., Ghude, S. D., Mahajan, A. S., Srinivas, R., Ali, K., Sahu, S. K., Parkhi, N., Surendran, D. E., Trimbake, H. K., & Tiwari, S. (2013). Evaluating population exposure to environmental pollutants during Deepavali fireworks displays using air quality measurements of the SAFAR network. *Chemosphere, 92*(1), 116–124. DOI: 10.1016/j.chemosphere.2013.02.043 PMID: 23561573

Bekele, N. A., Muffatto, M., & Ferrati, F. (2024). Desirability of consumer internet of things products: How emerging businesses address consumer desires to improve user experiences. *International Journal of Technology Marketing, 18*(1), 34–61. DOI: 10.1504/IJTMKT.2024.135668

Benson, P. K. (2024). Concept testing and selection—where and why will it fail. In *Innovation Anesthesiology* (pp. 131–138). Academic Press. DOI: 10.1016/B978-0-12-818381-6.00013-9

Beraldo, G., Oddi, A., & Rasconi, R. (2024). An Empirical Study of Grounding PPDDL Plans for AI-Driven Robots in Social Environment. *Frontiers in Artificial Intelligence and Applications, 392*, 4426–4433. DOI: 10.3233/FAIA241021

Berendes, D. M., de Mondesert, L., Kirby, A. E., Yakubu, H., Adomako, L., Michiel, J., Raj, S., Robb, K., Wang, Y., Doe, B., Ampofo, J., & Moe, C. L. (2020). Variation in E. coli concentrations in open drains across neighborhoods in Accra, Ghana: The influence of onsite sanitation coverage and interconnectedness of urban environments. *International Journal of Hygiene and Environmental Health, 224*, 113433. DOI: 10.1016/j.ijheh.2019.113433 PMID: 31978730

Bhardwaj, V., Anooja, A., Vermani, L. S., Sunita, , & Dhaliwal, B. K. (2024). Smart cities and the IoT: An in-depth analysis of global research trends and future directions. *Discover Internet of Things, 4*(1), 19. DOI: 10.1007/s43926-024-00076-3

Bhattacharjee, J., De, A., Kamila, B., & Mandal, A. (2025). Recent scenario of e-waste recycling: Chemical engineering. *International Journal of Chemical Reactor Engineering, 23*(6), 675–700. DOI: 10.1515/ijcre-2024-0058

Bhubaneswar Municipal Corporation. (2020). *Guidelines for eco-friendly festival practices* (BMC Circular No. 3823).

Bilici, F., & Özdemir, E. (2024). Consumer Perception of Planned Obsolescence: A Research on Smartphone Owners. *İşletme Araştırmaları Dergisi, 16*(1), 244-261.

Biyada, S., & Urbonavičius, J. (2025). Circularity in textile waste: Challenges and pathways to sustainability. *Cleaner Engineering and Technology, 24*, 100905. DOI: 10.1016/j.clet.2025.100905

Blahnik, V., & Schindelbeck, O. (2021). Smartphone imaging technology and its applications. *Advances in Optical Technologies, 10*(3), 145–232. DOI: 10.1515/aot-2021-0023

Blom, K. (2015). Drainage systems, an occluded source of sanitation related outbreaks. *Archives of Public Health = Archives Belges de Santé Publique, 73*(8), 8. Advance online publication. DOI: 10.1186/s13690-014-0056-6 PMID: 25722855

Bonifazi, G., Grosso, C., Palmieri, R., & Serranti, S. (2025). Current trends and challenges in construction and demolition waste recycling. *Current Opinion in Green and Sustainable Chemistry, 53*, 101032. DOI: 10.1016/j.cogsc.2025.101032

Bovea, M., Ibáñez-Forés, V., Gallardo, A., & Colomer-Mendoza, F. (2010). Environmental assessment of alternative municipal solid waste management strategies: A Spanish case study. *Waste Management (New York, N.Y.)*, *30*(11), 2383–2395. DOI: 10.1016/j.wasman.2010.03.001 PMID: 20381331

Bremenkamp, I., & Gallagher, M. J. S. (2024). Life cycle assessment methods for investigating novel food packaging systems. *Environmental Science. Advances*, *3*(10), 1337–1350. DOI: 10.1039/D3VA00380A

Bressanelli, G., Adrodegari, F., Pigosso, D. C., & Parida, V. (2022). Towards the smart circular economy paradigm: A definition, conceptualization, and research agenda. *Sustainability (Basel)*, *14*(9), 4960. DOI: 10.3390/su14094960

Brotosusilo, A., & Naldi, A. (2021, March). Policy on optimization of household waste and hazardous waste management based on community empowerment at the local level. In *IOP Conference Series: Earth and Environmental Science* (Vol. 716, No. 1, p. 012085). IOP Publishing. DOI: 10.1088/1755-1315/716/1/012085

Brziak, M., & Urban, K. (2025). Phenomenology meets self-regulated learning: Exploring the lived experience of learning. *Review of Education*, *13*(1), e70063. DOI: 10.1002/rev3.70063

Business Standard. (2024, October 25). Diwali fireworks and air pollution: How to protect your respiratory health. https://www.business-standard.com/india-news/diwali-fireworks-and-air-pollution-how-to-protect-your-respiratory-health-124102500784_1.html

Buttenheim, A. M. (2008). The sanitation environment in urban slums: Implications for child health. *Population and Environment*, *30*(1-2), 26–47. DOI: 10.1007/s11111-008-0074-9 PMID: 25825551

Butturi, M. A., Neri, A., Mercalli, F., & Gamberini, R. (2025). Sustainability-oriented innovation in the textile manufacturing industry: Pre-consumer waste recovery and circular patterns. *Environments (Basel, Switzerland)*, *12*(3), 82. DOI: 10.3390/environments12030082

Buys, T., Casteleijn, D., Heyns, T., & Untiedt, H. (2022). A reflexive lens on preparing and conducting semi-structured interviews with academic colleagues. *Qualitative Health Research*, *32*(13), 2030–2039. DOI: 10.1177/10497323221130832 PMID: 36194785

Camilleri, R., & Vella, A. J. (2010). Effect of fireworks on ambient air quality in Malta. *Atmospheric Environment*, *44*(35), 4521–4527. DOI: 10.1016/j.atmosenv.2010.07.057

Capoor, M. R., & Parida, A. (2021). Current perspectives of biomedical waste management in context of COVID-19. *Indian Journal of Medical Microbiology, 39*(2), 171–178. DOI: 10.1016/j.ijmmb.2021.03.003 PMID: 33766404

Carroccio, S. C., Scarfato, P., Bruno, E., Aprea, P., Dintcheva, N. T., & Filippone, G. (2021). Impact of nanoparticles on the environmental sustainability of polymer nanocomposites based on bioplastics or recycled plastics – A review of life-cycle assessment studies. *Journal of Cleaner Production, 335*, 130322. Advance online publication. DOI: 10.1016/j.jclepro.2021.130322

Cefis, E., Leoncini, R., & Marengo, L. (2024). Is innovation failure just a dead end? *Structural Change and Economic Dynamics, 70*, 481–493. DOI: 10.1016/j.strueco.2024.05.004

Central Pollution Control Board. (2020). *Guidelines for idol immersion.*

Centre for Science and Environment (CSE). (2017). *SFD Report Kannur*. Retrieved from SFD Promotion Initiative.

Chaudhari, M. M., & Shirude, S. B. (2025). Current E-library System: Challenges, Solutions and Open Issues. *Solutions and Open Issues (March 24, 2025).*

Chauhan, A., Jakhar, S. K., & Chauhan, C. (2020). The interplay of circular economy with industry 4.0 enabled smart city drivers of healthcare waste disposal. *Journal of Cleaner Production, 279*, 123854–123854. DOI: 10.1016/j.jclepro.2020.123854 PMID: 32863607

Cheela, V. R., Ranjan, V. P., Goel, S., John, M., & Dubey, B. (2021). Pathways to sustainable waste management in Indian Smart Cities. *Journal of Urban Management, 10*(4), 419–429. DOI: 10.1016/j.jum.2021.05.002

Chen, Z., King, B., & Suntikul, W. (2019). Festivalscapes and the visitor experience: An application of the stimulus–organism–response approach. *International Journal of Tourism Research, 21*(6), 758–771. DOI: 10.1002/jtr.2302

Chikowore, N. (2021). Factors influencing household waste management practices in Zimbabwe. *Journal of Material Cycles and Waste Management, 23*(1), 386–393. DOI: 10.1007/s10163-020-01129-9

Chioatto, E., & Sospiro, P. (2023). Transition from waste management to circular economy: The European Union roadmap. *Environment, Development and Sustainability, 25*(1), 249–276. DOI: 10.1007/s10668-021-02050-3

Chisholm, J. M., Zamani, R., Negm, A. M., Said, N., Abdel daiem, M. M., Dibaj, M., & Akrami, M. (2021). Sustainable waste management of medical waste in African developing countries: A narrative review. *Waste Management & Research, 39*(9), 1149–1163. DOI: 10.1177/0734242X211029175 PMID: 34218734

Christensen, D., Hjul-Nielsen, J., Moalem, R. M., & Johansen, B. (2021). Circular Economy in Denmark: Bornholm's Vision to Achieve 100 Percent Reuse and Recycling. In *Circular Economy: Recent Trends in Global Perspective* (pp. 385–424). Springer Nature Singapore. DOI: 10.1007/978-981-16-0913-8_13

Čiarnienė, R., & Vienažindienė, M. (2012). Lean manufacturing: Theory and practice. *Economics and Management, 17*(2), 359–366. DOI: 10.5755/j01.em.17.2.2205

Civilsdaily. (2024). *The Maha Kumbh and river pollution: Are we learning from past mistakes?* https://www.civilsdaily.com/the-maha-kumbh-and-river-pollution-are-we-learning-from-past-mistakes/

Clapp, J. (2021). *Waste trading among rich nations: Building a new theory of environmental regulation*. Routledge.

CockColours. (2024). *How to prepare natural colours for the festivities of Holi.* https://cockcolours.com/article/How-to-Prepare-Natural-Colours-for-the-Festivities-of-Holi

Cointreau, S. (2006). Occupational and environmental health issues of solid waste management. (Urban Papers, No. 2). The World Bank. https://www.worldbank.org/urban/

Collins, A. J., & Flynn, A. (2015). *The ecological footprint: New developments in policy and practice.* http://ci.nii.ac.jp/ncid/BB18739155

Collins, A., & Cooper, C. (2016). Measuring and managing the environmental impact of festivals: The contribution of the ecological footprint. *Journal of Sustainable Tourism, 25*(1), 148–162. DOI: 10.1080/09669582.2016.1189922

Corrocher, N., & Paganuzzi, S. (2025). Planned obsolescence and smartphone replacement: Empirical evidence on the Italian market. *Telecommunications Policy, 49*(8), 103022. DOI: 10.1016/j.telpol.2025.103022

Costa, N., Oliveira, J., Martins, A., & Gomes, A. (2024). The planned obsolescence of knowledge: information with an expiry date. In *ICERI2024 Proceedings* (pp. 5407-5411). IATED. DOI: 10.21125/iceri.2024.1321

Costa, A., Ferreira, A. C., Silva, Â. M. E., Ramos, J., & Ramos, B. (2025). Integration of Reverse Logistics and Continuous Improvement in Portuguese Industry: Perspectives from a Qualitative Survey. *Sustainability (Basel)*, *17*(9), 4056. Advance online publication. DOI: 10.3390/su17094056

Crystal, D. (2019). *English as a global language* (3rd ed.). Cambridge University Press.

Ćurić, M., Zafirovski, O., Spiridonov, V., Ćurić, M., Zafirovski, O., & Spiridonov, V. (2022). Air quality and health. *Essentials of medical meteorology*, 143-182.

Curly Tales. (2024, April 12). Maha Kumbh 2025 to boost spiritual tourism in Prayagraj; 162% YOY increase in flight, 187% surge in train bookings reported. https://curlytales.com/maha-kumbh-2025-boost-spiritual-tourism-in-prayagraj-162-y-o-y-increase-in-flight-187-surge-in-train-bookings-reported/

Czrnhak, T., Schreiber, D., & Mazzotti, K. G. (2025). A Comprehensive Analysis of Circular Fashion in Small Clothing Firms; [Uma Análise Compreensiva sobre Moda Circular em Pequenas Empresas]. *Revista de Administracao Contemporanea*, *29*(3). https://doi.org/DOI: 10.1590/1982-7849rac2025240239.por

Danwitz, L., & von Helversen, B. (2025). Observational learning of exploration-exploitation strategies in bandit tasks. *Cognition*, *259*, 106124. DOI: 10.1016/j.cognition.2025.106124 PMID: 40117983

Das, D. (2023, October 23). From culture to economy: How Kolkata's Durga Puja fuels economic growth. *LinkedIn*. https://www.linkedin.com/pulse/from-culture-economy-how-kolkatas-durga-puja-fuels-economic-dona-das

Datta, P., Mohi, G., & Chander, J. (2018). Biomedical waste management in India: Critical appraisal. *Journal of Laboratory Physicians*, *10*(1), 6–14. DOI: 10.4103/JLP.JLP_89_17 PMID: 29403196

De Koning, D. (2023). Green Ganesha Chaturthi: The ritualising and materialising of a green Hindu identity and the emerging of an alternative representation of Ganesha. *Religions*, *14*(1), 22. DOI: 10.3390/rel14010022

Deo, S. (2023). Methodology to implement lean manufacturing in cement industry. *ProQuest*. https://www.proquest.com/openview/8df736e5ec74d93920526b5bbe38063e/?pq-origsite=gscholar&cbl=18750

Dhiman, S., Seema, S., & Akhai, S. (2025). Performance evaluation of OPC 43 concrete incorporating micro silica and recycled ceramic tile waste as partial cement and coarse aggregate replacements. *Asian Journal of Civil Engineering*, *27*(1), 299–317. DOI: 10.1007/s42107-025-01504-7

Dhingra, S., Raut, R. D., Yadav, V. S., Cheikhrouhou, N., & Naik, B. K. R. (2025). Blockchain adoption challenges in the healthcare sector: A waste management perspective. *Operations Management Research : Advancing Practice Through Research, 18*(2), 518–536. DOI: 10.1007/s12063-023-00413-9 PMID: 40625400

Do, A. D., Ha, D. L., Pham, M. T., Khuat, M. A. T., Le, V. A. T., Nguyen, D. N. D., & La, T. Q. (2025). Impacts of e-RLSQ on repurchase intention in Vietnam's e-commerce market: The mediating role of customer satisfaction and trust. *International Journal of Information Management Data Insights, 5*(2), 100346. Advance online publication. DOI: 10.1016/j.jjimei.2025.100346

Dukovska-Popovska, I., Kaipia, R., Kjellsdotter Ivert, L., Dreyer, H. C., & Jonsdottir, H. (2025). End-of-life textile collection from consumers: A design science approach. *International Journal of Physical Distribution & Logistics Management, 55*(5), 482–504. DOI: 10.1108/IJPDLM-12-2023-0459

Ebrahimzadehsarvestani, E., Sarvestani, E. M., Safarzadeh, H., Enayati, M., & Di Maria, F. (2025). Implementing circular economy in healthcare waste: Technological innovations and process management. *Journal of Solid Waste Technology Management, 51*(1), 125–134. DOI: 10.5276/jswtm/iswmaw/51SI1/2025.014

EPA. (2021). *Resource Conservation and Recovery Act (RCRA) overview*. U.S. Environmental Protection Agency. https://www.epa.gov/rcra

Espejo-Herrera, N., Gràcia-Lavedan, E., Boldo, E., Aragonés, N., Pérez-Gómez, B., Pollán, M., & Villanueva, C. M. (2016). Colorectal cancer risk and nitrate exposure through drinking water and diet. *International Journal of Cancer, 139*(2), 334–346. DOI: 10.1002/ijc.30083 PMID: 26954527

European Commission. (2008). Directive 2008/98/EC on waste and repealing certain directives. https://eur-lex.europa.eu/legal-content/EN/TXT/?uri=celex%3A32008L0098

European Commission. (2020). A new circular economy action plan: For a cleaner and more competitive Europe. https://environment.ec.europa.eu/strategy/circular-economy-action-plan_en

Faiz, F., Ninduwezuor-Ehiobu, N., Adanma, U. M., & Solomon, N. O. (2024). AI-Powered waste management: Predictive modeling for sustainable landfill operations. Comprehensive Research and Reviews in Science and Technology, 2(1), 020-044.

Faizan, M. (2021). Solid Waste Management in India under Covid 19 Pandemic: Challenges and Solutions. *International Journal of Waste Resources, 11*, 1–4.

Fang, B., Yu, J., Chen, Z., Osman, A. I., Farghali, M., Ihara, I., Hamza, E. H., Rooney, D. W., & Yap, P.-S. (2023). Artificial intelligence for waste management in smart cities: A review. *Environmental Chemistry Letters*, *21*(4), 1959–1989. DOI: 10.1007/s10311-023-01604-3 PMID: 37362015

Fan, Z., Li, X., Gao, Q., & Li, S. (2025). Optimizing Reverse Logistics Network for Waste Electric Vehicle Batteries: The Impact Analysis of Chinese Government Subsidies and Penalties. *Sustainability (Basel)*, *17*(9), 3885. Advance online publication. DOI: 10.3390/su17093885

Farjana, M., Fahad, A. B., Alam, S. E., & Islam, M. M. (2023). An IoT- and cloud-based e-waste management system for resource reclamation with a data-driven decision-making process. *IoT*, *4*(3), 202–220. DOI: 10.3390/iot4030011

Fatimah, H., & Hafsah, U. M. (2020). Social Work's Program By Universitas Negeri Malang Spraying Desinfectant of Distribution of Covid-19 Trafficking Kunti District, Bungkal Ponorogo. *International Research-Based Education Journal*, *2*(2), 63–68. DOI: 10.17977/um043v2i2p63-68

Fatorachian, H., & Pawar, K. (2025). Enhancing waste management in cold supply chains via digital transformation. *Procedia Computer Science*, *253*, 167–176. DOI: 10.1016/j.procs.2025.01.080

Fazlollahtabar, H. (2024). Waste opportunity gap for energy productivity in new product design. *Energy Sources. Part B, Economics, Planning, and Policy*, *19*(1), 2371407. DOI: 10.1080/15567249.2024.2371407

Firstpost. (2024, April 13). Maha Kumbh 2025: Soaring prices, hotels in Prayagraj & flight bookings explained. https://www.firstpost.com/explainers/maha-kumbh-2025-soaring-prices-hotels-in-prayagraj-flight-train-bookings-13852657.html

Forastiere, F., Badaloni, C., de Hoogh, K., von Kraus, M. K., Martuzzi, M., Mitis, F., Palkovicova, L., Porta, D., Preiss, P., Ranzi, A., Perucci, C. A., & Briggs, D. (2011). Health impact assessment of waste management facilities in three European countries. *Environmental Health : A Global Access Science Source*, *10*(1), 53. Advance online publication. DOI: 10.1186/1476-069X-10-53 PMID: 21635784

Forti, V., Baldé, C. P., & Kuehr, R. (2022). *E-waste statistics: Global, regional, and national trends*. United Nations University.

Freisinger, E., & McCarthy, I. P. (2024). What fails and when? A process view of innovation failure. *Technovation*, *133*, 102995. DOI: 10.1016/j.technovation.2024.102995

Ganesan, P. (2023). Revolutionizing Robotics with AI, Machine Learning, and Deep Learning: A Deep Dive into Current Trends and Challenges Journal of Artificial Intelligence, Machine Learning and Data Science. *J Artif Intell Mach Learn & Data Sci*, *2023*(4), 1124–1128. DOI: 10.51219/JAIMLD/premkumar-ganesan/263

Ganguly, N. D., Tzanis, C. G., Philippopoulos, K., & Deligiorgi, D. (2019). Analysis of a severe air pollution episode in India during Diwali festival – A nationwide approach. *Atmósfera (México)*, *32*(3), 225–236. DOI: 10.20937/ATM.2019.32.03.05

Gardezi, A., Rathore, F. A., & Farooq, F. (2020). Sanitization Walk-Through Gates During COVID-19 Pandemic: Effective or A False Sense of Protection? *Annals of King Edward Medical University*, *26*(Special Issue), 259–261.

Gaur, T. S., Yadav, V., Prakash, S., & Panwar, A. (2025). Integration of industry 4.0 and circular economy for sustainable E-waste management. *Management of Environmental Quality*, *36*(5), 1304–1325. DOI: 10.1108/MEQ-07-2024-0277

Getz, D. (2010). The nature and scope of festival studies. *International Journal of Event Management Research*, *5*(1), 1–47. https://ijemr.org/wp-content/uploads/2014/10/Getz.pdf

Gharpure, R., Hunter, C. M., Schnall, A. H., Barrett, C. E., Kirby, A. E., Kunz, J., & Garcia-Williams, A. G. (2020). Knowledge and practices regarding safe household cleaning and disinfection for COVID-19 prevention—United States, May 2020. *Morbidity and Mortality Weekly Report*, *69*(23), 705709. PMID: 32525852

Ghei, D., & Sane, R. (2018). *Air pollution and fireworks: Evidence from Diwali.* Indian Statistical Institute.

Ghosh, B., Barman, H. C., & Padhy, P. K. (2024). Trends of air pollution variations during pre-Diwali, Diwali and post-Diwali periods and health risk assessment using HAQI in India. *Discover Environment*, *2*(1), 81. DOI: 10.1007/s44274-024-00082-w

Gibson, C., & Wong, C. (2011). Chapter 6. Greening rural festivals: Ecology, sustainability and human–nature relations. In *Multilingual Matters eBooks* (pp. 92–106). https://doi.org/DOI: 10.21832/9781845411688-009

Gina, M., Ofenloch, R., Schwebke, I., Hübner, N. O., Brüning, T., & Fartasch, M. (2025). The Effect of Alcohol-Based Virucidal Hand Sanitizers on Skin Barrier Function—A Randomised Experimental Study. *Contact Dermatitis*, *93*(2), 119–130. DOI: 10.1111/cod.14808 PMID: 40356051

Gold, J. R., & Gold, M. M. (2020). *Festival cities.* Routledge., DOI: 10.4324/9781003130802

Government of Maharashtra. (2022). *Sustainable Ganesh Idol Making Scheme – Department of Environment and Climate Change.*

Gracias, J. S., Parnell, G. S., Specking, E., Pohl, E. A., & Buchanan, R. K. (2023). *Smart Cities—A Structured Literature Review*. Smart Cities. DOI: 10.3390/smartcities6040080

Grand View Research. (2024). *India religious tourism market size, share & trends analysis report, 2024–2030.* https://www.grandviewresearch.com/industry-analysis/india-religious-tourism-market-report

Greven, F. E., Vonk, J. M., Fischer, P., Duijm, F., Vink, N. M., & Brunekreef, B. (2019). Air pollution during New Year's fireworks and daily mortality in the Netherlands. *Scientific Reports, 9*(1), 5735. Advance online publication. DOI: 10.1038/s41598-019-42080-6 PMID: 30952902

Grow Billion Trees. (2024). *Tree plantation on festivals.* https://growbilliontrees.com/collections/tree-plantation-on-festivals

Gulivindala, A. K., Bahubalendruni, M. R., P, M. B., & Eswaran, M. (2023). Mechanical disassembly sequence planning for end-of-life products to maximize recyclability. *Sadhana, 48*(3), 109. DOI: 10.1007/s12046-023-02160-w

Guo, X., & Yan, K. (2017). Estimation of obsolete cellular phones generation: A case study of China. *The Science of the Total Environment, 575*, 321–329. DOI: 10.1016/j.scitotenv.2016.10.054 PMID: 27744159

Gupta, A., Bherwani, H., Gautam, S., Anjum, S., Musugu, K., Kumar, N., Anshul, A., & Kumar, R. (2020). Air pollution aggravating COVID-19 lethality? Exploration in Asian cities using statistical models. *Environment, Development and Sustainability, 23*(4), 6408–6417. DOI: 10.1007/s10668-020-00878-9 PMID: 32837279

Gupta, N., & Yadav, K. K. (2019). Sustainable solid waste management in India. *Environmental Science and Pollution Research International, 26*(3), 2164–2175.

Gupta, N., Yadav, K. K., & Kumar, V. (2015). A review on current status of municipal solid waste management in India. *Journal of Environmental Sciences (China), 37*, 206–217. DOI: 10.1016/j.jes.2015.01.034 PMID: 26574106

Gurjot, A. P., Sharma, K. K., Reshma, & Bindle, A. (2022). Future Tendency Anticipation for Users in Blogging. *2022 2nd International Conference on Advance Computing and Innovative Technologies in Engineering, ICACITE 2022*, 1194 – 1198. DOI: 10.1109/ICACITE53722.2022.9823429

Habuer, N., Nakatani, J., & Moriguchi, Y. (2014). Time-series product and substance flow analyses of end-of-life electrical and electronic equipment in China. *Waste Management (New York, N.Y.)*, *34*(2), 489–497. DOI: 10.1016/j.wasman.2013.11.004 PMID: 24332400

Hamdoon, S., & Alabadee, S. (2025). The effect of strategic monitoring on lean manufacturing system: A case study of Northern Cement Company, Iraq. *International Journal of Social Science Management and Economics Research*, *3*(1), 59–75. DOI: 10.61421/IJSSMER.2025.3106

Han, Z., Pappas, E., Simmons, A., Fox, J., Donskey, C. J., & Deshpande, A. (2020). Environmental cleaning and disinfection of hospital rooms: A nationwide survey. *American Journal of Infection Control*, *49*(1), 34–39. DOI: 10.1016/j.ajic.2020.08.008 PMID: 32798634

Health Care Without Harm. (2025). *Waste management*. https://us.noharm.org/waste-management

Heljakka, K. (2022). On longevity and lost toys: sustainable approaches to toy design and contemporary play. In *Toys and Sustainability* (pp. 19–37). Springer Singapore. DOI: 10.1007/978-981-16-9673-2_2

Hennink, M., & Kaiser, B. N. (2022). Sample sizes for saturation in qualitative research: A systematic review of empirical tests. *Social Science & Medicine*, *292*, 114523. DOI: 10.1016/j.socscimed.2021.114523 PMID: 34785096

Hines, S. E., Brown, C. H., Oliver, M., Gucer, P., Frisch, M., Hogan, R., Roth, T., Chang, J., & McDiarmid, M. (2020). Cleaning and Disinfection Perceptions and Use Practices Among Elastomeric Respirator Users in Health care. *Workplace Health & Safety*, *68*(12), 572–582. DOI: 10.1177/2165079920938618 PMID: 32812845

Holidify. (n.d.). *Jagannath Rath Yatra, Puri overview*. https://www.holidify.com/places/puri/jagannath-rath-yatra-sightseeing-3321.html

Hoveidi, H., Ahmadi Pari, M., Pazoki, M., & Koulaeian, T. (2013). Industrial waste management with application of RIAM environmental assessment: A case study on Toos Industrial State, Mashhad. *Iranica Journal of Energy and Environment*, *4*(2), 142–149. DOI: 10.5829/idosi.ijee.2013.04.02.11

Hudomiet, P., & Willis, R. J. (2022). Computerization, obsolescence and the length of working life. *Labour Economics*, *77*, 102005. DOI: 10.1016/j.labeco.2021.102005 PMID: 37035254

Huffling, K., & Schenk, E. (2014). Environmental sustainability in the intensive care unit: Challenges and solutions. *Critical Care Nursing Quarterly, 37*(3), 235–250. DOI: 10.1097/CNQ.0000000000000028 PMID: 24896556

Hunka, A. D., Linder, M., & Habibi, S. (2021). Determinants of consumer demand for circular economy products. A case for reuse and remanufacturing for sustainable development. *Business Strategy and the Environment, 30*(1), 535–550. DOI: 10.1002/bse.2636

Husin, N., Ibrahim, N. A., Rahim, N. A., Abd Manaf, K., & Shafie, I. S. M. (2024). Exploring the Impact of Authentic Personality and Social Media Posting on Self-Interested Charitable Behavior among Millennials Online Community through the Lens of Social Cognitive Theory. *Information Management and Business Review, 16*(1), 39–48. DOI: 10.22610/imbr.v16i1(I)S.3724

Ibrahim, M., Kebede, M., & Mengiste, B. (2023). Healthcare waste segregation practice and associated factors among healthcare professionals working in public and private hospitals, Dire Dawa, Eastern Ethiopia. *Journal of Environmental and Public Health, 2023*(3), 1–7. DOI: 10.1155/2023/8015856 PMID: 36747502

Ihamäki, P., & Heljakka, K. (2019, November). Out of the box, into the cubes: Envisioning user experiences through a tool for gamification, toyification and playification. In *International Conference on ArtsIT, Interactivity and Game Creation* (pp. 49-68). Cham: Springer International Publishing.

Immerman, G. (2022, December 13). Lean manufacturing: The principles, wastes, benefits, and tools. *MachineMetrics.* https://www.machinemetrics.com/blog/lean-manufacturing

India Today. (2022, June 29). After two Covid-curtailed years, Puri Rath Yatra preparations in full swing. https://www.indiatoday.in/india-today-insight/story/after-two-covid-curtailed-years-puri-rath-yatra-preparations-in-full-swing-1968178-2022-06-29

Intauno, K., & Poshai, L. (2023). Examining the occupational safety and health hazards encountered by municipal waste collectors in murewa rural district council, Zimbabwe. [IJLAPP]. *International Journal of Law and Public Policy, 5*(2), 59–71. DOI: 10.36079/lamintang.ijlapp-0502.532

Ionescu, A.-M., & Cazan, C. (2024). Pharmaceutical Waste Management: A Comprehensive Analysis of Romanian Practices and Perspectives. *Sustainability (Basel), 16*(15), 6571. DOI: 10.3390/su16156571

Islam, M. M., Zaman, S., Ferdous, J., & Sarker, M. H. (2025). Policy gaps, actor dynamics, and implementation challenges in sustainable medical waste management in Bangladesh. *Journal of Environmental Management, 389*, 126203. DOI: 10.1016/j.jenvman.2025.126203 PMID: 40517630

Jakhar, R., Samek, L., & Styszko, K. (2023). A comprehensive study of the impact of waste fires on the environment and health. *Sustainability (Basel), 15*(19), 14241. DOI: 10.3390/su151914241

Javed, A. R., Zikria, Y. B., Rehman, S. U., Shahzad, F., & Jalil, Z. (2021). *Future Smart Cities: Requirements*. Emerging Technologies, Applications, Challenges, and Future Aspects.

Jemberie, M. A., Melesse, A. M., & Abate, B. (2023). Urban Drainage: The Challenges and Failure Assessment Using AHP, Addis Ababa, Ethiopia. *Water (Basel), 15*(5), 957. DOI: 10.3390/w15050957

Jerie, S., Shabani, T., Mudyazhezha, O. C., & Shabani, T. (2024). A review towards developing a hierarchical model for sustainable hospital solid waste management in rural areas of Zimbabwe. *Environmental Monitoring and Assessment, 196*(3), 308. DOI: 10.1007/s10661-024-12488-3 PMID: 38407739

Joshi, M., & Khanvilkar, V. (2018). Urban waste management during religious festivals: A study of Pune city. *Journal of Environmental Planning and Management, 61*(11), 1934–1950.

Kabir, S. A., Ali, F., Ahmed, R. L., & Sulaiman-Hill, R. (2025). Exploring the Use of AI in Qualitative Data Analysis: Comparing Manual Processing with Avidnote for Theme Generation. *International Journal of Qualitative Methods, 24*, 16094069251336810. DOI: 10.1177/16094069251336810

Kaithery, N., & Karunakaran, U. (2019). Study on attitude of household waste management in a rural area of Northern Kerala. *International Journal of Community Medicine and Public Health, 6*(5), 2095. DOI: 10.18203/2394-6040.ijcmph20191826

Kalonde, P. K., Austin, A. C., Mandevu, T., Banda, P. J., Banda, A., Stanton, M. C., & Zhou, M. (2023). Determinants of household waste disposal practices and implications for practical community interventions: Lessons from Lilongwe. *Environmental Research: Infrastructure and Sustainability, 3*(1). Advance online publication. DOI: 10.1088/ersi/3/1/015001

Kane, G. M., Bakker, C. A., & Balkenende, A. R. (2018). Towards design strategies for circular medical products. *Resources, Conservation and Recycling, 135*, 38–47. DOI: 10.1016/j.resconrec.2017.07.030

Kang, K., & Tan, B. Q. (2025). Multi-echelon reverse logistics network design in the context of circular economy: A Hong Kong case study. *Humanities & Social Sciences Communications*, *12*(1), 40. Advance online publication. DOI: 10.1057/s41599-024-04323-4

Kastanaki, E. (2025). Dynamic assessment of photovoltaic waste streams in the EU-27 countries under the circular economy principles of 'Reduce, Reuse and Recycle'. *Resources, Conservation and Recycling*, *214*, 108033. DOI: 10.1016/j.resconrec.2024.108033

Kastanaki, E., & Giannis, A. (2022). Forecasting quantities of critical raw materials in obsolete feature and smart phones in Greece: A path to circular economy. *Journal of Environmental Management*, *307*, 114566. DOI: 10.1016/j.jenvman.2022.114566 PMID: 35091243

Kaur, A., & Sharma, V. (2024). Nexus between climate change and mitigation approaches for sustainable development: A bibliometric review. *Global Knowledge, Memory and Communication*. Advance online publication. DOI: 10.1108/GKMC-11-2023-0418

Keck, M. E., & Sikkink, K. (1998). *Activists beyond borders: Advocacy networks in international politics*. Cornell University Press.

Kenny, C., & Priyadarshini, A. (2021). Review of current healthcare waste management methods and their effect on global health. *Health Care*, *9*(3), 284. DOI: 10.3390/healthcare9030284 PMID: 33807606

Khade, S., & Sawant, M. (2023). Assessment of festival-related pollution impacts on urban soil. *CMR Journal, 11*, 43–54. https://cmrj.in/011-May-23-Khade-Savant.pdf

Khajuria, A., Atienza, V. A., Chavanich, S., Henning, W., Islam, I., Kral, U., & Li, J. (2022). Accelerating circular economy solutions to achieve the 2030 agenda for sustainable development goals. *Circular Economy*, *1*(1), 100001. DOI: 10.1016/j.cec.2022.100001

Khang, A., & Akhai, S. (2025). E-Waste and Lithium-ion Battery Recycling Insights for Sustainable Transportation. In Khang, A. (Ed.), *Driving Green Transportation System Through Artificial Intelligence and Automation* (pp. 203–230). Springer., DOI: 10.1007/978-3-031-72617-0_11

Khan, R., & Kumar, A. (2024). A Hybrid Model Approach to Machine Learning Based Heart Disease Prediction. *2nd International Conference on Signal Processing, Communication, Power and Embedded Systems, SCOPES 2024*. DOI: 10.1109/SCOPES64467.2024.10991142

Kharat, M. G., Kapoor, S., Parhi, S., Kharat, M. G., & Pandey, S. (2025). Operationalizing sustainability in pharmaceuticals: Green supply chain metrics for circular economy. *Sustainable Futures : An Applied Journal of Technology, Environment and Society*, *9*, 100413. Advance online publication. DOI: 10.1016/j.sftr.2024.100413

Khoo, H. H. (2009). Life cycle impact assessment of various waste conversion technologies. *Waste Management (New York, N.Y.)*, *29*(6), 1892–1900. DOI: 10.1016/j.wasman.2008.12.020 PMID: 19157835

Kiehbadroudinezhad, M., Merabet, A., & Hosseinzadeh-Bandbafha, H. (2024). Landfill source of greenhouse gas emission. In *Advances and technology development in greenhouse gases: Emission, capture and conversion* (pp. 123–145). Elsevier. DOI: 10.1016/B978-0-443-19231-9.00023-5

Klabusayová, N. (2014). The possibilities for increasing production processes efficiency by utilizing the lean manufacturing principles. *Applied Mechanics and Materials*, *693*, 483–488. . DOI: 10.4028/www.scientific.net/AMM.693.483

Klima, V., Chadyšienė, R., Ivanec-Goranina, R., Jasaitis, D., & Vasiliauskienė, V. (2020). Assessment of air pollution with polychlorinated dibenzodioxins (PCDDs) and polychlorinated dibenzofuranes (PCDFs) in Lithuania. *Atmosphere (Basel)*, *11*(7), 759. DOI: 10.3390/atmos11070759

Kramer, A., Galabov, A. S., Sattar, S. A., Döhner, L., Pivert, A., Payan, C., Wolff, M. H., Yilmaz, A., & Steinmann, J. (2006). Virucidal activity of a new hand disinfectant with reduced ethanol content: Comparison with other alcohol-based formulations. *The Journal of Hospital Infection*, *62*(1), 98–106. DOI: 10.1016/j.jhin.2005.06.020 PMID: 16307826

Krikke, H. R., Sayers, M., & Smith, L. (2025). Should Recovery Strategies Be Reuse or Recycling Based? A Case Study on Discarded Electronics in Melbourne, Australia. *Logistics (Basel)*, *9*(2), 58. Advance online publication. DOI: 10.3390/logistics9020058

Kuligowski, K., Konkol, I., Świerczek, L., Chojnacka, K., Cenian, A., & Szufa, S. (2023). Evaluation of Kitchen Waste Recycling as Organic N-Fertiliser for Sustainable Agriculture under Cool and Warm Seasons. *Sustainability (Basel)*, *15*(10), 7997. DOI: 10.3390/su15107997

Kulkarni, B. N., & Anantharama, V. (2020). Repercussions of COVID-19 pandemic on municipal solid waste management: Challenges and opportunities. *The Science of the Total Environment*, *743*, 140693. DOI: 10.1016/j.scitotenv.2020.140693 PMID: 32663690

Kullman, S. M., & Chudyk, A. M. (2025). Participatory member checking: A novel approach for engaging participants in co-creating qualitative findings. *International Journal of Qualitative Methods*, *24*, 16094069251321211. DOI: 10.1177/16094069251321211

Kulshrestha, U. C. (2018). The social rituals and environmental degradation: How to tackle? *Current World Environment*, *13*(3), 285–287. DOI: 10.12944/CWE.13.3.01

Kulshrestha, U. C., Nageswara Rao, T., Azhaguvel, S., & Kulshrestha, M. J. (2004). Emissions and accumulation of metals in the atmosphere due to crackers and sparklers during Diwali festival in India. *Atmospheric Environment*, *38*(27), 4421–4425. DOI: 10.1016/j.atmosenv.2004.05.044

Kumar, A. (2024). A novel framework for waste management in smart city transformation with industry 4.0 technologies. *Research in globalization, 9*, 100234.

Kumar, A., Khan, R., & Deepika. (2025). Heart disease prediction using machine learning techniques. In *Sustainable Materials, Structures and IoT [SMSI-2024]*. CRC Press. DOI: 10.1201/9781003596776-30

Kumar, A., Singh, P. P., & Khanna, M. K. (2024a). Sustainable e-waste management and its effect on environment and human health. In *Integrated Waste Management: A Sustainable Approach from Waste to Wealth* (pp. 349–373). Springer Nature Singapore. DOI: 10.1007/978-981-97-0823-9_17

Kumar, P., Singh, S., Gacem, A., Yadav, K. K., Bhutto, J. K., Alreshidi, M. A., & Alam, M. W. (2024b). A review on e-waste contamination, toxicity, and sustainable clean-up approaches for its management. *Toxicology*, *508*, 153904. DOI: 10.1016/j.tox.2024.153904 PMID: 39106909

Kumar, R.. (2020). Eco-friendly festive practices in India. *Journal of Environmental Management*, *260*, 110143.

Kumar, R., Somrongthong, R., & Ahmed, J. (2016). Impact of waste management training intervention on knowledge, attitude and practices of teaching hospital workers in Pakistan. *Pakistan Journal of Medical Sciences*, *32*(3), 705–710. DOI: 10.12669/pjms.323.9903 PMID: 27375718

Kumar, R., & Tripathi, S. (2021). Sustainable celebrations: A case study of eco-friendly Ganeshotsav in Pune. *Journal of Environmental Management*.

Kumar, S. G., & Joseph, N. (2012). Drainage and sewerage system in urban India: Need for action. *Indian Journal of Occupational and Environmental Medicine*, *16*(3), 150–151. DOI: 10.4103/0019-5278.111767 PMID: 23776328

Kumar, S., Sinha, S., Ranjan, R., & Saurav, S. (2025). Optimization reverse logistics for waste plastic recycling in India: A genetic algorithm approach. *Innovative Infrastructure Solutions*, *10*(5), 167. Advance online publication. DOI: 10.1007/s41062-025-01969-0

Kumar, S., Smith, S. R., Fowler, G., Velis, C., Kumar, S. J., Arya, S., Rena, , Kumar, R., & Cheeseman, C. (2017). Challenges and opportunities associated with waste management in India. *Royal Society Open Science*, *4*(3), 4. DOI: 10.1098/rsos.160764 PMID: 28405362

Kwatra, S., Pandey, S., & Sharma, S. (2014). Understanding public knowledge and awareness on e-waste in an urban setting in India: A case study for Delhi. *Management of Environmental Quality*, *25*(6), 752–765. DOI: 10.1108/MEQ-12-2013-0139

Laeequddin, M., Kareem Abdul, W., Sahay, V., & Tiwari, A. K. (2022). Factors that influence the safe disposal behavior of E-waste by electronics consumers. *Sustainability (Basel)*, *14*(9), 4981. DOI: 10.3390/su14094981

Lakhouit, A. (2025). Revolutionizing urban solid waste management with AI and IoT: A review of smart solutions for waste collection, sorting, and recycling. *Results in Engineering*, *25*, 104018. DOI: 10.1016/j.rineng.2025.104018

Lechelt, S., Gorkovenko, K., & Speed, C. (2024). On Disused Connected Devices: Understanding Disuse,"Holding On," and Barriers to Circularity. *ACM Journal on Computing and Sustainable Societies*, *2*(2), 1–27. DOI: 10.1145/3651171

Lee, S. M., & Lee, D. (2022). Effective medical waste management for sustainable green healthcare. *International Journal of Environmental Research and Public Health*, *19*(22), 14820. DOI: 10.3390/ijerph192214820 PMID: 36429539

Leow, N. X., Lin, W. L., Yong, J. Y., Chong, S. C., Pek, C. K., & Feranita, F. (2026). Optimizing Remote Work for a Sustainable Workplace: Strategies to Promote Employee Green Behavior. In Yusliza, M. Y., Renwick, D. W. S., Yong, J. Y., & Kamil, N. L. M. (Eds.), *Employee Green Behavior* (pp. 97–117). Springer., DOI: 10.1007/978-981-95-1422-9_6

Lev, N., Negev, M., & Ayalon, O. (2023). Sometimes littering is acceptable—Understanding and addressing littering perceptions in natural settings. *Sustainability (Basel)*, *15*(18), 13784. DOI: 10.3390/su151813784

Lidgren, L., Raina, D. B., Tägil, M., & Tanner, K. E. (2020). Recycling implants: A sustainable solution for musculoskeletal research. *Acta Orthopaedica*, *91*(2), 125. DOI: 10.1080/17453674.2019.1706301 PMID: 31902268

Likotiko, E., Matsuda, Y., & Yasumoto, K. (2023). Garbage Content Estimation Using Internet of Things and Machine Learning. *IEEE Access : Practical Innovations, Open Solutions*, *11*, 13000–13012. DOI: 10.1109/ACCESS.2023.3242547

Limenh, L. W., Simegn, W., Bizuneh, G. K., Chanie, G. S., Seid, A. M., Mengesha, A. K., Beyna, A. T., Melese, M., Gela, Y. Y., Esubalew, D., Mitku, M. L., & Ayenew, W. (2025). Management of pharmaceutical waste from hospitals and community pharmacies in Bahir Dar and Gondar cities, Ethiopia. *BMC Health Services Research*, *25*(1), 379. DOI: 10.1186/s12913-025-12555-6 PMID: 40087596

Limna, P., Kraiwanit, T., & Siripipattanakul, S. (2023). The relationship between cyber security knowledge, awareness and behavioural choice protection among mobile banking users in Thailand. *International Journal of Computing Sciences Research*, *7*, 1133–1151. DOI: 10.25147/ijcsr.2017.001.1.123

Li, T., & Qiao, J. (2025). A novel q-rung orthopair fuzzy MAGDM method for healthcare waste treatment based on three-way decisions. *Pattern Recognition*, *157*, 110867. DOI: 10.1016/j.patcog.2024.110867

Liu, J., Chen, Y., Chao, S., Cao, H., & Zhang, A. (2019). Levels and health risks of PM2.5-bound toxic metals from firework/firecracker burning during festival periods in response to management strategies. *Ecotoxicology and Environmental Safety*, *171*, 406–413. DOI: 10.1016/j.ecoenv.2018.12.104 PMID: 30634092

Llatas, C., Quiñones, R., & Bizcocho, N. (2022). Environmental Impact Assessment of Construction Waste Recycling versus Disposal Scenarios Using an LCA-BIM Tool during the Design Stage. *Recycling*, *7*(6), 82. Advance online publication. DOI: 10.3390/recycling7060082

Longe, E. O., & Williams, A. (2006). A preliminary study of medical waste management in Lagos metropolis, Nigeria. *Journal of Environmental Health Science & Engineering*, *3*(2), 133–139. https://ijehse.tums.ac.ir/index.php/jehse/article/view/82

Lorang, S., Yang, Z., Zhang, H., Lü, F., & He, P. (2022). Achievements and policy trends of extended producer responsibility for plastic packaging waste in Europe. *Waste Disposal & Sustainable Energy*, *4*(2), 91–103. DOI: 10.1007/s42768-022-00098-z

Luo, H., Fu, M., Chen, X., Chen, Y., Li, Y., Liu, X., Su, X., & Chen, Y. (2020). Exploration of Infection Prevention and Control Practice of Designated Hospitals in During Corona Virus Disease 2019 Epidemic Period. *Journal of Clinical and Nursing Research*, *4*(4), 80–90. DOI: 10.26689/jcnr.v4i4.1347

Ly, N. H. T., Hong, T. T. K., & Giao, N. T. (2025). Impact of Informal Electronic Waste Recycling and Dumping Activities on Soil Ecosystems: A review from 2000 to 2024. *Journal of Applied Science & Environmental Management*, *29*(4), 1075–1085.

Maher, C., & Bedwei-Majdou, C. (2025). Semi-structured Qualitative Interview Guide: Process and Considerations for Doctoral Students. In *Qualitative Research Methods for Dissertation Research* (pp. 195-220). IGI Global Scientific Publishing.

Mahtta, R. (2015). *Green growth and air pollution in India* (18 pp.). The Energy and Resources Institute (TERI).

Mairizal, A. Q., Sembada, A. Y., Tse, K. M., & Rhamdhani, M. A. (2021). Electronic waste generation, economic values, distribution map, and possible recycling system in Indonesia. *Journal of Cleaner Production*, *293*, 126096. DOI: 10.1016/j.jclepro.2021.126096

Mair, J., & Whitford, M. (2013). An exploration of events research: Event topics, themes and emerging trends. *International Journal of Event and Festival Management*, *4*(1), 6–30. DOI: 10.1108/17582951311307485

Makov, T., & Fitzpatrick, C. (2021). Is repairability enough? big data insights into smartphone obsolescence and consumer interest in repair. *Journal of Cleaner Production*, *313*, 127561. DOI: 10.1016/j.jclepro.2021.127561

Malinauskaite, J., & Erdem, F. B. (2021). Planned obsolescence in the context of a holistic legal sphere and the circular economy. *Oxford Journal of Legal Studies*, *41*(3), 719–749. DOI: 10.1093/ojls/gqaa061 PMID: 34584495

Mamun, A., Yang, Q., Naznen, F., Aziz, N. A., & Masud, M. M. (2023). Modelling the mass adoption potential of food waste composting among rural Chinese farmers. *Heliyon*, *9*(8), e18998. DOI: 10.1016/j.heliyon.2023.e18998 PMID: 37609413

Mandadi, G. K., Asmatulu, R., Khan, W. S., & Asmatulu, E. (2020). Fast and affordable recycling approach to electronic waste above the melting point using induction heat combined with centrifugal forces. *Asia-Pacific Journal of Chemical Engineering*, *15*(4), e2483. DOI: 10.1002/apj.2483

Mane, S.R. (2021). A Systematic Overview of India's Smart City Mission.

Manimaran, S., Razalli, N. H., Manaf, Z. A., & Shahar, S. (2025). Challenges and strategies to reduce food waste in malaysian hospitals from the perspective of multidisciplinary professionals: A qualitative descriptive study. *BMC Health Services Research*, *25*(1), 224. DOI: 10.1186/s12913-025-12365-w PMID: 39930430

Mani, S., & Singh, S. (2016). Sustainable Municipal Solid Waste Management in India: A Policy Agenda☆. *Procedia Environmental Sciences*, *35*, 150–157. DOI: 10.1016/j.proenv.2016.07.064

Marciniuk-Kluska, A., & Kluska, M. (2025). Energy Recovery from Municipal Biodegradable Waste in a Circular Economy. *Energies*, *18*(9), 2210. DOI: 10.3390/en18092210

Marques, L., Borba, C., & Michael, J. (2020). Grasping the social dimensions of event experiences: Introducing the Event Social Interaction Scale (ESIS). *Event Management*, *25*(1), 9–26. DOI: 10.3727/152599520X15894679115448

Martínez-Torres, J., Gonzálvez, C., & Antón, N. (2024). School refusal profiles maintained by negative reinforcement and their relationship with self-perceived health. *Frontiers in Psychology*, *15*, 1340010. DOI: 10.3389/fpsyg.2024.1340010 PMID: 38562235

Martín, H. R. (2024). *How do we learn? A scientific approach to learning and teaching (evidence-based education)*. John Wiley & Sons.

Marzouk, M., & Azab, S. (2013). Environmental and economic impact assessment of construction and demolition waste disposal using system dynamics. *Resources, Conservation and Recycling*, *82*, 41–49. DOI: 10.1016/j.resconrec.2013.10.015

Masmali, M. (2021). Implementation of lean manufacturing in a cement industry. *Engineering Technology & Applied Science Research*, *11*(3), 7069–7074. DOI: 10.48084/etasr.4087

Matengo, T., Mwedzi, T., & Mbetu, J. (2025). Spatial and temporal patterns in water quality analysis near effluent dump-sites in Harare, Zimbabwe. *Journal of Water, Sanitation, and Hygiene for Development : a Journal of the International Water Association*, *15*(3), 206–214. DOI: 10.2166/washdev.2025.270

McDonnell, G. E. (2020). *Antisepsis, disinfection, and sterilization: types, action, and resistance*. John Wiley & Sons.

Menikpura, S. N. M., Gheewala, S. H., Bonnet, S., & Chiemchaisri, C. (2012). Evaluation of the effect of recycling on sustainability of municipal solid waste management in Thailand. *Waste and Biomass Valorization*, *4*(2), 237–257. DOI: 10.1007/s12649-012-9119-5

Milios, L., Davani, A. E., & Yu, Y. (2018). Sustainability Impact Assessment of increased plastic recycling and future pathways of plastic waste management in Sweden. *Recycling*, *3*(3), 33. Advance online publication. DOI: 10.3390/recycling3030033

Ministry of Environment, Forest and Climate Change. (MoEFCC) (2021). *Guidelines for implementation of extended producer responsibility (EPR).*

Ministry of Housing and Urban Affairs. (2024). Smart Cities Mission. *Evaluation.*

Mohajan, H. K. (2025). Waste management strategy to save environment and improve safety of humanity. *Frontiers in Management Science, 4*(2), 74–81. DOI: 10.56397/FMS.2025.03.05

Mohan, S. V., Hemalatha, M., Amulya, K., Velvizhi, G., Chiranjeevi, P., Sarkar, O., Kumar, A. N., Krishna, K. V., Modestra, J. A., Dahiya, S., Yeruva, D. K., Butti, S. K., Sravan, J. S., Chatterjee, S., & Kona, R. (2020). Decentralized Urban Farming Through Keyhole Garden: A Case Study with Circular Economy and Regenerative Perspective. *Materials Circular Economy, 2*(12), 12. Advance online publication. DOI: 10.1007/s42824-020-00011-1

MoHUA, NIUA, & Smart City. (2024). Smart Cities and Academia towards [+ Case Studies of Innovative Projects of Smart Cities Mission.]. *Action Research*, **•••**, 75.

Moneycontrol. (2024, April 11). Oh my God! Maha Kumbh drives 162% jump in flight bookings; hotels brimming with tourists. https://www.moneycontrol.com/travel/oh-my-god-maha-kumbh-drives-162-jump-in-flight-bookings-hotels-brimming-with-tourists-article-12908601.html

Montgomery, S. L. (2013). *Does science need a global language? English and the future of research.* University of Chicago Press. DOI: 10.7208/chicago/9780226010045.001.0001

Mouloudj, K., Bouarar, A. C., Hoti, F., Asanza, D. M., & Kabashi, D. (2025). Nurses' Intentions to Implement Sustainable Practices in Public Hospitals: An Empirical Investigation. In Carrasco-Garrido, C., Martínez Raya, A., & De-Pablos-Heredero, C. (Eds.), *Global Economic Challenges and Opportunities for the Workplace* (pp. 1–24). IGI Global., DOI: 10.4018/979-8-3693-6289-1.ch001

Mouloudj, K., Njoku, A., Asanza, D. M., Bouarar, A. C., Evans, M. A., Mouloudj, S., & Bouarar, A. (2023). Modeling predictors of medication waste reduction intention in Algeria: Extending the theory of planned behavior. *International Journal of Environmental Research and Public Health, 20*(16), 6584. DOI: 10.3390/ijerph20166584 PMID: 37623170

Muhammad. (2017). Implementation of lean manufacturing system to eliminate wastes on the production process of line assembling electronic car components with WRM and VSM method. *SciSpace.* https://typeset.io/papers/implementation-of-lean-manufacturing-system-to-eliminate-5023wabkhr

Mukherjee, R., & Chakraborty, A. (2022). Cultural campaigns and environmental awareness: A study of Green Durga Puja. *International Journal of Cultural Studies, 25*(3), 255–270.

Mukucha, P., Mushanyuri, B. E., & Jaravaza, D. C. (2021). Green sourcing: effects on supplier performance metrics in fast food restaurants in frontier markets. *Logistics, supply chain, sustainability and global challenges, 12*(1), 47-59.

Mukucha, P., & Jaravaza, D. C. (2025). Waste material recycling: The horrors of informal waste material picking in Zimbabwe. *Journal of Responsible Production and Consumption*, *2*(1), 235–253. DOI: 10.1108/JRPC-01-2024-0003

Mukucha, P., Jaravaza, D. C., & Nyengerai, S. (2023a). Circular economy of shopping bags in emerging markets: A demographic comparative analysis of propensity to reuse plastic bags versus cotton bags and paper bags. *Cogent Engineering*, *10*(1), 2176582. DOI: 10.1080/23311916.2023.2176582

Mukucha, P., Katsande, T., Dube, T., Jaravaza, D. C., & Chisango, S. (2025a). Disseminating Sustainable End-of-life Product Disposal Information Through Social Media Nudging. In *Sustainable Green Marketing Strategies for a Circular Economy: African Realities and Imperatives* (pp. 239–264). Springer Nature Switzerland. DOI: 10.1007/978-3-031-81627-7_12

Mukucha, P., Mwapenya, D., & Jaravaza, D. C. (2025b). The efficacy of food waste reduction strategies in an African frontier market foodservices industry. *Tourism (Zagreb)*, *73*(2), 215–226. DOI: 10.37741/t.73.2.2

Mukucha, P., Sixpence, S., Muzanenhamo, L., Muskwe, N., & Ukpere, W. I. (2023b). Green supply chain management: The role of sustainable packaging on corporate image and patronage intentions. *Annals of Spiru Haret University. Economic Series, 23*(4), 553–572.

Munguiko, C., Masereka, E. M., & Sang, C. (2025). Applying Learning Theories to Clinical Teaching in Contemporary Settings: A Conceptual Analysis. *South Asian Res J Med Sci*, *7*(2), 16–27. DOI: 10.36346/sarjms.2025.v07i02.001

Murthy, V., & Ramakrishna, S. (2022). A review on global E-waste management: Urban mining towards a sustainable future and circular economy. *Sustainability (Basel)*, *14*(2), 647. DOI: 10.3390/su14020647

Naeem, M., Ozuem, W., Howell, K., & Ranfagni, S. (2023). A step-by-step process of thematic analysis to develop a conceptual model in qualitative research. *International Journal of Qualitative Methods*, *22*, 16094069231205789. DOI: 10.1177/16094069231205789

Nasir, M., Salman, M., Ahmad, F., & Siddiquee, T. A. (2019). Critical Assessment of Smart City Mission - A Research of Gwalior Smart City Proposals. *International Journal of Innovative Technology and Exploring Engineering.*

Nawar, K. N., Mahbub, T., Tashfiq, R. A., & Rashid, T. U. (2024). Municipal Solid Waste Collection, Transportation, and Segregation. In *Environmental Engineering and Waste Management: Recent Trends and Perspectives* (pp. 29–71). Springer Nature Switzerland. DOI: 10.1007/978-3-031-58441-1_2

Nelda. (2024). *Festivals plantation.* https://nelda.org.in/festivals-plantation/

Nesmachnow, S., Rossit, D., & Moreno-Bernal, P. (2025). A Literature Review of Recent Advances on Innovative Computational Tools for Waste Management in Smart Cities. *Urban Science (Basel, Switzerland), 9*(1), 16. DOI: 10.3390/urbansci9010016

NEXT IAS. (n.d.). *Environmental pollution.* https://www.nextias.com/blog/environmental-pollution

Nexteel. (2024). *Balancing tradition and environmental sustainability: Water challenges at the Kumbh Mela.* https://nexteel.in/balancing-tradition-and-environmental-sustainability-water-challenges-at-the-kumbh-mela/

Nguyen, T. T. T., Malek, L., Umberger, W. J., & O'Connor, P. J. (2022). Household food waste disposal behavior is driven by perceived personal benefits, recycling habits, and ability to compost. *Journal of Cleaner Production, 379*(1), 134636. DOI: 10.1016/j.jclepro.2022.134636

Nicolet, J., Mueller, Y., Paruta, P., Boucher, J., & Senn, N. (2022). What is the carbon footprint of primary care practices? A retrospective life-cycle analysis in Switzerland. *Environmental Health : A Global Access Science Source, 21*(1), 3. DOI: 10.1186/s12940-021-00814-y PMID: 34980135

Nivas, A.H. (2019). A Critique on Need of Private Sector Participation to Enhance Solid Waste Management Practices in India.

Noliya, S., Gupta, A., & Kumar, V. (2025). E-waste management through Blockchain Technology: Supply chain led environmental sustainability. *EDPACS*, 1-11.

Nowakowski, P. (2019). Investigating the reasons for storage of WEEE by residents–A potential for removal from households. *Waste Management (New York, N.Y.), 87*, 192–203. DOI: 10.1016/j.wasman.2019.02.008 PMID: 31109518

Ojadi, J. O., Owulade, O. A., Odionu, C. S., & Onukwulu, E. C. (2025). AI-Driven Optimization of Water Usage and Waste Management in Smart Cities for Environmental Sustainability. *Engineering and Technology Journal., 10*(3), 4284–4306. DOI: 10.47191/etj/v10i03.36

Oluka, A. (2025). Phenomenological Research Strategy: Descriptive and Interpretive Approaches. *F1000 Research, 14*, 725. DOI: 10.12688/f1000research.166273.1

Onwudiegwu, C. A., Aigberua, A. O., & Elegbede, I. O. (2024). Uncontrolled Bush Burning and Air Quality. In *Air Pollutants in the Context of One Health: Fundamentals, Sources, and Impacts* (pp. 369–394). Springer Nature Switzerland. DOI: 10.1007/698_2024_1136

Onyeaka, H., Tamasiga, P., Nwauzoma, U. M., Miri, T., Juliet, U. C., Nwaiwu, O., & Akinsemolu, A. A. (2023). Using artificial intelligence to tackle food waste and enhance the circular economy: Maximising resource efficiency and minimising environmental impact: A review. *Sustainability (Basel), 15*(13), 10482. DOI: 10.3390/su151310482

Oraee, A., Pohl, L., Geurts, D., & Reichel, M. (2024). Overcoming premature smartphone obsolescence amongst young adults. *Cleaner and Responsible Consumption, 12*, 100174. DOI: 10.1016/j.clrc.2024.100174

Oswald, D. (2021). Homeowner vulnerability in residential buildings with flammable cladding. *Safety Science, 136*, 105185. DOI: 10.1016/j.ssci.2021.105185

Oyenuga, M. O., Singh, R., Apata, S. B., Khan, S., & Kumar, V. (2025). Circular Economy and Waste Reduction Technologies. In *Smart Technologies for Climate Change and Net Zero Policies: Practical Approaches Towards Sustainability* (pp. 117–129). Springer Nature Switzerland. DOI: 10.1007/978-3-031-92221-3_10

Ozbay, G., Jones, M., Gadde, M., Isah, S., & Attarwala, T. (2021). Design and operation of effective landfills with minimal effects on the environment and human health. *Journal of Environmental and Public Health, 2021*(1), 6921607. DOI: 10.1155/2021/6921607 PMID: 34531916

Özden, M. (2024). The Importance of Ecofeminism in Sustainable Development. *Reconstructing Feminism through Cyberfeminism, 275*, 88.

Paleologos, E. K., O'Kelly, B. C., Tang, C. S., Cornell, K., Rodríguez-Chueca, J., Abuel-Naga, H., & Goli, V. S. N. S. (2020). Post COVID-19 Water and Wastewater Management to Protect Public Health and Geoenvironment. *Environmental Geotechnics, 40*, 1–15. DOI: 10.1680/jenge.20.00067

Parida, L., & Patel, T. N. (2023). Systemic impact of heavy metals and their role in cancer development: A review. *Environmental Monitoring and Assessment, 195*(6), 766. DOI: 10.1007/s10661-023-11399-z PMID: 37249740

Patel, M.. (2021). Circular economy in organic waste management. *Resources, Conservation and Recycling, 168*, 105291. DOI: 10.1016/j.resconrec.2021.105291

Paulchamy, D., Alwar, E.B., Anbarasu, K., Hemalatha, R., Lavanya, R., & Manasa, K.M. (2018). IOT Based Waste Management in Smart City.

Pilone, V., di Santo, N., & Sisto, R. (2023). Factors affecting food waste: A bibliometric review on the household behaviors. *PLoS One, 18*(7), e0289323. DOI: 10.1371/journal.pone.0289323 PMID: 37506105

Podlasek, A., Vaverková, M. D., Jakimiuk, A., & Koda, E. (2024). A comprehensive investigation of geoenvironmental pollution and health effects from municipal solid waste landfills. *Environmental Geochemistry and Health, 46*(3), 97. DOI: 10.1007/s10653-024-01852-4 PMID: 38393507

Poetz, M. K., & Schreier, M. (2012). The value of crowdsourcing: Can users really compete with professionals in generating new product ideas? *Journal of Product Innovation Management, 29*(2), 245–256. DOI: 10.1111/j.1540-5885.2011.00893.x

Power, J. R., Tanner, D., & Buckley, J. (2025). Self-efficacy development in undergraduate engineering education. *European Journal of Engineering Education, 50*(1), 1–25. DOI: 10.1080/03043797.2025.2468350

Prabhu, N. S., & Majhi, R. (2023). Disposal of obsolete mobile phones: A review on replacement, disposal methods, in-use lifespan, reuse and recycling. *Waste Management & Research, 41*(1), 18–36. DOI: 10.1177/0734242X221105429 PMID: 35722875

Prasad, D., Alizadeh, T., & Dowling, R. (2023). Smart city planning and the challenges of informality in India. *Dialogues in Human Geography.*

Prasath, C. A. (2025). Automated Waste Classification and Segregation Using YOLOv5 and Robotic Arm. In Frontiers in Mathematical and Computational Research (Vol. 1, Issue 1). wwwfrontiermcr.com

Press Information Bureau. (2024, February 10). *India records 2,509 million domestic tourist visits in 2023.* https://pib.gov.in/PressReleasePage.aspx?PRID=2024XXXX

Proske, M., & Jaeger-Erben, M. (2019). Decreasing obsolescence with modular smartphones?–An interdisciplinary perspective on lifecycles. *Journal of Cleaner Production, 223*, 57–66. DOI: 10.1016/j.jclepro.2019.03.116

Pulido-Ocegueda, J. C., Santibáñez-Aguilar, J. E., Ramírez-Márquez, C., & Ponce-Ortega, J. M. (2025). Strategic development and economic impact of e-waste management in Mexico: A quantitative and perspective analysis on collection, recycling, and material recovery systems. *Journal of Material Cycles and Waste Management, 27*(5), 1–21. DOI: 10.1007/s10163-025-02344-y

Putra, A. N. (2024). *The Influence Of Observational Learning On Consumer Purchase Intention Is Mediated By Perceived Message Authenticity And Consumer Attitudes* (Doctoral dissertation, Universitas Muhammadiyah Surakarta).

Qasim, M., Ahmad, W., Yoshida, M., Gould, M., & Yasir, M. (2020). Analysis of the worldwide corona virus (COVID-19) pandemic trend; a modelling study to predict its spread. *medRxiv*. https://doi.org/DOI: 10.1101/2020.03.30.20048215

Rahman, N. A. A., Sharif, S. M., & Esa, M. M. (2013). Lean manufacturing case study with Kanban system implementation. *Procedia Economics and Finance, 7*, 174–180. DOI: 10.1016/S2212-5671(13)00232-3

Rai, P. (2020). Disinfection of Endoscopy and Reusability of Accessories. *Journal of Digestive Endoscopy, 11*(1), 61–66. DOI: 10.1055/s-0040-1712238 PMID: 40477028

Ramteke, S., & Sahu, B. L. (2019). Health care waste management: A review. *International Journal of Trend in Scientific Research and Development, 3*(5), 1235–1240. DOI: 10.31142/ijtsrd23109

Rana, N. P., Luthra, S., Mangla, S. K., Islam, R., Roderick, S., & Dwivedi, Y. K. (2018). Barriers to the Development of Smart Cities in Indian Context. *Information Systems Frontiers, 21*(3), 503–525. DOI: 10.1007/s10796-018-9873-4

Rashid, S., & Malik, S. H. (2023). Transition from a Linear to a Circular Economy. In *Renewable energy in circular economy* (pp. 1–20). Springer International Publishing. DOI: 10.1007/978-3-031-42220-1_1

Rastogi, N., Singh, A., & Satish, R. (2018). Characteristics of submicron particles coming from a big firecracker burning event: Implications to atmospheric pollution. *Atmospheric Pollution Research, 10*(2), 629–634. DOI: 10.1016/j.apr.2018.11.002

Rauf, A., Puluhulawa, F. U., & Ahmad, A. (2024). Ideal Arrangements for Fines to Enhance Legal Awareness and Minimize Waste Effectively in Society. *Estudiante Law Journal, 6*(3), 593–606. DOI: 10.33756/eslaj.v6i3.28916

Raut, S., Hossain, N. U. I., Kouhizadeh, M., & Fazio, S. A. (2025). Application of artificial intelligence in circular economy: A critical analysis of the current research. *Sustainable Futures : An Applied Journal of Technology, Environment and Society, 9*, 100784. Advance online publication. DOI: 10.1016/j.sftr.2025.100784

Renou, S., Thomas, J., Aoustin, E., & Pons, M. (2007). Influence of impact assessment methods in wastewater treatment LCA. *Journal of Cleaner Production, 16*(10), 1098–1105. DOI: 10.1016/j.jclepro.2007.06.003

Ren, Y., Liu, S., Yang, L., Shan, J., Liu, Y., Li, H., Chen, L.-H., & Chen, H. (2020). Practice and exploration of infection prevention and control measures based on risk management of surgical patients during the epidemic of corona virus disease 2019 (COVID-19). *American Journal of Infection Control, 49*(2), 151–157. DOI: 10.1016/j.ajic.2020.07.023 PMID: 32702389

Reshma, & Kant Sharma, K. (2021). Impact of Solar cells Materials on Energy Efficiency using Nanotechnology. *Proceedings - 2nd International Conference on Smart Electronics and Communication, ICOSEC 2021*. DOI: 10.1109/ICOSEC51865.2021.9591626

Reshma, & Singh Attwal, K. P. (2024). A Review on Medicinal Plant identification through Leaf Image. In M. H. (Ed.), *Proceedings - 2024 6th International Conference on Computational Intelligence and Communication Technologies, CCICT 2024* (pp. 27 – 33). Institute of Electrical and Electronics Engineers Inc. DOI: 10.1109/CCICT62777.2024.00015

Reuters. (2025, January 15). *Kumbh Mela 2025 projected to boost Indian economy by up to $35 billion*. https://www.reuters.com/article/kumbh-mela-2025-economic-impact-idUSKBNXXXX

Reyes-Santias, F., Portela-Romero, M., Cinza-Sanjurjo, S., Cordova-Arevalo, O., & González-Juanatey, J. R. (2024). Estimated technological obsolescence of computed tomography equipment. *Academic Radiology, 31*(3), 951–955. DOI: 10.1016/j.acra.2023.07.006 PMID: 37541825

Richards, G., & Palmer, R. (2010). *Eventful cities: Cultural management and urban revitalization*. http://ci.nii.ac.jp/ncid/BB04121970

Rittl, L. G. F., Zaman, A., & de Oliveira, F. H. (2025). Digital transformation in waste management: Disruptive innovation and digital governance for zero-waste cities in the global south as keys to future sustainable development. *Sustainability (Basel), 17*(4), 1608. DOI: 10.3390/su17041608

Rivas, C. (2012). Coding and analysing qualitative data. *Researching society and culture, 3*(2012), 367-392.

Rodrigues, S. P., Gomes, L. de C., Peres, F. A. P., Correa, R. G. de F., & Baierle, I. C. (2025). A Framework for Leveraging Digital Technologies in Reverse Logistics Actions: A Systematic Literature Review. *Logistics (Basel)*, *9*(2), 54. Advance online publication. DOI: 10.3390/logistics9020054

Rodriguez, D. (2024, August 23). Lean tools, principles, and applications. *Invensis Learning Blog*. https://www.invensislearning.com/blog/lean-methodology-tools-principles-applications/

Roes, A. L., Marsili, E., Nieuwlaar, E., & Patel, M. K. (2007). Environmental and cost assessment of a polypropylene nanocomposite. *Journal of Polymers and the Environment*, *15*(3), 212–226. DOI: 10.1007/s10924-007-0064-5

Roy, P., Mohanty, A. K., Dick, P., & Misra, M. (2023). A Review on the Challenges and Choices for Food Waste Valorization: Environmental and Economic Impacts. *ACS Environmental Au*, *3*(2), 58–75. DOI: 10.1021/acsenvironau.2c00050 PMID: 36941850

Rutala, W. A., Gergen, M. F., Sickbert-Bennett, E. E., & Weber, D. J. (2020). Comparative evaluation of the microbicidal activity of low-temperature sterilization technologies to steam sterilization. *Infection Control and Hospital Epidemiology*, *41*(4), 391–395. DOI: 10.1017/ice.2020.2 PMID: 32098638

Sabeet, . (2017). Sustainable Solid Waste Management in India. International Journal of Scientific Research in Science and Technology, 3, 1317-1324 .

Saha, A. (2020). The Socio-political Relevance of the Indian Smart City Mission:A Critical Analysis.

Saha, K., Farhanj, Z., & Kumar, V. (2025). A systematic review of circular economy literature in healthcare: Transitioning from a 'post-waste' approach to sustainability. *Journal of Cleaner Production*, *505*, 145427. Advance online publication. DOI: 10.1016/j.jclepro.2025.145427

Saha, K., Patel, B., & Paladini, S. (2025). The role of leadership and cultural barriers in the adoption of lean six sigma in clinical pharmacy practice and medicine waste reduction. The case of NHS-UK. *International Journal of Quality & Reliability Management*, *42*(3), 809–836. DOI: 10.1108/IJQRM-02-2024-0069

Sahoo, S., Rathod, W., Vardikar, H., Biswal, M., Mohanty, S., & Nayak, S. (2024). Biomedical waste plastic: Bacteria, disinfection and recycling technologies—a comprehensive review. *International Journal of Environmental Science and Technology*, *21*(1), 1141–1158. DOI: 10.1007/s13762-023-04975-w PMID: 37360566

Saini, H. (2024). Impact of solid waste dumping on soil quality and its potential risk on human health and environment. *Environmental Monitoring and Assessment, 196*(2). https://www.researchgate.net/publication/382532973 PMID: 38183498

Sakthivelmurugan, E., Senthilkumar, G., & Karthick, K. N. (2022). Analysis of the impact of circular economy over linear economy in the paper processing industry. *Materials Today: Proceedings, 66*, 1446–1452. DOI: 10.1016/j.matpr.2022.05.449

Salvia, G., Zimmermann, N., Willan, C., Hale, J., Gitau, H., Muindi, K., Gichana, E., & Davies, M. (2021). The wicked problem of waste management: An attention-based analysis of stakeholder behaviors. *Journal of Cleaner Production, 326*, 129200. DOI: 10.1016/j.jclepro.2021.129200 PMID: 34866810

Sarada, B. V., Vijay, R., Johnson, R., Rao, T. N., & Padmanabham, G. (2020). Fight against COVID-19: ARCI's technologies for disinfection. *Transactions of the Indian National Academy of Engineering : an International Journal of Engineering and Technology, 5*(2), 349–354. DOI: 10.1007/s41403-020-00153-3 PMID: 38624431

Sarc, R., Curtis, A., Kandlbauer, L., Khodier, K., Lorber, K. E., & Pomberger, R. (2019). Digitalisation and intelligent robotics in value chain of circular economy oriented waste management – A review. In *Waste Management* (Vol. 95, pp. 476–492). Elsevier Ltd., DOI: 10.1016/j.wasman.2019.06.035

Sarfo, J. O., Debrah, T., Gbordzoe, N. I., Afful, W. T., & Obeng, P. (2021). Qualitative research designs, sample size and saturation: Is enough always enough. *Journal of Advocacy. Research in Education, 8*(3), 60–65.

Sari, T. K., & Rahmani, P. H. (2024). The Role of Positive Reinforcement on Students in English Language Learning: A Skinnerian behaviorist: English. *JOEY: Journal of English Ibrahimy, 3*(1), 1–5. DOI: 10.35316/joey.2024.v3i1.1-5

Sarkar, S., Khillare, P. S., Jyethi, D. S., Hasan, A., & Parween, M. (2010). Chemical speciation of respirable suspended particulate matter during a major firework festival in India. *Journal of Hazardous Materials, 184*(1–3), 321–330. DOI: 10.1016/j.jhazmat.2010.08.039 PMID: 20817345

Sarkhoshkalat, M. M., Afkham, A., Bonyadi Manesh, M., & Sarkhosh, M. (2024). Circular Economy and the Recycling of E-Waste. In *New Technologies for Energy Transition Based on Sustainable Development Goals: Factors Contributing to Global Warming* (pp. 319–354). Springer Nature Singapore. DOI: 10.1007/978-981-97-2527-4_16

Sati, A., & Sudhakar, K. (2018). Smart Dustbins: An Approach for Sustainable Solid Waste Management in Smart Cities.

Schade, A., Melzer, M., Zimmermann, S., Schwarz, T., Stoewe, K., & Kuhn, H. (2024). Plastic waste recycling— A chemical recycling perspective. *ACS Sustainable Chemistry & Engineering*, *12*(33), 12270–12288. DOI: 10.1021/acssuschemeng.4c02551

Schafer, R., & Phillippi, J. C. (2025). Updating and advancing member-checking methods: Use of video and asynchronous technology to optimize participant engagement. *International Journal of Qualitative Methods*, *24*, 16094069251315395. DOI: 10.1177/16094069251315395 PMID: 40933861

Schanes, K., Dobernig, K., & Gözet, B. (2018). Food waste matters - A systematic review of household food waste practices and their policy implications. *Journal of Cleaner Production*, *182*, 978–991. DOI: 10.1016/j.jclepro.2018.02.030

Schellenberg, T., Subramanian, V., Ganeshan, G., Tompkins, D., & Pradeep, R. (2020). Wastewater Discharge Standards in the Evolving Context of Urban Sustainability–The Case of India. *Frontiers in Environmental Science*, *8*, 30. Advance online publication. DOI: 10.3389/fenvs.2020.00030

Scottish Environment Protection Agency. (n.d.). *Air pollution impacts on the environment*. https://www.environment.gov.scot/our-environment/air/air-pollution-impacts-on-the-environment/

Sedláková, S. (2016). Separation of Waste - One of the Ways to Reduce the Cost of Waste Management Municipality.

Seif, R., Salem, F. Z., & Allam, N. K. (2023). E-waste recycled materials as efficient catalysts for renewable energy technologies and better environmental sustainability. *Environment, Development and Sustainability*, *26*(3), 1. DOI: 10.1007/s10668-023-02925-7 PMID: 36691418

Sepetis, A., Parlavatzas, I., Zaza, P. N., Platis, C., Fotios, R., & Nikolaou, I. E. (2024). The Role of Organizational Behavior to Sustainable Health Care: The Case of Greece. *Environmental Health Insights*, *18*, 1–15. DOI: 10.1177/11786302241298788 PMID: 39641108

Serpe, A., Purchase, D., Bisschop, L., Chatterjee, D., De Gioannis, G., Garelick, H., Kumar, A., Peijnenburg, W. J. G. M., Piro, V. M. I., Cera, M., Shevah, Y., & Verbeek, S. (2025). 2002–2022: 20 years of e-waste regulation in the European Union and the worldwide trends in legislation and innovation technologies for a circular economy. *RSC Sustainability*, *3*(3), 1039–1083. DOI: 10.1039/D4SU00548A

Shabani, T., Jerie, S., Mutekwa, T. V., & Shabani, T. (2024). Electronic waste: 21st century scenario in Zimbabwe—a review. *Circular Economy and Sustainability*, *4*(2), 1269–1284. DOI: 10.1007/s43615-024-00344-y

Shafipour, M., Forghani, M. A., & Sadeghi, Z. (2025). Measurement of Reverse Logistics in Incomplete Production Process Mode, Variable Demand and Return Rate, with Different Degrees of Quality in Order to Increase Value: A Case Study of Kerman Bahonar Copper Complex. *Tasmimgiri va Tahqiq dar Amaliyyat, 10*(2), 214 – 237. DOI: 10.22105/dmor.2025.484648.1884

Shah, N. P.Nirali P Shah. (2024). Navigating challenges in new product development: Strategies for reducing failure rates in the medical device industry. *World Journal of Advanced Research and Reviews, 22*(3), 786–795. DOI: 10.30574/wjarr.2024.22.3.1740

Sharma, A. (2020). Logo.png AI and Robotics for Reducing Waste in the Food Supply Chain: Systematic Literature Review, Theoretical Framework and Research Agenda.

Sharma, A., Singh, P. V., & Musunur, L. P. (2020). Artificial intelligence and robotics for reducing waste in the food supply chain: Systematic literature review, theoretical framework, and research agenda. *Working Paper, Date of current version Nov. 2, 2020.*

Sharma, S. (2020). How To Reduce Corona Spread. *Journal of Medical Case Reports and Reviews, 3*(5), 658–662.

Sharma, V., Jamwal, A., Agrawal, R., & Pratap, S. (2025). A review on digital transformation in healthcare waste management: Applications, research trends and implications. *Waste Management & Research, 43*(6), 828–849. DOI: 10.1177/0734242X241285420 PMID: 39352741

Sharpley, R., & Stone, P. (2014). *Contemporary tourist experience: Concepts and consequences*. Routledge. DOI: 10.4324/9780203139110

Shin, S.-K., Um, N., Kim, Y.-J., Cho, N.-H., & Jeon, T.-W. (2020). New policy framework with plastic waste control plan for effective plastic waste management. *Sustainability (Basel), 12*(15), 6049. DOI: 10.3390/su12156049

Shrivastava, S. R., Bobhate, P. S., Singh, S., Quazi, S., & Jawade, S. (2025). Minimizing the dermatological consequences of exposure to e-waste. *Indian Dermatology Online Journal, 16*(5), 882–884. DOI: 10.4103/idoj.idoj_128_25 PMID: 40847670

Shukla, S., Halli, P., Khalid, M. K., & Lundström, M. (2022). Waste pharmaceutical blister packages as a source of secondary aluminum. *Journal of the Minerals Metals & Materials Society, 74*(2), 612–621. DOI: 10.1007/s11837-021-05038-6

Singh, A., Goel, A., Chauhan, A., & Singh, S. K. (2025). Sustainability of electronic product manufacturing through e-waste management and reverse logistics. *Sustainable Futures : An Applied Journal of Technology, Environment and Society, 9*, 100490. Advance online publication. DOI: 10.1016/j.sftr.2025.100490

Singh, A., Pant, P., & Pope, F. D. (2019). Air quality during and after festivals: Aerosol concentrations, composition and health effects. *Atmospheric Research, 227*, 220–232. DOI: 10.1016/j.atmosres.2019.05.012

Singh, N. (2021). *Religion, ecology, and sustainability: Engaging faith leaders for environmental transformation*. Sage Publications.

Singh, R. (2024). Ecosystem dynamics of Prayagraj. In *Environmental impact of ritual practices in India* (pp. 79–88). ResearchGate., https://www.researchgate.net/publication/392315941

Singh, T., Solanki, A., Sharma, S. K., Nayyar, A., & Paul, A. (2022). A Decade Review on Smart Cities: Paradigms, Challenges and Opportunities. *IEEE Access : Practical Innovations, Open Solutions, 10*, 68319–68364. DOI: 10.1109/ACCESS.2022.3184710

Singh, U., & Upadhyay, S. P. (2022). Fractured smart cities: Missing links in India's smart city mission. *Environment and Planning. B, Urban Analytics and City Science, 50*(7), 1790–1805. DOI: 10.1177/23998083221144321

Smaller Footprint. (2024). *Festivals and flowers*. https://smallerfootprint.co/blogs/good-steps/festivals-and-flowers

Sonkusare, S., & Shastri, Y. (2025). A Systemic Evaluation of Reuse, Recycling, and Reduction Strategies in the Circular Economy for Global Sustainability Using an Integrated Planetary Model. *Circular Economy and Sustainability, 5*(3), 2009–2032. DOI: 10.1007/s43615-025-00513-7

South Tourism. (n.d.). *Brahmotsavam Festival – Tirupati*. https://www.southtourism.in/andrapradesh/festivals/brahmotsavam.php

Spångberg, J., Tidåker, P., & Jönsson, H. (2014). Environmental impact of recycling nutrients in human excreta to agriculture compared with enhanced wastewater treatment. *The Science of the Total Environment, 493*, 209–219. DOI: 10.1016/j.scitotenv.2014.05.123 PMID: 24946033

Springer. (2024). *Festival-related pollution in the River Ganges*. https://link.springer.com/content/pdf/10.1007/s13201-024-02227-x.pdf

Starikovskiy, A. Y., & Usmanova, D. R. (2020). Medical mask with plasma sterilizing layer. *arXiv preprint arXiv:2004.00807*. Available at: https://arxiv.org/abs/2004.00807

Stephen, S. A. (2024). *Relationships between social identity, self-efficacy, other-efficacy, relation-inferred-self-efficacy and performance in sport* (Doctoral dissertation, Heriot-Watt University).

Subbanarasimha, R. P., Venumuddala, V. R., Prakash, A., & Chaudhuri, B. (2025). Embracing Complexity in Smart City Services: A Case of Waste Management. *Journal of Innovation & Knowledge*, *10*(2), 100661. DOI: 10.1016/j.jik.2025.100661

Swachh Bharat Mission. (2022). *Best practices compendium*. Ministry of Housing and Urban Affairs, Government of India.

Tafadzwa, Z., & Steven, J. (2025). A critical review of innovative strategies for the sustainable management of solid waste generated in the health institutions of Zimbabwe. *Environmental Monitoring and Assessment*, *197*(5), 611. DOI: 10.1007/s10661-025-14043-0 PMID: 40301144

Taiwo, A. (2011). Composting as a Sustainable Waste Management Technique in Developing Countries. *Journal of Environmental Science and Technology*, *4*(2), 93–102. Advance online publication. DOI: 10.3923/jest.2011.93.102

Takunda, S., & Steven, J. (2023). Medical solid waste management status in Zimbabwe. *Journal of Material Cycles and Waste Management*, *25*(2), 717–732. DOI: 10.1007/s10163-022-01578-4 PMID: 36686405

Tang, L., Murphree, M., & Breznitz, D. (2016). Structured uncertainty: A pilot study on innovation in China's mobile phone handset industry. *The Journal of Technology Transfer*, *41*(5), 1168–1194. DOI: 10.1007/s10961-015-9432-9

Tan, J., Guo, S., Ma, Y., Duan, J., Cheng, Y., He, K., & Yang, F. (2011). Characteristics of particulate PAHs during a typical haze episode in Guangzhou, China. *Atmospheric Research*, *102*(1–2), 91–98. DOI: 10.1016/j.atmosres.2011.06.012

Tao, X., & Ucbasaran, D. (2025). How does failure normalization foster product innovativeness in new product development? The role of passion and learning. *Journal of Product Innovation Management*, *42*(2), 338–364. DOI: 10.1111/jpim.12755

Tardy, C. (2021). Academic publishing and English dominance: Challenges for scholars in multilingual contexts. *Journal of English for Academic Purposes*, *53*, 101034. DOI: 10.1016/j.jeap.2021.101034

Taslimi, M., Batta, R., & Kwon, C. (2020). Medical waste collection considering transportation and storage risk. *Computers & Operations Research, 120*, 104966. DOI: 10.1016/j.cor.2020.104966

Tathavadekar, V. P., & Mahankale, N. R. (n.d.). Neuromorphic Computing-Enabled Digital Twin Framework for Sustainable IT Supply Chain Integration in Smart Urban Ecosystems. https://fdtue.cultechpub.com/index.php/fdtue

TERI. (2020). *Sustainable celebrations: A guidebook for eco-friendly festivals*. The Energy and Resources Institute.

Thakur, V., Parida, D., & Raj, V. (2022). Sustainable municipal solid waste management (MSWM) in the smart cities in Indian context. *International Journal of Productivity and Performance Management.*

Thane, P. P. S. (2023). *Effects of Ganapati festival*. https://www.ppsthane.com/blog/effects-of-ganapati-festival

The Hindu. (2020, February 15). *Why India needs to clean its air*. https://www.thehindu.com/sci-tech/energy-and-environment/why-india-needs-to-clean-its-air/article69400021.ece

The Indian Express. (2024). *From gardens to gulal: Celebrating Holi with sustainable herbal colours*. https://indianexpress.com/article/lifestyle/art-and-culture/from-gardens-to-gulal-celebrating-holi-with-sustainable-herbal-colours-9885601/

Tiedemann, T., Keppner, M., Runge, T., Vögele, T., Wittmaier, M., & Wolff, S. (2021). Concept of a robotic system for autonomous coarse waste recycling. *Proceedings of the 18th International Conference on Informatics in Control, Automation and Robotics, ICINCO 2021*, 493–500. DOI: 10.5220/0010584004930500

Times of India. (2023a, October 2). *Six lakh devotees attend grand Brahmotsavams at Tirumala*. https://timesofindia.indiatimes.com/city/vijayawada/six-lakh-devotees-attend-grand-brahmotsavams-at-tirumala/articleshow/114195251.cms

Times of India. (2023b, October 25). *Kolkata airport sees record footfall during Durga Puja 2023*. https://timesofindia.indiatimes.com/city/kolkata/kolkata-airport-sees-record-footfall-during-durga-puja-2023/articleshow/114229931.cms

Times of India. (2024). *Unique Diwali festive decor ideas with sustainable materials*. https://timesofindia.indiatimes.com/life-style/home-garden/unique-diwali-festive-decor-ideas-with-sustainable-materials/articleshow/114726870.cms

Torrejón-Ramos, M., Medina-Salgado, M. S., & Ortiz-de-Urbina-Criado, M. (2025). A scoping review on reducing food waste and loss in hospitals. *Gaceta Sanitaria, 39*, 102462. DOI: 10.1016/j.gaceta.2025.102462 PMID: 39985827

Tourki, T. (2010). *Implementation of lean within the cement industry* [Doctoral dissertation, De Montfort University]. https://hdl.handle.net/2086/4562

Travel and Tour World. (2024, April 17). *India tourism rolls out free pavilion access for international visitors at Kerala's iconic Thrissur Pooram.* https://www.travelandtourworld.com/news/article/india-tourism-rolls-out-free-pavilion-access-for-international-visitors-at-keralas-iconic-thrissur-pooram/

Ugandhar, T. (2024). Eco-conscious celebrations: Bonalu festival's impact on environmental sustainability and public health in Telangana, India. *ResearchGate.* https://www.researchgate.net/publication/383467369

Unegg, M. C., Steininger, K. W., Ramsauer, C., & Rivera-Aguilar, M. (2023). Assessing the environmental impact of waste management: A comparative study of CO2 emissions with a focus on recycling and incineration. *Journal of Cleaner Production, 415*, 137745. Advance online publication. DOI: 10.1016/j.jclepro.2023.137745

UNEP & ISWA. (2024). Global Waste Management Outlook 2024 - Beyond An Age of Waste: Turning rubbish into a resource.

UNEP. (2020). BASEL: Waste management an essential public service in the fight to beat [The United Nations Environment Programme] [UNEP] [and The Basel Convention.]. *COVID*, •••, 19.

United Nations Environment Programme (UNEP). (2020). Waste management: An essential public service in the fight against. *COVID*, •••, 19.

United Nations Environment Programme (UNEP). (2023). *E-waste management and the circular economy.* UNEP-IETC.

Uysal, İ., Altın, F. G., & Özcan, F. (2026). Predicting medical waste amounts and costs using machine learning and XAI for sustainable waste management in healthcare. *International Journal of Environmental Science and Technology, 23*(1), 21. DOI: 10.1007/s13762-025-06895-3

Vadodaria, D. (2014). Optimizing time with increasing productivity using lean manufacturing and overall equipment effectiveness. *SciSpace.* https://typeset.io/papers/optimizing-time-with-increasing-productivity-using-lean-1r3ccdxre1

Vail, J. P. (2022). The need for a sustainability pledge: Fighting planned obsolescence. *Geo. Wash. J. Energy & Env't L., 13*, 1.

van Baar, A., Knoote, F., & Consultancy, D. (2024). A fair price for human rights due diligence. *Bonn: Fairtrade International/Amsterdam: Vrije Universiteit Amsterdam. Retrieved, 30.*

Van Straten, B., Dankelman, J., Van Der Eijk, A., & Horeman, T. (2021). A circular healthcare economy; a feasibility study to reduce surgical stainless steel waste. *Sustainable Production and Consumption*, *27*, 169–175. DOI: 10.1016/j.spc.2020.10.030

Vishwakarma, A., Kanaujia, K., & Hait, S. (2023). Global scenario of E-waste generation: trends and future predictions. In *Global e-waste management strategies and future implications* (pp. 13–30). Elsevier. DOI: 10.1016/B978-0-323-99919-9.00013-1

Vukicevic, A. M., Petrovic, M., Jurisevic, N., Djapan, M., Knezevic, N., Novakovic, A., & Jovanovic, K. (2025). Versatile waste sorting in small batch and flexible manufacturing industries using deep learning techniques. *Scientific Reports*, *15*(1), 3756. Advance online publication. DOI: 10.1038/s41598-025-87226-x PMID: 39885307

Walia, J., Sharma, R., & Akhai, S. (2025). Stubble-reinforced materials for sustainable construction. In Akhai, S. (Ed.), *Mitigating unsustainable practices in construction and architecture* (pp. 99–118). IGI Global., DOI: 10.4018/979-8-3373-2555-2.ch004

Wang, J., Shen, J., Ye, D., Yan, X., Zhang, Y., Yang, W., Li, X., Wang, J., Zhang, L., & Pan, L. (2020). Disinfection technology of hospital wastes and wastewater: Suggestions for disinfection strategy during coronavirus Disease 2019 (COVID-19) pandemic in China. *Environmental Pollution*, *262*, 114665. DOI: 10.1016/j.envpol.2020.114665 PMID: 32443202

Wang, S., Yu, R., Shen, H., Wang, S., Hu, Q., Cui, J., Yan, Y., Huang, H., & Hu, G. (2019). Chemical characteristics, sources, and formation mechanisms of PM2.5 before and during the Spring Festival in a coastal city in Southeast China. *Environmental Pollution*, *251*, 442–452. DOI: 10.1016/j.envpol.2019.04.050 PMID: 31103004

WHO. (2014). Safe management of wastes from health-care activities. https://iris.who.int/bitstream/handle/10665/85349/9789241548564_eng.pdf

WHO. (2024b). Safe management of wastes from health-care activities. https://iris.who.int/bitstream/handle/10665/85349/9789241548564_eng.pdf

Wijewickrama, M. K. C. S., Chileshe, N., Rameezdeen, R., & Ochoa, J. J. (2025). Integrating quality into the reverse logistics in construction: An empirical study in South Australia. *Built Environment Project and Asset Management*, *15*(3), 483–499. DOI: 10.1108/BEPAM-12-2023-0241

Wijnands, A. P. G., Koppen, A., de Vries, I., de Lange, D. W., & Rietjens, S. J. (2025). Poison center surveillance of occupational incidents with hazardous materials (2016–2023): Insights for risk mitigation and incident preparedness. *International Journal of Environmental Research and Public Health, 22*(2), 158. DOI: 10.3390/ijerph22020158 PMID: 40003384

Wilson, D. C., Velis, C., & Cheeseman, C. (2006). Role of informal sector recycling in waste management in developing countries. *Habitat International, 30*(4), 797–808. DOI: 10.1016/j.habitatint.2005.09.005

World Health Organization (WHO). (2020). *Coronavirus disease (COVID-19) outbreak situation.*

World Health Organization (WHO). (2022). *Children and digital dumpsites: E-waste exposure and child health.* WHO.

World Health Organization. (2024a). Health-care waste. https://www.who.int/news-room/fact-sheets/detail/health-care-waste

World Health Organization. (n.d.). *Air pollution.* https://www.who.int/india/health-topics/air-pollution

Wu, P., & Low, S. P. (2011). Lean management and low carbon emissions in precast concrete factories in Singapore. *Journal of Architectural Engineering, 18*(2), 176–186. DOI: 10.1061/(ASCE)AE.1943-5568.0000053

Wu, W., Li, L., Chen, H., Xu, M., & Yuan, Y. (2022). Farmers' Preference for Participating in Rural Solid Waste Management: A Case Study from Shaanxi Province, China. *International Journal of Environmental Research and Public Health, 19*(21), 14440. DOI: 10.3390/ijerph192114440 PMID: 36361319

Xu, C., & Yan, M. (2024). Product life cycle lengths and real activities management. *Journal of Accounting, Auditing & Finance, 39*(4), 980–1007. DOI: 10.1177/0148558X221108970

Yadav, S. K., Rajagopal, K., Priya, A. K., & Sharma, G. D. (2020). Smart Waste Management and Energy Extraction from Waste in Indian Smart Cities –. *RE:view.*

Yaspal, B., Jauhar, S. K., Kamble, S., Belhadi, A., & Tiwari, S. (2023). A data-driven digital transformation approach for reverse logistics optimization in a medical waste management system. *Journal of Cleaner Production, 430*, 139703. DOI: 10.1016/j.jclepro.2023.139703

Ye, G., Lin, H., Chen, L., Wang, S., Zeng, Z., Wang, W., & Yang, Z. (2020). Environmental contamination of the SARS-CoV-2 in healthcare premises: an urgent call for protection for healthcare workers. *medRxiv*. https://doi.org/DOI: 10.1101/2020.03.11.20034546

Yin, J., Gao, Y., & Xu, H. (2014). Survey and analysis of consumers' behaviour of waste mobile phone recycling in China. *Journal of Cleaner Production*, *65*, 517–525. DOI: 10.1016/j.jclepro.2013.10.006

Yu, L., Moriguchi, Y., Nakatani, J., Zhang, Q., Li, F., He, W., & Li, G. (2019). Environmental Impact Assessment on the recycling of waste LCD panels. *ACS Sustainable Chemistry & Engineering*, *7*(6), 6360–6368. DOI: 10.1021/acssuschemeng.9b00119

Yuvaraja, H. (2020). Challenges in development and implementation of Indian smart city mission. International Journal of Advance Research. *Ideas and Innovations in Technology*, *6*, 534–540.

Zaman, S.. (2018). Water quality of the River Ganga in and around the city of Kolkata during and after Goddess Durga immersion. *Parana Journal of Science and Education*, *4*(9), 1–7.

Zambujal-Oliveira, J., Franco, A., & Fernandes, B. (2025). Aligning Financing Strategies With Circular Business Principles: A Multicriteria Decision Framework. *Business Strategy and the Environment*, *34*(4), 4252–4273. DOI: 10.1002/bse.4199

Zhang, L., Xuan, Y., Li, Z., Gao, P., & Zheng, Y. (2024). How to obtain customer requirements for each stage of the product life cycle from online reviews: Using mobile phones as an example. *Journal of Retailing and Consumer Services*, *80*, 103928. DOI: 10.1016/j.jretconser.2024.103928

Zhang, M. (2023). Burn to harvest, burn to sabotage: Between fire and water on a sugar plantation in Madagascar. *American Anthropologist*, *125*(1), 100–111. DOI: 10.1111/aman.13811

Zhang, X., Shen, H., Li, T., & Zhang, L. (2020). The effects of fireworks discharge on atmospheric PM2.5 concentration in the Chinese Lunar New Year. *International Journal of Environmental Research and Public Health*, *17*(24), 9333. DOI: 10.3390/ijerph17249333 PMID: 33322228

Zhilyaev, D., Cimpan, C., Cao, Z., Liu, G., Askegaard, S., & Wenzel, H. (2021). The living, the dead, and the obsolete: A characterization of lifetime and stock of ICT products in Denmark. *Resources, Conservation and Recycling*, *164*, 105117. DOI: 10.1016/j.resconrec.2020.105117

Zhuang, E., Chen, H. H., Kolesnik, O., & Hines, S. E. (2023). Tolerability, user acceptance and preference for a novel reusable respirator among health care workers. *American Journal of Infection Control, 51*(7), 821–826. DOI: 10.1016/j.ajic.2022.09.006 PMID: 36122632

Zimba, A., Mukupa, G., & Chama, V. (2024). On emergent mobile phone-based social engineering cyberattacks in developing countries: The case of the Zambian ICT sector. *African Journal of Science, Technology, Innovation and Development, 16*(6), 774-791. 1248.

Zocco, F., Sleath, D., & Rahimifard, S. (2025). Towards a Thermodynamical Deep-Learning-Vision-Based Flexible Robotic Cell for Circular Healthcare. *Circular Economy and Sustainability*, *5*(4), 3187–3209. Advance online publication. DOI: 10.1007/s43615-025-00532-4 PMID: 40881316

Zulfa, V., & Nugraheni, P. L. (2020, May). Effectiveness of community empowerment in waste management program to create sustainable tourism in Karawang, West Java. In *IOP Conference Series: Earth and Environmental Science* (Vol. 485, No. 1, p. 012087). IOP Publishing. DOI: 10.1088/1755-1315/485/1/012087

About the Contributors

Muhammad Rizwan Yaseen, Ph.D. Current Position: Associate Professor & Head of Economics Department Government College University Faisalabad, Pakistan Research Interests: Microeconomics, Managerial Economics, Sustainability, Competitiveness, Economic Policy Education: Ph.D. in Economics and Management Sciences, AgroParisTech, France (2013) Master's in Economics and Management Sciences, University of Montpellier, France (2009) B.Sc. Honors (Agricultural Economics), University of Agriculture, Faisalabad, Pakistan (2005) Selected Publications: Yaseen, M. R., Ali, Q., & Khan, M. T. I. (2024). Attaining global peace and tourism competitiveness: Linkages between tourism, security measures, and economic stability. Kybernetes, Impact Factor: 2.4. Yaseen, M. R., et al. (2023). Climate change vulnerability, adaptation, and feedback hypothesis: A comparison of lower-middle, upper-middle, and high-income countries. Sustainability, Impact Factor: 3.9. Yaseen, M. R., et al. (2022). The nexus between research and development, renewable energy, and environmental quality: Evidence from developed and developing countries. Renewable Energy, Impact Factor: 8.7. Teaching Expertise: Courses: Microeconomics, Managerial Economics, Advanced Economic Theory, Environmental Economics Innovative Methods: Flipped classrooms, simulation-based learning, data analysis (Stata/SPSS) Leadership & Supervision: Supervised 8 PhD and 43 MPhil theses on topics like sustainability, globalization, and socio-economic development. Led curriculum redesign and AACSB accreditation efforts as Head of Economics Department. Commitment to Diversity: Promoted inclusive classrooms and mentorship for underrepresented students. Proposed initiatives: Faculty diversity training, global exchange programs, and need-based scholarships. Contact: Email: rizwany2001@yahoo.com | rizwanyaseen@gcuf.edu.pk

Qamar Ali is lecturer in Department of Economics, the Virtual University of Faisalabad Pakistan.

***

Abhishek Allapur is a postgraduate student pursuing an MBA with a specialization in Operations and Marketing at [Woxsen University], India. His academic interests include supply chain management, digital transformation, and sustainable business practices. He has worked on projects related to e-waste management, AI-driven innovation, and circular economy frameworks. He also completed an internship with Seva Satkar Foundation, where he focused on marketing research for community development initiatives. Currently, he has contributed as a co-author to a book chapter on AI-driven cognitive robotics for sustainable e-waste management.

A.C. Bouarar is a Professor in the College of Economic at University of Medea (Algeria). His main research interests include online learning, digital transformation, modern apps, entrepreneurship, telemedicine, volunteering, save energy, green practices. He received his Bachelor and Master in business sciences from University of Medea, and Doctor of Accounting from University of Algiers 3. From 2018 to 2019 he served as Department Chair. He is the head of a research project on digital transformation (2022-2026) at Medea University. Prof. Dr. Bouarar is a Speaker and participant in many foreign conferences (such as the United Kingdom, the Czech Republic, the United States, Serbia, Indonesia, Turkey, etc.). He has published about 40 papers in journals.

Susanta Das is a seasoned educator with a proven record of teaching and mentoring diverse students, working at multiple institutions, and providing service to the various initiatives of organizations. He received his Ph.D. and M.A. degrees from Western Michigan University (WMU), USA, and M.Sc. degree from Banaras Hindu University (BHU), India all in Physics. He continued his research as a Marie-Curie post-doctoral fellow at Stockholm University, Stockholm, Sweden on beam diagnostics for the DESIREE (Double ElectroStatic Ion Ring ExpEriment) in the project DITANET (Diagnostic Techniques for particle Accelerators – a Marie-Curie initial training NETwork), at the Indian Institute of Science Education and Research-Kolkata, India on high-pressure physics as a Project Scientist-B, and at the University of Electro-Communications, Tokyo, Japan on ion-surface interactions as a Post-doctoral fellow. Throughout his career, Dr. Das worked and collaborated with many researchers, Ph.D., master, and visiting students from different countries. He further visited many countries to discuss research and an international conference participant (UK, Italy, Germany, Greece, Belgium, Bulgaria, Romania, Brazil etc.). He worked at several institutions before assuming his current role at Ajeenkya D Y Patil University. He received the Marie-Curie post-doctoral fellowship in Sweden,

Science Academies' Summer Research Fellowship in India, Gowen Frostic Doctoral Fellowship, Department Graduate Research and Creative Scholar Award by WMU, and Leo R. Parpart Doctoral Fellowship by Dept. of Physics, WMU, among many others, throughout his academic journey. He co-authored several research articles in WoS/Scopus indexed international journals, conference proceedings, and scholarly book chapters and co-edited several books published by renowned international publishing houses (IGI-Global, Springer Nature, Nova Science Publishers, Wiley, Scrivener-Wiley, Notion Press). At present, Dr. Das continues to delve into cutting-edge fields, with a keen interest in nanotechnology, UN SDGs, and data science. His multifaceted contributions, spanning teaching, research, and administrative leadership, showcase a dedicated professional who is instrumental in advancing the vision and mission of the institutions he serves.

Rui Manuel Teixeira Santos Dias is a Postdoctoral Fellow in Finance at the State University of Feira de Santana (Brazil), in the Department of Exact Sciences. He has a PhD in Management from the University of Évora (Portugal). He obtained his Diploma of Advanced Studies (DEA) in Doctoral Studies at the University of Extremadura (Spain), in the scientific area of Financial Economics.

Ginu George is passionate about delivering high-quality education and research for over 10 years in the field of human resources management, organisational behaviour, business communication and research methodology for undergraduate and postgraduate students. Holds a PhD in Commerce (HRM & OB) and an MPhil in Commerce (Marketing) from Christ University, Bangalore, India. Dr Ginu have more than 10 Scopus-indexed publications and a prolific record of presenting research papers at international conferences, showcasing a commitment to advancing knowledge." Additionally, Dr Ginu George is currently guiding two Ph.D. scholars, providing mentorship and guidance to the next generation of researchers. Beyond academia, also comes with 3 years of practical experience as HR contributing to employee engagement, and talent acquisition. Through the combined experience as an educator and a practitioner, Dr Ginu aims to bridge the gap between theory and practice, and to contribute to the advancement of knowledge and skills in the field of HR.

Basima Kzar Hasan is an Assistant Professor in the Department of Economic Studies at the Center for Basra and Arab Gulf Studies, University of Basra, Basra, Iraq. Her academic and research interests center on sustainable development, green finance, economic development, and the green economy. She has published several scholarly articles and has actively participated in numerous local and international conferences. Dr. Hasan's research primarily focuses on analyzing the relationship between fiscal policies, development, and economic growth, with particular attention to sustainability-oriented economic strategies.

Flaka Hoti holds a Master's degree in Healthcare Management and bring extensive experience in research focused on public health, nursing, and elderly care. My research addresses key aspects of autonomy, privacy, and care quality, aiming to improve standards and practices in nursing and institutional care. My expertise in healthcare management allows me to contribute valuable insights toward sustainable, evidence-based solutions for the well-being of vulnerable populations.

Mohammad Irfan is presently working as a Professor at the Faculty School of Business and Commerce, Chandigarh University, Lucknow, Unnao, India. Dr. Irfan has done his Ph.D. at the Central University of Haryana. He is MBA (Finance), M.Com (Account and Law), and MA (Economics). He has qualified UGC-JRF/SRF/NET in Management and Commerce. Dr. Irfan certified NSEs (NCFM) and BSEs certification. He has experience of more than sixteen years in the area of SAPM, Artificial Intelligence, Machine Learning, Blockchain, Cryptocurrency, Financial Engineering, Fintech, Green Finance, and Alternative Finance. He has to his credit more than 101+ Scopus-indexed articles, including The Journal of Economic Cooperation and Development (Q2), International Journal of Business Excellence (IJBEX), International Journal of Economics and Management (IJEM), Montenegrin Journal of Economics (Q2), Cogent Business & Management (Taylor & Francis) (Q2), Indian Journal of Finance (IJF), and Journal of Islamic Monetary Economics and Finance (JIMF). His citations reached 702+ along with 14 H-index. Dr. Irfan has published 11 books in Springer, IGI Global Publication (Scopus indexed).

Pawan Kumar, PhD, is a prominent academic and researcher with over 20 years of expertise in business studies. A professor at Graphic Era Deemed to be University, Dehradun, India, he has supervised six PhD scholars and published over 60 research papers, including 51 in SCOPUS/ 06 in Web of Science, and 14 books, including eight indexed in Scopus. His works, featured by prominent publishers like Emerald, IGI Global, Nova Science, and Cambridge Scholars, have gained recognition. Dr. Kumar earned his PhD in Commerce from Kurukshetra University in 2007 and holds an M.Phil. and M.Com. Degrees. Specializing in finance, banking, and securities markets, he has authored books on green finance, AI, and fintech. His innovative research includes five patents and has served as a guest editor for Scopus-indexed journals. In 2022, he received the prestigious Researcher Award from Chandigarh University. Dr. Kumar's impactful work on sustainable finance, blockchain, and fintech is globally recognized. His active participation in high-impact conferences and faculty development programs reflects his commitment to advancing research, teaching, and innovation in commerce and management.

Sanjeev Kumar Is a Professor in Lovely Professional University Punjab India. He received Undergraduate in Hotel Management from Osmania University, Postgraduate in Tourism Management from Madurai Kamaraj University. He completed Doctor of Philosophy in Hospitality Management in 2019 from Amity University Jaipur India. His research areas lies in food Service Industry, his research focuses on Alcoholic beverages, Event management and Sustainable Management Practices, Metaverse, AI Machine Learning and Artificial Intelligence. He published more than 80 research papers, articles and chapters in Scopus Indexed, UGC Approved and peer reviewed Journals and books. Dr. Sanjeev Kumar participated and acted as resource person in various National and International conferences, seminars, session chair, research workshops and industry talks and his work has been widely cited. Dr. Sanjeev also edited more than 20 books with various International Publishers such as IGI Global, Cambridge Scholars, Emerald, CABI, Bentham Science, Taylor & Francis. Dr Sanjeev is an active researcher and has published numerous research articles in prestigious National and International journals and books.

Kritikka Kunte is a third-year undergraduate student pursuing a Bachelor of Technology in Artificial Intelligence and Data Science at the School of Engineering, Ajeenkya DY Patil University, Pune, Maharashtra, India. With a strong academic orientation toward data-driven problem solving, she has developed a focused interest in the domains of data analytics and data science, supported by both theoretical foundations and applied project experience. Throughout her academic journey, Kritikka has cultivated proficiency in core programming and analytical competencies, including Python, SQL, and Java, along with essential data science methodologies such as data preprocessing, exploratory data analysis (EDA), and machine learning techniques. She is proficient in industry-relevant tools and platforms, including Advanced Microsoft Excel, Matplotlib, Tableau, and Power BI, to derive insights from complex datasets and present them in a structured and interpretable manner. Her practical exposure includes the successful completion of projects such as Ocean Salinity Analysis, which involved analyzing environmental datasets to identify patterns and trends, and Speech Recognition and Translation, where she explored applied machine learning concepts in natural language processing. These projects reflect her ability to translate conceptual knowledge into functional, real-world solutions. Kritikka has further strengthened her technical profile through professional certifications, including the AWS re/Start Graduate credential from Amazon Web Services, a certification in Software Development, and Python training certified by

Google. Her engagement in competitive and collaborative academic environments is demonstrated by her participation in the Smart India Hackathon (SIH) and by securing Second Runner-Up in an inter-collegiate Tableau competition, highlighting her analytical aptitude and visualization skills. With a growing interest in advanced transportation systems and data-centric engineering solutions, Kritikka brings a multidisciplinary perspective to the GHC Hyperloop Project Competition hosted at IIT Madras. She aspires to contribute meaningfully to innovative, technology-driven initiatives that integrate artificial intelligence, data science, and sustainable engineering to address complex global challenges.

Kamel Mouloudj is a Professor at the College of Economics at the University of Medea in Algeria, specializing in research areas such as individual behavior prediction (including customers, employees, and students), marketing, waste management, sustainability, and green start-ups. He holds a Doctorate in Marketing from the University of Algiers 3 and has over 20 years of teaching experience. With more than 50 published works, Professor Mouloudj is also serving as the Chair of the Doctoral Training Committee in the Department of Commercial Sciences from 2019 to 2026. Additionally, he has actively participated in over 20 local and international conferences, contributing his expertise to a wide range of academic discussions.

Paul Mukucha is the chairman of the department of marketing at Bindura University of Science Education.

Krutika Patre, is an aspiring programmer and technology enthusiast, currently pursuing Engineering in Artificial Intelligence and Data Science. With a passion for learning, she is building experience in a variety of programming languages and technologies, including C, C++, Python, Data Structures (DS), HTML, CSS, and JavaScript, DBMS (My SQL). After learning this programming language, she has created interest in Full Stack Development. She had also attended an Arduino-Based Automation course at Dolphin Labs to improve my software knowledge. There, she learned how to work with sensors, microcontrollers, and circuit design. She has written review paper on Sustainable engineering on build environment. She is doing researching and learning many topics related to AI, Policy and governance and related to my interest. She has taken part in hackathons, workshops, and tech events to explore many fields. She is also working with a group on Hyperloop. Being part of this project has really helped her grow, especially in areas like teamwork, innovation, and systems thinking.

Afzalur Rahman is a distinguished Professor of Commerce at Woxsen University, Hyderabad, with over 14 years of expertise in Finance and FinTech. He holds a PhD in Accounts from Patna University (2011) and an Executive MBA in

Finance from VM University (2008). His teaching encompasses advanced courses such as Quantitative Finance with Python, Machine Learning in Equity Research, Financial Modeling, and Advanced FinTech. Dr. Rahman has developed algorithmic trading applications on platforms like Kite-Zerodha and Alice Blue, fine-tuned a 7B-parameter LLM for financial predictions, and authored 33 publications on cryptocurrency, blockchain, and AI in sustainable finance. He has published books on strategic financial management and guided two PhD scholars. Proficient in Python, R, cloud technologies, and SAP FICO, he has led international certifications, MOUs, and conferences, fostering industry-academia collaborations.

Sunil Sankathala is pursuing B. Tech. in Computer Science and Engineering at Ajeenkya D Y Patil University, with a diverse range of interests that bridge technology, sustainability, and innovation. His academic focus lies in computer science fundamentals like programming, data structures, algorithms, and software development. He actively supplements this with hands-on experience in web development and robotics. Beyond the core curriculum, He has a strong research inclination in emerging and impactful fields. His current research interests include Artificial Intelligence in Healthcare, bioplastics, nano bioplastics, e-waste management, and solid waste management—areas where he believes technology can make a meaningful difference in solving real-world problems. He has been publishing research articles and book chapters in reputed national and international journals and books. He is also part of an ambitious Hyperloop project, which involves interdisciplinary collaboration and cutting-edge engineering aimed at revolutionizing transportation. Working on this project has enhanced his skills in teamwork, innovation, and systems thinking. In the tech domain, he has developed proficiency in web development using technologies such as HTML, CSS, JavaScript, and React. His interests in robotics further reflect building smart, responsive systems that interact with the physical world. Throughout his course, he has helped strengthen problem-solving and collaborative skills. He is proficient in languages such as Python, C, C++, MySQL, Ubuntu, Machine Learning and familiar with frameworks and tools like React, Git. He actively participates in hackathons, coding competitions, and tech events to challenge himself and learn from peers. He is also a quick learner and highly adaptable, with an eagerness to stay updated with the latest trends in the tech world. He aspires to become a skilled software engineer and contribute to building innovative solutions that have a positive impact on society. The long-term goal is to work in a dynamic environment where knowledge, grow professionally, and make meaningful contributions to the field of technology. Driven by a spirit of exploration and sustainability, he aims to contribute to technological solutions that address both human and environmental needs. The long-term goal is to become a versatile engineer and researcher, working at the intersection of computer science, environmental innovation, and human well-being.

Mamoona Sheikh is working as a lecturer at The university of Faisalabad . Meanwhile had a focus on research .

Sujit Singh is an academic and researcher specializing in operations management, supply chain strategies, and agile practices in higher education with over 24 years of Experiences. He holds a PhD in Manufacturing Management from University of Malaya, Kuala Lumpur, Malaysia and M.Tech. from Indian Institute of Technology, Delhi. He has authored several book chapters and research papers focusing on case study–based approaches, emphasizing practical applications in supply chain, logistics, procurement and sustainable manufacturing. With a keen interest in interdisciplinary work, he explores the intersection of technology, management, and organizational excellence. He is currently working an associate professor at Woxsen University Hyderabad. His academic vision centers on integrating theory with practice to drive sustainable organizational growth.

Ale Udaya Sri is pursuing an MBA with a dual specialization in Finance and Data Science & AI at Woxsen University. Her academic interests include capital investment decisions, financial analysis, and data-driven decision-making.She has worked on projects focused on capital investment decisions and hold finance-related certifications. She aims to integrate academic knowledge with practical application and contribute meaningfully to the fields of management and finance.

Angadi Kavya Sri is currently pursuing an MBA in General Management with a specialisation in Operations and Finance at Woxsen University. Her academic interests include supply chain management, pricing and revenue management, project management with agile practices, and financial management. She has worked on projects related to dealership revenue enhancement and capstone studies in operations, and hold a certification in Project Management with an Agile focus. She aims to bridge academic learning with practical application and contribute meaningfully to the fields of management and finance.

Antigona Ukëhaxhaj, had finished Medical Faculty at the University of Prishtina, where she also got a Master science degree in Public Health. During the work in National Institute of Public Health of Kosovo, she is specialised in Hygiene. Being related and interested in Public Health she participated in many workshops, conferences and projects in Public Health area in local,regional and international level. PhD study she enrolled in school of Public Health "Adrija Stampar"in Zagreb and finished at University of "Kirij i Metodij"of Skopje; FYROM. She is

Professor Asociated at the University of Gjakova and also Dean in Medical Faculty in University of Gjakova. She works from 1999 at Centre for Laboratory Testing at National Institute of Public Health too. Her expertise includes water quality monitoring, environmental pollution and public health. She was local coordinator for project "Supply of project management, air quality information management, behavior change and communication services in supporting Outreach and Behavior Change", which was founded from Millenium Challenge Cooperation (2019-2021), and also coordinator of Project from WHO.UVW and UNDP "Healthier Kosovo 1" and "Healthier Kosovo 2" for research in indoor air quality in schools and households in Kosovo. He has also led many projects at the University level through Erasmus + programs, the American Embassy and other international partners. Monitoring and quality of drinking water in Kosovo, food quality and safety, prevention of intra-hospital infections, environmental pollution with special emphasis on outdoor and indoor air pollution, where her profile is also completed by a large number of publications and projects, list in the field of Public Health. From 2005-2015, she was a teaching assistant at the University of Pristina. She has worked at the University of Gjakova since 2010 as an assistant, then in 2011 as a lecturer, head of the midwifery program, deputy dean and from 2021-2023, she has been the dean of the Faculty of Medicine at the "Fehmi Agani" University of Gjakova. 11th ECLSS Conferences on Language and Social Sciences, March 2021:" Teaching during social distance in the era of pandemic" Best research paper, and in 2025 she was chosen as "Researcher of the Year" in University of Gjakova "Fehmi Agani". She is currently the head of the laboratory at the Centre for Environment and Health at NIPH/Pristina, Kosovo.

Aryan Vasa is a passionate and self-driven Bachelor of Technology (B.Tech.) student specializing in Computer Science Engineering at Ajeenkya DY Patil University, Pune. Currently in his third year, Aryan is committed to building a solid foundation in core technical domains, including web development, data science, and artificial intelligence. With an inquisitive mindset and a strong eagerness to explore diverse fields, he strives to balance his academic journey with hands-on learning and project development. Originally from Jamnagar, Gujarat, and having completed his schooling in Rajkot under the ICSE board, Aryan has always been drawn toward technology, logic, and creativity. He is actively engaged in learning full-stack development while also keeping a close eye on the evolving job market, particularly the shift toward AI and generative technologies. His academic curiosity extends to research in advanced topics such as Generative AI and Digital Synthetic Outputs, for which he is currently writing a comprehensive review article. Aryan has also been a dedicated athlete, having played basketball for over ten years. This experience has helped him develop teamwork, discipline, and a growth mindset. In addition, he enjoys writing analytical and review-based content, merging his technical knowledge with strong communication skills. Committed to personal growth, Aryan maintains a balanced lifestyle that includes regular workouts, general fitness, and a focus on healthy eating. While continuously learning, he aims to contribute meaningfully to India's tech innovation landscape and become a versatile engineer capable of solving real-world problems through thoughtful design and smart solutions. Whether developing web applications, studying emerging technologies, or sharing knowledge through articles and mentorship, Aryan Vasa stands out as a motivated learner and a future-ready technologist.

Index

A

C

D

E

G

H

K

www.ingramcontent.com/pod-product-compliance
Lightning Source LLC
LaVergne TN
LVHW060004031026
849161LV00025B/1415